Kickin' Bot

An Illustrated Guide to Building Combat Robots

Grant Imahara

WILEY

Wiley Publishing, Inc.

Kickin' Bot: An Illustrated Guide to Building Combat Robots

Published by
Wiley Publishing, Inc.
909 Third Avenue
New York, NY 10022
www.wiley.com

Library of Congress Cataloging-in-Publication Data:

Imahara, Grant, 1970-
  Kickin' bot : an illustrated guide to building combat robots / Grant
Imahara.
    p. cm.
  ISBN 0-7645-4113-7 (PAPER/WEBSITE)
  1. Robots—Design and construction—Amateurs' manuals. I. Title.
  TJ211.15.I43 2003
  796.15—dc22

2003020215

ISBN: 0-764-54113-7

Printed in the United States of America

10 9 8 7 6 5 4 3 2 1

1MA/RU/RR/QT/IN

Published by Wiley Publishing, Inc., Indianapolis, Indiana
Published simultaneously in Canada

For general information on our other products and services please contact our Customer Care Department within the United States at (800) 762-2974, outside the United States at (317) 572-3993 or fax (317) 572-4002.

Credits

Executive Editor
Chris Webb

Development Editor
Eileen Bien Calabro

Production Editor
Pamela M. Hanley

Senior Production Editor
Fred Bernardi

Editorial Manager
Kathryn A. Malm

Vice President & Executive Group Publisher
Richard Swadley

Vice President and Executive Publisher
Bob Ipsen

Vice President and Publisher
Joseph B. Wikert

Executive Editorial Director
Mary Bednarek

Project Coordinator
Gina Snyder

Graphics and Production Specialists
Jennifer Heleine
Sean Decker
Carrie Foster
LeAndra Hosier
Brent Savage
Rashell Smith
Mary Gillot Virgin

Quality Control Technicians
John Tyler Connoley
John Greenough
Susan Moritz
Charles Spencer

Permissions Editor
Carmen Krikorian

Cover Designer
Anthony Bunyan

Book Designer
Kathie Schnorr

Proofreading and Indexing
Laura L. Bowman
Tom Dinse

Photos Provided by
BattleBots Inc.

Dedication

It was January of 2002. The ILM Model Shop had finished most of its work on *Star Wars: Episode I*, and things were winding down. It was between fighting robot seasons, since BattleBots ran in May and November. I got an e-mail from our community outreach coordinator, Dawn Yamada, who knew that I was a veteran robot builder and BattleBots competitor. She said that a struggling high-school robot team in Richmond, California was looking for volunteer mentors. The school was on the way home, halfway between ILM in Marin and my loft in Oakland, and I was intrigued. They were a rookie team entering a national robotics competition for high schools called FIRST (For Inspiration and Recognition of Science and Technology). As a way to build relationships with industry, the competition encourages industry mentoring from local engineering firms. The Richmond team had no mentors and little support from the faculty. What's more, every team gets only 6 weeks to design and build a robot, which is a pretty tight schedule, even with generous funding. I found out later that they were already into their second week, with little more than a few vague designs.

Richmond High School isn't in the best area in Northern California, far from the comfortable confines of Marin, although it's actually very much like the neighborhood where I grew up in Los Angeles, near Washington and La Brea (for those who know). Their automotive technology program had long been abandoned, and there was no metal shop. The wood shop had antiquated tools with blades that were inappropriate for metal work. Funding, to say the least, was very limited.

In contrast, my robot Deadblow cost in excess of $20,000, and I built it using exotic processes such as laser cutting, CNC (computer numerical control) machining, abrasive waterjet, and wire EDM (electrical discharge machining). It was constructed using the finest materials available: titanium armor, a 2024 aluminum frame, and an S7 tool steel hammer head. However, I was committed to following the rules of the FIRST competition, which encouraged the mentors and engineers to let the students do most (if not all) of the work themselves. This ruled out the vast resources of ILM, which included the aforementioned laser cutter and CNC machine. Basically, I had to rethink my whole approach to robot building.

We had no tools and very little money, so we started from scratch. We went to The Home Depot and bought a bandsaw, a jigsaw, a drill press and a belt sander. With this set of power tools, a handful of materials, and a dozen students, we set out to build a robot. We didn't even have our own room. We had to borrow a classroom and clean it up at the end of every day, hauling all the tools back into a closet.

I designed Deadblow on the computer, prototyped the frame with lasercut acrylic, and then machined the parts with the CNC machine. I had the titanium pieces cut by waterjet, and the S7 tool steel head got a slot cut all the way down the neck by wire EDM. For our robot Robzilla, we designed everything on graph paper, and used cardboard and Popsicle sticks to figure out the mechanics. We prototyped it with plastic plumbing pipe and plywood from the hardware store.

Thanks to the undying enthusiasm and energy of the students and the school's program coordinator, and after more than a few late nights (which I have accepted as par for the course when building a robot), we made it to the regional event in San Jose, California, competing with teams two and three times our size.

While we placed in the middle of the group of 50, the students were happy with getting the robot to work and about their learning experience. At the completion of the competition, they were surprised with an award for their enthusiasm both on and off the playing field, and for "overcoming difficult circumstances" to make it to competition. The next year, their robot MechPlow placed sixth overall and made it to the semifinal round.

I dedicate this book to the students of Richmond High School FIRST Team 841, also known as the *BioMechs*, who helped me rediscover robot building with limited resources. They share the spirit of robot builders everywhere, the desire to design and build mechanical wonders.

About the Author

Grant Imahara is an animatronics engineer and modelmaker for George Lucas's Industrial Light & Magic in Marin County, California. He specializes in electronics and radio control at the ILM Model Shop, and has credits on numerous movies, including *Jurassic Park: The Lost World*, *Star Wars: Episode I - The Phantom Menace*, *Galaxy Quest*, *AI: Artificial Intelligence*, *Star Wars: Episode II – Attack of the Clones*, *Terminator 3: Rise of the Machines*, and most recently, *Matrix: Reloaded* and *Revolutions*.

He has installed electronics in R2-D2 units for *Star Wars Episodes I* and *II*, replacing the halogen light source and rotating color wheel (for the sparkly lights) with a custom microcontroller-based LED circuit that was originally created to make the pulsating lights for the main engines of the Protector, from *Galaxy Quest*. He also upgraded all of the radio equipment and speed controls to modern standards. Along with R2-D2 Crew Chief Don Bies and Nelson Hall, he is one of only three official R2-D2 operators in the United States.

Grant developed a custom circuit to cycle the Energizer Bunny's arm beats and ears at a constant rate. He performed all electronics installation and radio programming on the current generation of Bunnies. He later became the Bunny's driver and the crew supervisor on numerous commercials.

Grant has a bachelor of science degree in electrical engineering from the University of Southern California. He picked up his mechanical skills from the machinists at the ILM Model Shop, many of whom date back to *Howard the Duck* (1986).

For fun, Grant competes in BattleBots with his robot Deadblow, which set a record for most number of hits in the first season of the show. Grant lives in a loft in Oakland, California, where he works on his robot and pursues other projects in his spare time.

Contents at a Glance

· ·

Contents

Foreword

Robotic combat is a sport, but to many of us who have competed, it has become much more.

Combat robots are things that you invent, you get dirty with, and you *build*. You obsess over them. You wake up in the middle of the night thinking about them. Then you get to compete with them and see what happens. You will never know exactly what the machine will really do until you fight with it. It becomes a chess game in your head trying to anticipate what an opponent will do when you don't have the faintest idea what the opponent is or what it is capable of. And this is before you even have an idea what your own robot is. What a wonderful challenge! What an adventure!

I think the biggest impact all this has had on me personally, and the thing that is the most valuable aspect of the sport, is that it is such a wonderful teacher. Your desire for glory and destruction has an insidious and unavoidable side effect: you *learn*. You *really* learn. When you put all that work and thought into designing these beasts, and then spend every spare second building, tearing down, and building again, getting ripped apart and tearing down and building again, and again—you stretch your brain. And there are no multiple-choice questions here. You kick butt or you're toast.

In my case, I have been able to take what I have learned here and use it every day in my business, and I can tell you it has refined and honed my abilities over such a wide range that I would not trade what I have experienced in the sport for anything. Heck, I even have my own TV show as a result of it!

What if you are average Joe with limited resources and no experience with engineering a robot? Well, that's another bonus of the sport—*anyone* can put together a machine and get in and fight. You don't have to be a rock star or a genius, in fact you can build a pretty simple robot and win some rounds if you are lucky and you stick with it.

My robot, Blendo, suffered its first defeat by just such a robot. The first time we went against this robot we destroyed him in about 30 seconds with one massive hit. This was a simple box with a motor in it, and we pretty much ruined it. The second time around the builder came with the same design, but it was bigger and he had mounted all the insides of the robot on really long rubber mounts to the point where they drooped out of the bottom of the thing when he picked it up. Blendo pummeled it, hammered it over and over and threw it all over the place, but it held together until Blendo ended up stuck in a corner. My opponent had beaten the notorious Blendo. This guy started out with a pretty basic robot, and had limited resources at his disposal, but he learned and persevered, and he took down one of the most successful robots in the sport. He learned, he *adapted*, and he had fun doing it.

The sport is pretty much like life: If you work hard, if you adapt, if you keep at it, you can pretty much do anything you want to, no matter who you are. This lesson is pretty important, especially for young people. Their robots will evolve and improve in response to the events in the ring; most importantly, the *builders* do as well. They will learn, they will adapt.

What you'll learn is this: creative mechanical engineering, design, electronics, physics, chemistry, combat strategy, physical coordination and control, even PR, finance, and team management. And it will consume you like no other hobby. Then you go in front of a bunch of your peers and you have a few minutes to pull together the result of your passion and months or even years of hard work. The adrenaline rush of all this during battle has to easily be equivalent to that coursing through the veins of an Olympic sprinter as he or she approaches the starting line. You don't get that sitting at a desk. You don't get that tinkering with your car or gardening. You don't get all that any other way than by putting your brain and your body on the line under pressure in robotic combat.

So if you are reading this book because you want to get into the sport, good for you! Realize that you are about to get an unparalleled education, and your life may change, in all likelihood for the better.

Jamie Hyneman

M5 Industries Inc.

Creator of Blendo

Host of Discovery Channel's *Mythbusters* series

Acknowledgments

As with any huge project (robots included), there is a team of people who contributed to make this book a reality. Thanks to the following people:

Aaron Haye, Rhino 3D modeler, for creating all of the outstanding renderings that grace the pages of this book.

Dr. Jason Dante Bardis, veteran robot builder, for making sure that I didn't say anything too dangerous or stupid.

Eileen Bien Calabro, development editor for Wiley, for helping me stay on top of my deadlines.

Bob Mimlitch and Tony Norman of IFI Robotics for their technical assistance with the IFI Control System.

Jascha Little of Team Mechanicus (The Judge), John Reid of Team Hurtz (KillerHurtz, TerrorHurtz, and Beta), and Zander Rose of Team Inertia Labs (Toro, T-Minus and Matador) for their technical input on the Pneumatics Appendix.

Norm Domholt and Rich Reed of NPC Robotics for their technical assistance on the motor section.

John Duncan for his technical assistance in the internal combustion engine section.

Fon Davis for additional CAD modeling and technical support.

Chris Webb, executive editor for Wiley, who convinced me that this was a good and possible idea for a book.

Ioanna Stergiades for supporting me while I was with thinking, eating, sleeping, and writing about robot building.

Thanks to my mom, Carolyn Imahara, for believing in me and for supporting me without question, no matter how strange or ridiculous my situation may have seemed.

Introduction

Welcome to the exciting sport of robot combat. It's unique among sports because it relies just as much on your brains as it does on your strength. Best of all, anyone can play, and anyone can become a champion.

You may ask, "I've never done this before. Can I do this?" Here's a little test to find out if you can and should.

- You have an interest in tinkering or knowing how things work.
- You played with Lego bricks as a kid (especially if you *still* do).
- You want to join an enthusiastic community of other tinkerers, engineers, artists, schoolteachers, special effects wizards, and dozens of other professions.
- You're looking for fun.

If at least one of the above applies to you, congratulations! You've got what it takes.

Your next question is probably, "How do I get started?" That's where *Kickin' Bot* comes in. I remember how it was in the beginning. It was hard to find information. I had to scrape together bits of information from all over the place — books, hardware stores, paintball retailers, scuba shops, the machinists and fabricators that I've worked with, and just about anyone or anyplace that might have had some part or technique that I could use to build a fighting robot.

Since then, the robot-building community has exploded with new ideas, and all kinds of information is now available both online and in the bookstores. I've compiled all of my scraps of information from books and product literature, as well as countless Web sites and online forums into this book. In addition, I've poured in everything that I've learned from the ILM Model Shop, where we specialize in making things quickly that perform reliably for the movies and TV, and added all the hard lessons that I've learned from the various upgrades and rebuilds of my personal competition robot, Deadblow.

About the Book

The chapters in this book fall into five categories:

- **Basic Design Techniques.** Developing and refining robot ideas into a realistic designs.
- **Starting with the Tools.** The tools for fabrication.
- **Working with Metal.** Techniques for cutting and finishing metal.
- **Parts, System by System.** Extensive research into the most popular parts and their strengths and weaknesses. Assembly details for each system are included in the applicable chapter.
- **Testing and Troubleshooting.** Making sure it all works once you've got everything together, and building driving skill.

Each chapter includes background information and hands-on advice about the topic at hand (everything from selecting materials and cutting metal to assembling the parts and testing), and many include *projects*, step-by-step instructions for basic procedures that result in a functioning robot.

The projects follow in order (from 1 to 9), and are structured around a fully designed drivetrain— a sort of *training* bot to learn how to work with metal. I didn't want the beginning robot builder to get stuck in design purgatory and end up with a *vaporbot* (a robot that is designed, but never built). My goal is to give you the skills you need to construct something.

I've based the projects around paper templates that tell you where to drill the holes and how to cut certain pieces. You will print out the templates and paste them to the metal or plastic to mark your holes, as described in the projects. These templates, along with complete plans and construction pictures, are available for free download from the Kickin' Bot Web site at www. kickinbot.com. The Wiley Web site also has links to evaluate the most recent releases of various CAD software packages that you can use to create drill/fabrication templates for your own robot designs. For more information about the projects, flip to "Project 1: Introduction to the Project Robot" in Chapter 2, "Designing the Robot."

Using This Book

How you use this book is up to you. The normal flow of building is research, design, prototype, test, build, test again, revise, and practice. However, since the emphasis here is on building skills by getting you started making things, we're going to jump directly into fabrication to get the beginning robot builder dirty as soon as possible. More advanced readers may want to skip directly to specific chapters for technical info, while beginners should proceed in a linear fashion.

You may be wondering why we're focusing on the drivetrain instead of weapons. The essence of every robot combat event is that if you can't move anymore, you're disabled, and you automatically lose. Often, it's a war of attrition, and some robots are as destructive to themselves as they are to other robots. If you have a hearty drivetrain, then you can either knock out your opponents, or they will knock themselves out. Play it safe with this system, and use battle-proven techniques. You can save all your innovative, never-before-seen ideas for the weapon and overall design. Remember, even if you have the most kick-ass weapon in the world, and it's still fully functional at the end of a match, if you can't move, and your opponent still can, you *lose*.

It may seem daunting at first, but you'll soon discover that building a robot is a series of little steps. I will be focusing specifically on combat robots and the materials, parts, and techniques used to build them. You won't see any chapters on autonomous control, sensors, or do-it-yourself electronics. Those are for noncombat robots, and there are several other excellent books that already exist on those subjects.

One last thing: Building and competing with combat robots is not a moneymaking venture. It's not a vehicle for fame. If you're looking for either of those things, then you're in the wrong place, and disappointment will inevitably follow. Robot combat is about feeling the satisfaction of building something with your own two hands, experiencing the joy of watching your creation perform like you designed it to in thrilling combat, contributing to the cooperative spirit of the robot-building community, and most of all, having fun!

That having been said, read on, be safe, and go build good robots!

Getting Started

With many projects, getting started is the hardest part. Not so with combat robots. You're probably already brimming with ideas about weapons and strategies. Before you get too far down the path of deciding what material to use for the armor, and which drive motor to buy, consider the points in this chapter and make sure that you're going in the right direction. We will answer some of the questions you may have about the process, including how much this will cost, and how long it will take, and give you a basic idea of what you're getting yourself into. As I mentioned in the introduction, it's a bit different than it appears on TV.

Getting Your Ideas on Paper

Before building anything, you've got to start with a picture or description of your idea. Most builders already have an idea in their heads for a specific weapon. Maybe you have more than one. Write them all down. Perhaps it's a new twist on something you've seen out there, or a combination of ideas. In the case of my robot Deadblow, inspiration came from a hammer robot called Thor that I'd seen at the Second Annual Robot Wars (San Francisco, 1995). The robot had genuine personality. It seemed excited as it fired its hammer, bucking wildly around the arena, while also commanding fear and respect with its deadly amount of power. Whatever your inspiration, it's important to put your plan down in writing. You don't have to create a dimensioned drawing or a fully rendered CAD (computer-aided design) model. Cocktail napkins and the backs of envelopes will do.

Tip
One of the more challenging aspects of the sport is coming up with a catchy name that hasn't been used before. You can use www.nameprotect.com to see if your choice has already been taken.

Pick Your Battles

The next step in the process is to select a competition to enter. The main reason you need to choose your competition so early in the process is that you've got to see what the weight limits are and what weapons are allowed. Before spending weeks and months pondering a weapon, it's a good idea to make sure that it's legal. The rules are different for every competition. What might be allowed at one event may be strictly prohibited at another (flame throwers, for example). Also, the rules change slightly from season to season. You should start by locating competitions that are available to enter. Then weigh the cost of hotels, entry fees, and other expenses. Finally, revise your ideas based on the rulebook and weight limits.

Finding the Events

Robot combat has gained widespread popularity in recent years through television coverage. At one time, there were three major televised combat robot events: BattleBots, Robot Wars, and Robotica. BattleBots was easily the largest of the three events in the United States, with hundreds of competitors each season vying for the coveted spots in the televised rounds. Since then, however, things have changed a bit, and BattleBots and Robotica have gone off the air, while Robot Wars continues both domestically and in England. More information on Robot Wars can be found at their Web site at www.robotwars.com. BattleBots is currently negotiating with other networks and exploring the possibility of hosting another live event. Updates and information can be found at their Web site: www.battlebots.com. Unfortunately, Robotica is pretty much a part of robot history now.

Where will you fight? Most of the builders in the sizeable U.S. fighting robot community have turned to untelevised regional combat events. The *Robot Fighting League* (or RFL) comprises over 20 large and small regional events, banded together under one central organizational structure. Events are hosted in states all across the country, including California, Arizona, Nevada, Minnesota, North Carolina, and Florida. There is a unified rule set that can be scaled to the size of each event's arena. More information is available at their Web site, www.botleague.com, as well as a schedule of local competitions.

Competition Costs

In deciding what competition you want to enter, logistics is usually the determining factor, since all competitors are responsible for their own travel and accommodations, as well as any entry fees and shipping costs. I'm sure you'd love to compete in a regional event in Los Angeles, but if you live in Miami, then you've got to consider airfare to the West Coast, shipping for the robot, and food and lodging during the competition, all of which could add up to more than the cost of your robot. I would target local events first. You don't have to travel far, and it will give you a taste of the combat experience.

Check the Rules

Don't just read the rules. Know them. Having a good grasp of the rules will save you from wasting precious time and effort on something that's illegal. Don't develop your design without knowing what's allowed and what's not.

Follow the rules. They're there to ensure fairness, and more importantly, to protect you, other competitors, and the audience from weapon systems that exceed the arena's capabilities, which could cause an injury. This isn't a sandlot sparring match behind the school at 3:00. Break the rules and you won't be allowed to compete.

Questionable Weapons

You'll find that what's allowed or prohibited is pretty clearly spelled out in some rulebooks, while others are a little vague. If you have an idea that's not specifically disallowed, but seems borderline, it's best to contact the competition's organizers and get a ruling from the source. Most, if not all, organizers are open to exploring fresh, new ideas that may not fit into established guidelines, as long as they can be done safely and don't cause any harm to the builders, the arena, or the audience. Remember that the organizers will ultimately determine if you can compete or not, because they're liable if something goes wrong.

This is not to say that you should bother them with *every* idea that you have, only *borderline* cases. If the rulebook says, "no explosives," don't e-mail them and ask if you can use explosives. Likewise, if a weapon clearly falls within established parameters, you won't need approval.

How Long Will It Take?

One thing that first-time builders often underestimate is how long it takes to build a robot. It will take a while before you will be able to compete. You can't go to "Joe's Used Robot Lot" and purchase last year's model with low-low financing (although a few robots have been bought and sold on eBay). These things are built from scratch, and as a rule of thumb, you will always have less time than you need, due to those pesky day jobs that cut into valuable robot-building time.

You should consider robot building a journey. Like any journey, it doesn't happen overnight. It will take *weeks* and *months* to finish. And then, as any veteran robot builder will tell you, it's only finished until *after* competition. Then, the process of repair begins, and inevitably, the robot evolves into something else, as you have the opportunity to go back and fix those few small things that were bugging you during the competition, and try new and better parts and designs.

The most important goal for time management is to try to finish before the competition. You must allow time for testing and fixing the robot (from your testing) before the event. Let me say from personal experience that finishing your robot in the pits *sucks* and is to be avoided. (A description of my rookie year experience can be found in Chapter 20, "Going to a Competition.")

In truth, fabrication time depends on your access to tools and your experience. The more you build, the more familiar you become with the tools, and the faster you get. You shouldn't be in a rush when building the robot, however, because rushing and cutting corners lead to accidents and injury.

You also need time to practice driving. Driving is what wins and loses matches, because it allows you to bring your weapon into contact with your opponent, while escaping theirs. Without any practice, how can you expect to drive well under pressure?

How Much Will It Cost?

Building a competitive robot isn't going to be cheap. If you do it cheaply, you may be putting yourself at a disadvantage compared to others who have more resources than you. As with any competitive sport (like auto racing, for example), higher-performance parts cost more. Another rule of thumb is that bigger robots are generally more expensive. A bigger robot means bigger motors and speed controls, more batteries, and more materials overall, all of which inflate the cost. If you have no experience and little money, then it's best to start small, and avoid the superheavyweight category until you've built up your skills and have more disposable income.

I would set aside a minimum of $3,000 for parts and materials to build a competitive lightweight (60-pound) robot. Add to that another $1,500 for each weight class you step up. Also, this does not include the cost of purchasing power tools and other equipment you'll need to construct the robot. Of course, these estimates could vary wildly, depending on what type of armor you use, if you already have a competition-legal radio system, and whether you have a weapon system that uses pneumatics, hydraulics, or a very large and expensive electric motor. Bear in mind that there is no limit on the amount of money you can spend on your robot, and some teams have spent $40,000 or more on a single robot (many accomplish this by enlisting sponsors).

Tip   The Law of Good, Fast, Cheap: pick any two. Prefabricated custom parts for robot combat can save you a lot of time engineering a motor bracket, or a bearing mount. Unfortunately, as the law predicts, they tend to be expensive.

You may be intimidated facing a big team with a lot of resources. It's true that not everyone has a large chunk of money to dedicate to this sport. That's okay. Here's the catch: Simply having expensive parts doesn't guarantee a victory. An expensive robot can face defeat just as easily as an average one if it's not thoroughly tested, or if the driver hasn't had enough practice time.

Other Tips to Keep the Process Moving

Here, you can find some general organizational techniques for keeping the design and building process moving smoothly, as well as some tips to keep you out of trouble that are often over-looked (even by veteran builders), like keeping track of your weight and ordering parts early.

Keep Track of Your Weight

The number one thing that makes builders scramble at the end of the build process is having an overweight robot. (You'll hear me mention this numerous times throughout the book.) In order to prevent (or at least minimize) the scramble, you've got to start your design with the weight limit in mind, and be aware of it at all times. Be as stingy as possible with your weight, and you won't end up horribly over the limit at the end. Keep a running list of your parts and their weight. You should be able to tally the weights of all your parts and armor at any time, which will help you make decisions about lightening other parts.

Remember that your weight is a fixed limit, and you've got to stick to it. Each item that you add to the robot's design eats into that budget. Novice designers and casual observers often wonder why you don't just put a saw on every side or a hammer on the front and back. These things require weight. Extra weapons mean less armor, or battery capacity, or motor power, and the best robot designs balance these categories to avoid weaknesses.

Chapter 4, "Selecting Materials," tells you all about how to calculate the weight of any given part, and the effect that choice of material has on the weight of a part. Since weight is such an important part of design, most (if not all) of the parts mentioned in this book also include their weight as part of the discussion.

Leave a Paper Trail

All construction diagrams and notes should be dated, so you can track your changes. It may seem simple enough to keep track of now, but things can become impossibly confusing as you get further along with construction. Designs often change from the original concept, and most of the time, more than once. If you move a bracket, for instance, referring to an old diagram could cause you to drill a bunch of holes in the wrong place, possibly making the part unusable. I usually have a dimensioned sketch or CAD drawing for each part that I make notes on. All that confusion can easily be avoided by taking a few seconds to put a date on a piece of paper.

Order Materials Early

Order your materials early in the process. They will take at least a few days to arrive. It's murder when you can't continue working on a part because you forgot to order something that you need. Waiting around for parts, especially when you're on a tight deadline, is a drag.

Before placing an order, read the manufacturer's catalog thoroughly, and make a list of parts. Make sure you've got all the correct (and complete) part numbers. Also have a backup choice if (gasp!) your part is out of stock, which happens occasionally, even with huge suppliers like McMaster-Carr. Keep the catalog nearby for reference when you call your order in. Being prepared will help you feel much less like an idiot if there are complications. Keep all of your parts lists in one place. All you need is an envelope or folder in which to collect them separately from your other notes. You'll surely need to refer to them again during the build process to get more of something that you've run out of (screws, for instance) or to source parts for another robot in the future.

Appendix D describes the Web sites of various online vendors for robot parts, while Appendix E lists the contact information for mail-order companies that have catalogs that are useful in robot building.

Make a Master To-Do List

Sit down and compose a list of things to do for every system in the robot. Sometimes it's hard to track your progress, and the task may seem daunting. As I mentioned earlier, it's a pretty long process. It will take more than a few weekends to complete. Having a running checklist of steps to perform gives you a little morale boost each time you complete a task and cross it off, helping to keep you motivated. Updating your lists can also help identify and prioritize the items that need more attention than others.

Document Your Progress with Pictures

You should take pictures of your robot as you built it. This will be important later on, if you decide to build a Web site. Right now, you may not feel that you'll want to have a Web site for the robot, but you should give yourself that option. Not only will you be able to share your experiences with other builders, but you can also direct potential sponsors to the site, so *they* can see all of the work that you've done.

It's always more interesting when you can see what went into the finished product. Try to capture at least the beginning and completion of each step. You don't have to get bogged down with documenting every little detail (although I'm sure some builders would appreciate it). Just keep a camera nearby so you can snap some shots of your work in progress here and there, when the opportunity presents itself.

Wrapping Up

This chapter has given you the basic rundown of the process of designing and building a combat robot. Things may still seem a bit vague at this point, but hopefully, you now have a better idea of what will probably be required of you, in terms of both time and money. If you're willing to accept the challenge, you will be rewarded with the feeling of accomplishment from having built something yourself.

Designing the Robot

R obot design is like a big jigsaw puzzle, where some pieces belong together, others kind of fit, and the rest don't fit together at all. Your job is to take the idea you have in your head, make it into a workable design, and select the parts that you need and decide where they should go, all without making the jigsaw pieces fit together with a hammer (both figuratively *and* literally).

This chapter will show you how to develop the overall plan for your robot by introducing some design rules for you to apply to your concept and revise accordingly. We get down and dirty with the specifics and the details of actual features that will save you time in the pits (like making your batteries easily accessible), reduce your robot's vulnerabilities (by moving fragile items inboard), and help you come up with a better overall design. With the information in this chapter, I'll try to point out common design mistakes, and offer better alternatives. Finally, I'll introduce some techniques to help you select parts and figure out how to fit them into the robot's frame and then we'll get started on your project robot.

Basic Design Rules

Below is a list of some of the basic rules for combat robot design that should be considered for every robot regardless of the type of weapon or strategy, including budgeting your weight, designing for the specific contest you will be entering, keeping it simple, and planning to get flipped over.

Start with the Weight

As mentioned in Chapter 1, "Getting Started," weight is the number one design issue that's overlooked by both veteran and rookie builders alike. Everything from drive and weapons to armor and batteries should be scaled to the weight limit of the robot. By now, you've selected the competition that you want to enter and are familiar with the different weight classes and their limits.

Note When selecting the weight class, remember that larger robots generally cost more to build, require more people to maintain and move around, and are more expensive to ship.

You should start by dividing your overall weight limit among the different systems according to the following rule of thumb: 30 percent for drive, 30 percent for weapons, 25 percent for armor and frame, and 15 percent for batteries and electrical. These percentages are a starting point, and you can vary them according to your design, as long as you don't exceed your maximum weight. However, the best designs are balanced so that you don't end up with no armor, or a really wimpy drive system.

Make a list of parts for each different system. If the sum of the weights on your list exceeds your budget for that system, then you know you've got to scale down. Keep updating your list during the design process. It will help you decide whether you can use a part, or whether it's too heavy.

Cross-Reference Weight calculations are discussed in detail in Chapter 4, "Selecting Materials."

Match Your Abilities to the Contest

What type of competition will you be entering? What type of environments will you face? Some competitions feature water and flame, while others have active weapons in the arena. These issues affect your robot's required *ground clearance* (distance from the bottom of the frame to the ground) and your tire selection. Another important consideration is whether there will be arena weapons or other "house" robots to contend with. Is it more of a head-to-head battle, or is there an obstacle course? Some weapon systems rely on speed rather than precise maneuvering, which may put you at a disadvantage on an obstacle course.

As mentioned in Chapter 1, each competition has its own set of rules that spell out the regulations and requirements for each robot. Before you can fight in a single match, you've got to pass safety. The safety inspectors will be scrutinizing your robot to make sure that you've followed all the rules. Failing inspection means scrambling in the pits to fix whatever's wrong, or else you don't get to play. Making huge modifications at the last minute can be incredibly stressful. Avoiding this unpleasantness is as simple as following the rules.

Keep It Simple

Strive to make your designs simple and efficient. The more complicated the system, the more things that can go wrong. Also keep in mind that there's no way to defeat every kind of robot. The number of different designs out there is a reflection of the diversity of the robot-building community. Instead, concentrate on removing weaknesses from your robot. You can mentally run your frame/armor prototype through the different weapons listed in Chapter 18, "Choose Your Weapon," and try to predict what aspects of your design might be exploited by the various strategies.

Get Used to Change

Rarely, if ever, do robot designs make it all the way from concept to completion without changes. Adjusting the design is an important part of the process. For example, you may have to make the robot's frame larger to accommodate your weapon's air tanks. Just keep in mind that every change requires you to update the weight of your robot.

In the design and building process, there are decisions. A lot of them. There are at least a half dozen ways to accomplish any given task on the robot. During prototyping, you may discover that there's a better way to implement your weapon. Don't be afraid to revise your idea.

Plan to Get Flipped Over

You will eventually be flipped or tipped over. It's just a matter of time. Many robot weapon systems are designed specifically for this purpose. Also, combat situations are fast and unpredictable, and even a random combination of events can flip a robot. Good designs should include a way to deal with this. The two most popular solutions are invertible drive and self-righting mechanisms.

An *invertible drive* means that your robot still has the ability to move around if it's flipped over. The best designs also allow the weapon to be used in this position. A *self-righting mechanism* is any means built into the robot that will allow you to flip yourself back over if you become inverted.

Although an invertible drive is easier to implement, self-righting is preferred, because although you can still drive, your weapon usually becomes useless when you're inverted. If you're in a close match that's left to a judge's decision, spending time on your back with a useless weapon will count against you in a big way.

Important Design Issues

There are several specific design issues that you need to keep in mind as you develop your design and position individual parts within the robot's frame.

Low Center of Gravity

Every particle of the robot is subject to the earth's gravity, and contributes to the total weight. If you take into account the mass distribution of the entire robot, it's possible to find a single balance point that behaves as if the total weight is concentrated there. This point is called the *center of gravity*. A low, flat object will have a lower center of gravity than a tall skinny object. The higher in the air the center of gravity is, the more likely you are to tip over. Keeping it low will improve your ability to maneuver and make it harder for your opponents to flip you over.

Weight Distribution

Some weapon designs require a certain amount of weight to be in a certain place for maximum performance. For example, when a hammer robot fires, the tendency is for the front of the robot to jump up. By positioning heavy items such as batteries towards the front of the robot, you can minimize the power lost by jumping, and transfer more of that energy to your opponent.

See Chapter 18 for the weight distribution requirements of other types of weapon systems.

Size of the Frame

As you increase the frame size, you also increase the weight dramatically, so it's better to keep the robot as small as possible. The weight you save can be used in other systems. Of course, packing everything into a tight space also has its disadvantages. It makes it harder to repair and maintain the robot, since it will be harder to get tools into the space.

Wheel Placement

Several issues are related to wheel number and placement. Generally speaking, most wheeled robots fall into two categories: two or four wheels. Two-wheeled robots usually have two driven wheels near the center of the robot (relative to front/back) along with caster wheels (described in Chapter 12, "Let's Get Rolling") that can swivel in any direction. Four-wheeled robots usually have a wheel in each of four corners. The difference between the two strategies is that two-wheeled robots consume less power when turning, but are less likely to drive in a straight line. Four-wheeled robots *slip-steer*, which means that all wheels have to slip to the side while the robot turns, burning a lot of power in the process. However, they inherently drive straighter and are more precise at positioning. Another issue is that four-wheel drive systems allow you to usually have at least one wheel on the ground if you're lifted, so that you have some means of escape. Two-wheeled robots are usually at the mercy of the lifter.

Wheel selection and placement issues are also discussed in Chapter 12.

Move Fragile Items Inboard

Protect fragile parts like the radio by moving them away from the edges of the frame. This way, even if the armor is pierced, an opponent will need deep penetration to do any damage to sensitive components. You may also consider secondary armor for batteries and electronics.

Easily Replaceable Batteries

Things can move pretty quickly at a competition, especially as you advance to later tournament rounds. The time between matches can become quite short. You won't have long enough to charge depleted batteries, and if it takes too long to swap batteries, you could end up in a bad situation. You can avoid this by having multiple sets of batteries and making them easy to change.

See Chapter 15, "Choosing Batteries," for more information about mounting and using batteries.

Ease of Maintenance/Repair

Some parts of the robot may require maintenance. Tightening chains, for example, may require access to a tensioning screw. Leaving yourself room for a tool or drilling an access hole will make minor maintenance between matches much easier. Remember, you've got to move quickly in the pits. You won't be under the gun all the time, but it always seems to happen that you have something important to fix when you have the least amount of time.

Shock Isolation

It's rough in the arena. Robots get tossed around and bashed repeatedly during a typical match. You can help minimize the damaging effects of this pummeling by shock isolating critical components. Shock isolating means protecting a component from these sudden jolts by mounting it with a shock-absorbing material like neoprene or natural rubber.

See Chapter 4 for more information about neoprene and other shock-absorbing options.

Leave Yourself Access

The following items should be made easily accessible in your design. Accessibility of these items will allow you to prepare the robot for combat, and quickly and easily activate it once you're in the arena, so you don't hold up the competition. Most competitions have a maximum activation time, which is intended to keep things moving along. Besides, it can be very unnerving trying to perform a lengthy and complicated series of procedures with everyone watching you (and waiting). All externally accessible controls and adjustments should be out of the path of the weapon.

- Master power switch (see Chapter 16, "Wiring the Electrical System")
- Secondary power switch (see Chapter16, "Wiring the Electrical System")
- Radio RESET switch and Tether/Program port, for IFI radios only (see Chapter13, "Choosing Your Control System," and Appendix B, "IFI System Programming and Troubleshooting")
- Internal combustion engine fuel tank fill spout (if applicable)
- Internal combustion engine starting mechanism (if applicable)
- Pneumatic air tank cutoff valve (if applicable, see Appendix C, "Pneumatics")
- Pneumatic system purge valve (if applicable, see Appendix C, "Pneumatics")
- Pneumatic air tank fill nipple (if applicable, see Appendix C, "Pneumatics")
- Hydraulic bypass/purge valve (if applicable)
- Hydraulic pressure test point (if applicable)

Cross-Reference See Chapter 17, "The First Test Drive," and Chapter 20, "Going to a Competition," for information on putting together a pre-flight checklist to help prepare for battle. This list is intended to help you perform whatever lengthy and complicated procedures you have to do outside of the arena, right before your match. This will keep your final activation time short and ensure that you didn't miss any important steps.

Developing Your Design

Now you will begin the process of transforming your rough ideas into a workable plan. Listed below are design strategies for refining you basic weapon idea and selecting which parts to use in the robot for drive and control systems. I'll also show you some techniques for figuring out how to fit everything into your frame, as well as getting an idea of the size of your robot.

Weapon Selection

All weapon systems have their inherent strengths and weaknesses, as described in Chapter 18. Perhaps you want to build the ultimate flipper, or a powerful horizontal spinner. Whatever your weapon, you should keep the following points in mind.

Be Realistic

You will be limited by cost, weight, and complexity. For example, specifying a 40-pound flywheel for a 60-pound robot is unrealistic because it leaves only 20 pounds for frame, armor, drive system, batteries, and other electronics, not to mention the motor that would be required to spin a flywheel of that size. Some materials are more costly than others or require special tools to work with, as described in Chapter 4. Finally, some weapon systems use technologies that require experience and research to be able to implement. When selecting your weapon, it's important to be realistic about your ability to meet all of these requirements.

Spend It All in One Place

I suggest spending your weight budget for the weapon system on making a single really effective weapon, rather than several mediocre weapons. It's rarely the case that you'll have enough weight to build in several powerful weapons at the same time. Also, having multiple weapons may mean driving the robot backwards in order to deploy a rear-mounted weapon, which can become confusing for the driver. While it's true that multiple weapons add versatility, they also add complexity.

Designing Only for the Weapon

Some builders prefer to design primarily around the weapon, making all other systems secondary in priority. This was the design approach I took for the first version of my robot Deadblow. The result was a powerful weapon, but a pitifully weak frame and 1/8" thick armor, which is laughable for a middleweight combat robot by today's standards. The weapon had the most attention, while the armor was almost an afterthought. As I learned, the danger in overbuilding the weapon results in weaknesses in other areas, such as too little armor or drive power. In this sport, balanced designs tend to fare the best in the arena.

Parts Selection

Parts help determine how fast your robot will be, and how it will handle in the arena. Every part should be evaluated for weight. Usually, if it's light and powerful, then it's expensive. Before committing to actually purchasing your parts, you should use mockups or CAD (see below) to try out your selections with the overall design to make sure they work.

Cookbook Design

Some builders prefer the *cookbook* approach, based on what other builders have used. Also called *cut and paste engineering*, this is a perfectly valid strategy as long as you are aware of the weaknesses of the systems that you're duplicating. Also, doing some research in the different weight classes is an excellent way to develop your ideas for weapons and defensive capabilities.

Designing with What's Available

For many builders, cost is the limiting factor that dictates the design. They find themselves scrounging from surplus stores and designing around what's available. Not all good motors need to cost hundreds of dollars. Maybe you have scrap aluminum. The first version of Mouser Mecha-Catbot used an aluminum satellite dish for the internal support structure. Blendo and Ziggo both used Chinese cooking woks for their shells. This type of design saves money for those builders on a limited budget, but requires them to be a bit more clever. It often results in non-ideal weight distributions and other compromises.

Parts Selection for Major Systems

The following components can be selected based on the information in their respective chapters:

- **Motors** - horsepower, built-in gearbox, mounting (see Chapter 9, "Selecting Drive Motors")

- **Other mechanical items (roller chain, sprockets, bearing blocks)** - used to gear down motors and/or distribute drive power to multiple wheels (see Chapter 10, "Mechanical Building Blocks," and Chapter 11, "Working with Roller Chain and Sprockets")

- **Wheels** - diameter and composition (see Chapter 12, "Let's Get Rolling")

- **Control system** - technology choices (see Chapter 13)

- **Speed controls** - power-handling capability (see Chapter 14, "Speed Controls")

- **Batteries** - capacity and chemistry (see Chapter 15)

- **Weapons** - materials (see Chapter 4), technology such as electric motors, pneumatics, and hydraulics (see Chapter 18 and Appendix C)

Laying Out the Internal Components

Having selected the parts that you want to use, the next step is to try and position them in the smallest frame that is feasible using the techniques listed below, including figuring out the design on the computer, shuffling around paper cutouts, and manipulating the actual parts and full-size mockups.

Digital Modeling

Although smaller robots mean less weight, it's a lot harder to fit all the things to do the job into a smaller package. This is where a *computer-aided design* (CAD) program can help out dramatically in figuring out the best placement for all of the motors, batteries, and electronics inside your robot's frame. Unfortunately, this process can become time consuming because you've got to create all the models on the computer yourself from available measurements before you can use them. However, you can use the computer to help you figure out dimensions, and print out exact diagrams to use in construction later.

Paper Shapes

Maybe you aren't computer-savvy, or don't have any CAD software. A quick and simple way to get a tentative parts layout is to cut the shapes of all your parts out of paper and push them around until you find a good arrangement. However, since this is only a 2-D representation, you will need to examine both top and side views of your parts to check for clearance issues, like a part poking through the bottom of the frame.

Make a Pile of Parts

If you're sure of your parts choices and have the means to do so, then buy all the parts for your robot at one time and lay them out on a table. This is the most direct way to figure out the layout because you're working with the exact dimensions for all the parts, and you can easily twist and push them around until you find an arrangement that works well for your needs. Perhaps you don't have the cash to buy all your parts at once, or maybe you're waiting for some to arrive. You can make a *mockup*, or placeholder for the part that's about the right size, with paper or cardboard and some tape.

Recalculate the Weight

If you adjust the frame size even slightly, you must recalculate the weight. Adding 1 or 2 inches to the length of a robot may have a dramatic effect on the weight. The same goes for changing the thickness of an armor plate. For example, if you take a 24" x 24" aluminum baseplate and change it from 1/8" thick to 3/16" thick (only making it 1/16" thicker), you add 3.6 pounds.

Cross-Reference   Weight calculations are discussed in greater detail in Chapter 4.

Designing the Frame and Armor

Now that you have a basic frame size that will fit all of your parts, the next step is to design the supporting structure. By supporting your armor with a well-designed frame, you can maximize its strength, while poor frame design can undermine a piece of armor, and cause it to fail before it should.

Note   I highly advocate making a cardboard mockup of the robot full size, so that you can get a feel for how big it's going to be. It's one thing to design on paper or on the computer, but it's often hard to judge exact size from a computer screen.

Frame and Armor Materials

There are a variety of different materials that can be used for combat robot armor. Not all of them are good, and some have become more popular than others. Aluminum and steel have become popular for frame components, while the most widely used materials for armor are aluminum, Lexan, steel, and titanium.

Cross-Reference The strength, weight, and cost of various materials are compared in Chapter 4.

Frame Types

The most popular design approach is a tubular steel or aluminum frame to which various armor plates are attached. Plates can be easily replaced during a competition, or removed and individually straightened. This strategy relies on a solid frame with a lot of reinforcement (see below). Other competitors use an integrated frame, where the armor actually forms part of the frame, and various armor pieces help support each other. Again, armor plates can be removed and repaired individually. A third option is a mono-body welded aluminum shell, which can either serve as its own cohesive structure, or be coupled to a tubular steel frame by shock absorbers for high impact resistance. Mono-body shells may be difficult to repair at competitions where access to welding is limited.

Full Armor

Most robots have 100 percent armor coverage all around the frame, including the bottom. This is usually a good idea because some competitions have saws that pop out of the floor, while others have challenges that feature water or fire. Protecting vital components while keeping out the elements is a worthy design goal.

Project 1: Introduction to the Project Robot

Now that you've gotten started and know a little about basic design, it's time to introduce you to the project robot that you will be building through the rest of the projects, as well as an explanation of the design choices I made. I felt that it was important to demonstrate how all of these techniques work by actually building a working robot. In keeping with the theme of the book, I used only tools that I could purchase at local hardware stores or through mail order, exactly as described in the chapters.

To allow you to jump directly into becoming familiar with the tools, I completed the design process ahead of time. While working on the design, I struggled with what kind of robot to demonstrate. Should it be a hammer, like my robot Deadblow, or a spinner, or maybe a lifter, perhaps? I came to the realization that a weapon wasn't essential in the project robot, because the real goal is to give you practice using the tools with the techniques described in the chapters, not to make a robot that you can take to competition. Besides, most builders have their own ideas about what weapons they want to put on their robots, and you're encouraged to be creative and find one that suits you (see Chapter 18).

How Much Will It Cost?

Since most beginning builders will be using this project to get their feet wet, I didn't want to commit them to building a huge superheavyweight robot, and consequently spending tens of thousands of dollars. However, I also didn't go the cheapest possible route, because I wanted to give you a realistic building experience. This is an expensive sport, and you should have an idea of what you're getting yourself into. Fortunately, all of the parts that you buy for these projects can be used in a competition robot. After you're done, you can strip the project robot and re-use all of the drive components and electronics for your own design.

The actual cost of the materials, including two sets of batteries, a charger, and the radio control system is a little over $2,000, while the cost of tools is about $1,500. The good news is that, as mentioned earlier, all of it except for about $250 is completely reusable in a robot of your own design. That $250 covers the price of the Lexan and aluminum armor plates. Even so, you could recycle those items into scrap pieces to make parts for another robot.

Note A complete listing of all the parts, materials, and tools used to build the project robot, as well as the list of vendors, is available at the *Kickin' Bot* Web site at www.kickinbot.com.

How Long Will It Take?

This is not the case for all robots, of course, but thanks to pre-fabricated parts, I was able to build the project robot from scratch in five days, including some delivery delays (due to holidays) and design errors. Some of the design errors are explained in the projects to show you how to deal with them and recover gracefully.

The projects are divided into logical breaks that correspond with the chapters. It's designed so that the parts and techniques that you read about in the chapter will then be demonstrated by the project. To get the maximum benefit, the novice builder should read along with the book and only try the project after finishing the accompanying chapter. This means that while it will probably only take a few days of actual building, you'll spend longer reading up on the project before actually doing it.

Caution It is essential before trying any of the projects that you review the power-tool safety guidelines in Chapter 5, "Cutting Metal." The first power-tool project is right after Chapter 6, "Shaping and Finishing Metal," so if you're going in order, the projects will conveniently start after that safety discussion.

To save time, I've also based the projects around paper templates that tell you where to drill the holes and how to cut certain pieces. You will print out the templates and paste them to the metal or plastic to mark your holes, as described in the projects. These templates, along with complete plans and construction pictures, are available for free download from the Kickin' Bot Web site at www.kickinbot.com. The Wiley Web site also has links to evaluate the most recent releases of various CAD software packages that you can use to create drill/fabrication templates for your own robot designs.

Robot Design Specifics

As mentioned earlier, the project robot is small. It's basically a 15" x 14" x 3.5" box with four wheels. I started by scaling the armor, drive, and electronics components for a 60-pound (lightweight class) robot. However, since this robot has no weapon, there were about 30 pounds left over, which is 50 percent of the total weight. The difference between this ratio and my recommended ratio of 30 percent for the weapon system will be offset by the larger frame that would be required to fit all of the weapon components.

One of the design goals of this robot was to get as many of the sensitive components (speed controls, receiver, and so on) as far away from the edge of the robot as possible, as shown in the parts layout in Figure 2.1 below. That way, any penetrating weapons would have to travel quite a ways to get at them. While it might have been more direct to mount components in the space above the battery, or pass wires across the top, I decided to keep the space open, so that the battery can be changed easily.

FIGURE 2.1: Parts laid out on the plan.

The Armor

The 3/16" aluminum base is thicker than really necessary for this robot, since many robots in the 60-pound weight class get away with 1/8" thick bases. However, I want to give you practice cutting thicker plates. Besides, if your competition features saws popping up out of the floor, a 1/8" plate can get cut through pretty easily, potentially damaging a battery or speed control.

The sides are made of 1/2" thick Lexan, which gives you enough thickness to create a screw thread into the side of the piece for mounting to the base. Since most attacks will come from the sides, I opted for a thinner piece of 1/4" Lexan for the top armor to keep the weight down. The strengths and weaknesses of these materials as well as others are discussed in Chapter 4.

The Drive System

The drive system is pretty basic. On each side of the robot, one wheel will be directly driven by a DeWalt drill motor, while the other wheel will be driven by a roller chain and sprocket system connected to the same motor. Although there are many other drive system options (as discussed in Chapter 12), this arrangement will give you an introduction to using roller chain and sprockets.

Since this will be the first fabrication project for many builders, I wanted to minimize frustration by incorporating solutions that I knew were easy to implement, and were going to work well. That's why I selected most of the drivetrain components from Team Delta. Engineer Dan Danknick of Team Delta designs and fabricates custom parts for the combat robot community. He sells the solutions that he designed for his own champion robots online at www.teamdelta.com.

The project robot uses a Team Delta RCM500 DeWalt motor mount for each drill motor. I opted for prefabricated custom RCM200 bearing blocks (also from Team Delta) because I didn't want you to get bogged down and spend a lot of time trying to line up individual bearings for each wheel. In addition, they are designed to integrate seamlessly with the motor mounts, and have a weight of only 8.2 ounces.

The wheels are Colson caster wheels. They are battle-proven to have excellent traction in a number of different arenas. Also, I didn't want to force you to make your own wheel hubs. These wheels were purchased preassembled from Team Delta ready for mounting, with keyways already cut into the hubs. Chapter 12 lists various wheel and hub choices, as well as drive layout considerations.

The Electronics

The control system for this robot is a Futaba 9CAP radio-control package. The accompanying receiver will have its own battery, with a separate on/off switch. Various control systems are covered in Chapter 13, while radio-control programming issues are explored in great detail in Appendix A, "Advanced R/C Programming."

The IFIrobotics Victor 883 speed controls were selected for this project because they are affordable and easy to obtain. They are also single-channel speed controls, so you will get some practice in manually enabling the elevon mixing on the radio as described in Chapter 13. All of the major speed control choices are listed in Chapter 14.

The battery will be a single 3.6 Ahr NiCad Battlepack, made by Robotic Power Solutions. You've also got to purchase a spare battery, a charger, and a power supply for the charger. Battery selection and technology are discussed in Chapter 15.

A West Marine power switch was chosen for this robot's master power switch because of its ease of mounting and shock resistance. Power switches, wire, and connectors and techniques for using them are discussed in Chapter 16.

The Result

In the end, the skeptical observer may wonder why you went to all that trouble to build what appears to be an "unattractive" box with four wheels that is essentially no more than a radio-controlled car. Remind them that this is a training robot, to introduce you to all of the different tools that you will need to master to build your own robots. Besides, this box with four wheels (shown in Figure 2.2) can carry a full-grown adult across the room at 5 to 7 feet per second, and withstand high inertia shocks without blinking an eye. Can your radio-controlled car do that?

FIGURE 2.2: The finished project robot.

Wrapping Up

This chapter has presented some of the most important flaws to avoid and critical features to keep when developing your combat robot idea into a full-blown design. By consulting the individual chapters described in the *parts selection* section above, you can get more detailed information to help you evaluate potential part choices and make educated decisions. You should try to avoid getting too attached to one particular design. It's important to experiment with different arrangements and continue to revise the design as you go along.

As I mentioned briefly in the introduction, this book is structured around helping a beginning robot builder develop the necessary skills for constructing his or her own combat robot, and the next chapter will begin the discussion of the tools.

Tools of the Trade

Tools are here to make your robot-building life easier. The more tools you buy, the easier your job will be, but if you're on a budget, you don't need to buy all of them at once, and you can still get the job done with a minimal set of tools. While some builders prefer to buy as many tools as they can all at once, others go out and buy tools as they need them. It's up to you.

This chapter contains some basic descriptions of the essential tools that are used in the building process. Other tools that have more specific applications, such as power tools for cutting metal, and tools for sanding and shaping, are discussed in their own chapters (see Chapter 5, "Cutting Metal," and Chapter 6, "Shaping and Finishing Metal," respectively).

Safety and Protection Equipment

Safety equipment is the first topic here because it's extremely important. You want the building process to be challenging yet enjoyable. A minor injury can put a damper on your efforts, while a major injury can be a life-changing event. While safety equipment cannot guarantee that you won't have an injury in the shop, it can go a long way in minimizing your risk.

- *Safety glasses* (see Figure 3.1) protect your eyes from flying bits of metal, wood, or anything else. They are usually made out of high-impact plastic, and have side shields, although they are still open on the top and bottom.

Tip

One of the biggest problems with eye protection is that it never seems to be where you're working. It's always on the table in the other room. Even with multiple pairs, they all seem to magically end up in the same place. The way to overcome this is very simple: Get an eyeglass strap from the drugstore and carry them around your neck. This does two good things: It keeps them within easy reach, and it reminds you to use them by hanging in your field of view.

Tip

Keep your glasses clean. In the process of building, you'll use all kinds of oils and cutting fluids that sometimes splash around and get on your safety glasses. Take a look at the lenses. If they're smeared with junk, then use a little Windex and clean them off. When you can see better, it won't feel like the glasses are "getting in the way."

■ *Goggles* (see Figure 3.1) offer more complete eye protection than safety glasses since they usually wrap around the sides and fit closely to the face, creating a tight seal all the way around the eyes.

Caution

Eye protection should be worn at all times during metal cutting and finishing operations.

■ A *face shield* (see Figure 3.1) takes protection to the next step, and incorporates a clear plastic guard that swings down in front of your whole face. Some power-tool operations, such as using a cutoff wheel in a rotary tool, require them for maximum protection.

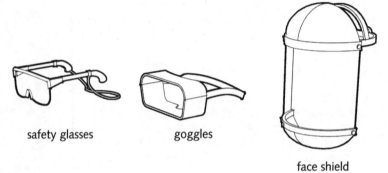

safety glasses   goggles

face shield

FIGURE 3.1: Safety glasses, goggles, and face shield.

■ *Ear protection* (see Figure 3.2) is useful for blocking out especially noisy power-tool operations, which helps you concentrate. It can come in the form of *earmuffs* or disposable foam *earplugs*. The earmuffs cover the whole ear and come as a pair with a headband joining them. Disposable foam earplugs are usually sold in bulk packages.

■ *Mechanic's gloves* (see Figure 3.2) are often required for handling metal pieces that are too hot to touch. They are heat insulated, but also close fitting, not bulky, so that they are less likely to catch or get sucked into a power tool. I recommend the Mechanix brand. They can be obtained from specialty auto supply shops or Eastwood, a mail-order auto supplier (www.eastwoodcompany.com).

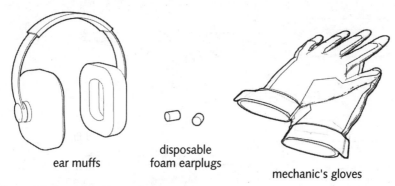

FIGURE **3.2: Ear protection and gloves.**

Hand Tools

Many of the hand tools in this section will be familiar to the weekend mechanic or handyman. They should be readily available at hardware stores. I've broken down hand tools into several different types, sorted by function.

Tip

Keep your hand tools clean. Wipe off dirt and grease. They'll be less likely to slip and cause an injury when you're using them.

Screwdrivers

Screwdrivers are used to turn machine screws. The two major types of screwdrivers are flathead and Phillips, as shown in Figure 3.3. It's important to keep pressure straight down along the shaft. If you don't have enough downward pressure, the blades may slip out of the screw head and cause damage to your part, injure you, or mangle the screw head, making it impossible to tighten or loosen the screw. The screwdriver should be kept perpendicular to the screw head. Make sure to use the correct-sized screwdriver for the screw. Too large a screwdriver simply won't fit. Too small a screwdriver may damage the tool as well as the screw.

- *Flathead screwdrivers* have a single flat blade at the end of a handle.

Caution

Flathead screwdrivers are neither chisels nor pry bars. Using them this way can damage the tip as well as cause the tool to shatter.

- *Phillips screwdrivers* have a pair of blades in the form of a cross at the end for Phillips screws.

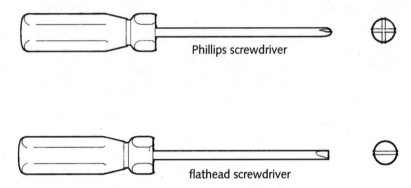

FIGURE **3.3: Flathead and Phillips screwdrivers.**

Wrenches

Wrenches are used like a long lever arm to tighten nuts and hex-head bolts. They give you the necessary leverage to make sure everything is held snug.

- *Allen wrenches* (see Figure 3.4) are used to turn cap screws (see Chapter 8, "Fasteners – Holding It All Together." They are pieces of hexagonal steel that are L- or T-shaped, or are a part of a foldout kit.

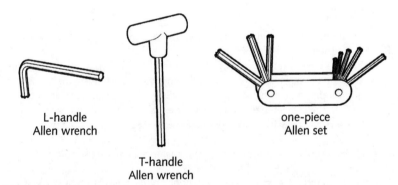

FIGURE **3.4: Allen wrenches.**

- *Combination wrenches* (see Figure 3.5) are used to tighten hex-head bolts and nuts. They combine two types of wrenches: open-end and box-end. The open-end wrench looks like a "C" and allows you to slide the wrench horizontally onto the bolt head, while the box-end side completely encloses the bolt head and must be applied from the top.

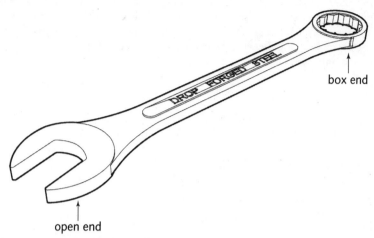

open end

FIGURE 3.5: Combination wrench.

- *Socket wrenches* (see Figure 3.6) are used to quickly tighten hex-head bolts and nuts with a back-and-forth motion. The ratcheting mechanism in the head has a one-way stroke that ratchets in one direction and pulls in the other. It's quicker than a combination wrench because you don't have to remove the wrench from the bolt head to set up for the next stroke. Three major sizes that are common in mechanic's tool sets are $\frac{1}{4}$" , $\frac{3}{8}$", and $\frac{1}{2}$". You can get by with just a $\frac{3}{8}$" set for almost all robot jobs.

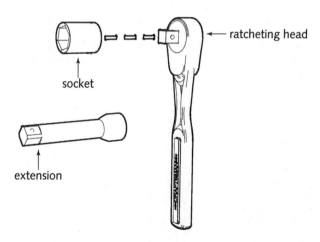

FIGURE 3.6: Socket wrench and accessories.

■ *Adjustable (crescent) wrenches* are used to tighten hex-head bolts and nuts. They have a moveable jaw that can be set to many different sizes, and they are often used in the pits in place of a full set of combination wrenches. Make sure that the jaws are fully tightened before turning the wrench, or you'll end up damaging the bolt or nut. Pulling force should only be applied to the fixed side, as shown in Figure 3.7.

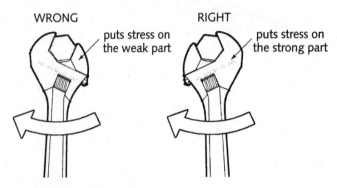

FIGURE 3.7: Adjustable wrench usage.

 All wrenches should always be pulled towards you. Avoid pushing whenever possible. You should consider which way your hand will go if a stubborn nut suddenly releases, or if the wrench slips. By always pulling the wrench, you can prevent crushed and bloody knuckles caused by slamming into something in the path behind the wrench.

Pliers

Pliers are used to grasp or pinch things and hold them securely. Unlike a clamp, which holds something securely to a surface, they are portable and use hand power to maintain tension on the object. The different types of pliers are shown in Figure 3.8.

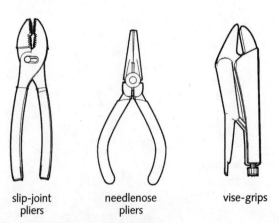

slip-joint
pliers

needlenose
pliers

vise-grips

FIGURE 3.8: Various types of pliers.

- *Combination (slip-joint) pliers* have a large joint in the middle that allows the pliers to be set to several positions, each with a different clamping capacity.

- *Needlenose pliers* have a pointed tip, and are useful for holding small items, and inserting and removing pins and clips.

- *Vise-grips* are locking pliers. The squeezing force of this device when locked can be adjusted by a thumbscrew in the handle. They come in many different styles.

 Note Never use a pair of pliers to hold a nut. It may seem quicker, but you will mangle the sides of the nut where torque needs to be applied. The proper tool for holding a nut is a wrench.

Clamping Devices

Most metal-cutting operations require that the work be firmly held down for the most safety and best results. Clamps are used to hold things securely to a workbench or to each other. They are also used to make sure that bench-top power tools are held to the worktable. You should have at least two to four clamps handy in the shop. Various types of clamps are shown in Figure 3.9.

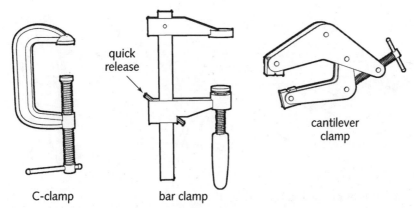

FIGURE **3.9: C-clamp, bar clamp, and cantilever clamp.**

- *C-clamps* are the standard for holding anything down in the shop. They have a C-shaped body and a large screw thread that's used to tighten down the clamp.

- *Bar clamps* have a larger capacity than C-clamps of comparable size. They have a quick-release latch that allows you to loosen and readjust the size of the clamp more quickly than a C-clamp, where you have to continue to turn the handle until you reach the desired size.

- *Cantilever clamps* (Kant-Twist clamps) have the handle parallel to the work, as opposed to C-clamps, where the handle is perpendicular to the work. This lower profile can be an advantage if space is limited. The jaws also remain parallel in any position, and they won't transmit the twisting motion that happens when you tighten a C-clamp.

Tip If the material you're clamping is easily scratched, you can protect the finish by using a scrap of wood between the part and the clamp.

- *Bench vises* (shown in Figure 3.10) are large, heavy vises that bolt to your workbench for securing items that you need to cut or file.

Tip You can make "soft jaws" for the bench vise to prevent the textured, hardened jaws from marring your parts as you tighten the vise down. These can be made by bending some thin aluminum strips to cover the jaws of the vise. Also, there are third-party aftermarket soft jaws that are magnetic so they're easily removed if you don't need them. You can also cut aluminum 90-degree-angle stock pieces.

- *Drill press vises* are small, portable vises that are intended to be clamped to the drill press table for holding small work pieces in place for drilling. Some vises have flanges along the sides with boltholes for clamping, while others have straight sides (see Figure 3.10) and should be clamped down on the vise body.

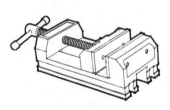
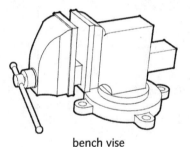

drill press vise   bench vise

FIGURE 3.10: Drill press vise and bench vise.

Marking and Distance Measuring

Before you can do any cutting or drilling, you need to mark your material as described in Chapter 5. The tools here will help you do the job accurately.

- A *felt-tip marker* is useful for marking aluminum pieces. Markers made by Sharpie work best on aluminum. The extra-fine tip should be used for the best accuracy.

- A steel *tape measure* (see Figure 3.11) at least 12 feet long is great for long pieces of stock, while a 12" × 1/2" *steel rule* (also see Figure 3.11) made by General or Starrett with either 1/64" or 1/100" markings should be used for fine measurements.

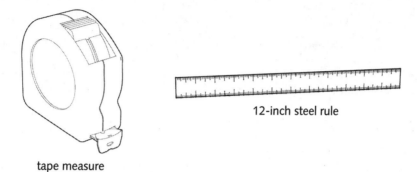

12-inch steel rule

tape measure

FIGURE 3.11: Tape measure and 12-inch steel rule.

■ *Squares* give you a true 90-degree angle for making straight edges. A *machinist square* is useful for making precision marks on small pieces. A carpenter's *combination square* can be used for medium-size jobs, and a 16" × 24" *framing square* is great for marking large sheets. All three of these types of squares are shown in Figure 3.12.

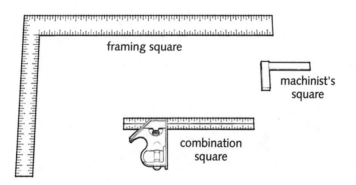

framing square

machinist's square

combination square

FIGURE 3.12: Various squares.

■ *Angle finders* and *protractors* (shown in Figure 3.13) are both used for indicating angles. While a protractor is usually a plastic half-circle with the angle ticks marked out along the circumference, the angle finder is a steel tool that adds an arm that can be set to any angle on the half-circle, making the marking job easier.

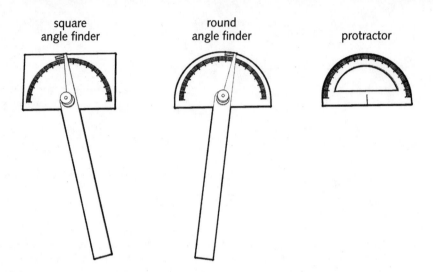

FIGURE 3.13: Angle finders and protractor.

- *Calipers* are used to make high precision measurements, often to 0.001" accuracy. They come in both dial and digital readout versions. *Micrometers* offer higher accuracy than calipers. Both are shown in Figure 3.14.

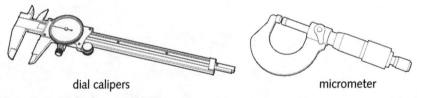

FIGURE 3.14: Dial calipers and micrometer.

Drilling and Tapping Tools

The tools listed here are used in marking and drilling holes and creating threads (tapping) for fasteners to screw into. There are several types of specialty drill bits for different jobs, some of which, such as the Uni-bit and hole saw, are mentioned here. They are described in more detail in Chapter 7, "Drilling and Tapping Holes," along with their correct application techniques.

- High-speed steel (HSS) two-flute spiral twist *drill bits* are generally used for making holes in aluminum and steel. A *stepped drill bit* (also called a Uni-bit) has an increasing radius over several steps, and is used for making holes in thin materials. Both tools are shown in Figure 3.15.

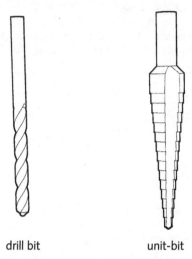

drill bit   unit-bit

FIGURE **3.15: Drill bit and Uni-bit.**

- *Hole saws* (see Figure 3.16) are used to make big holes, and usually have a pilot drill bit to keep the saw centered over the target position. Some models have an interchangeable design that allows you to use many different saw sizes with the same arbor and pilot bit.

- *Countersinks* (also see Figure 3.16) are used to put an angle (chamfer) on the inside of a hole, so that you can use flathead screws (as described in Chapter 8) or to help guide a larger drill bit into the center for enlarging the hole.

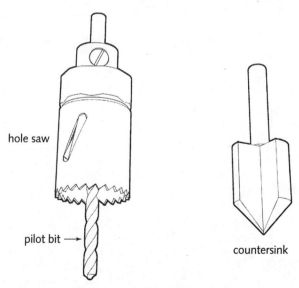

hole saw

pilot bit →

countersink

FIGURE **3.16: Hole saw and countersink.**

■ An *automatic center punch*, as shown in Figure 3.16, is used to mark the positions for holes. It's called "automatic" because it's spring-loaded, so that all you have to do is push straight down on the mark. After a certain amount of pressure, the center punch fires and leaves a little divot on the mark to guide the drill bit and keep it from wandering. Manual center punches are also available.

FIGURE **3.17: Automatic center punch.**

Tip

The center punch is good for making lineup marks on complicated assemblies. If you want to make sure you reassemble something correctly, put matching dots on each side of two pieces that go together. By varying the number (one, two, or three dots next to each other), you can get the correct alignment every time.

■ Spiral point *taps* are used for threading holes for screws. A *tap wrench* holds the tap and allows you to turn it to make the thread. These tools are illustrated in Figure 3.18.

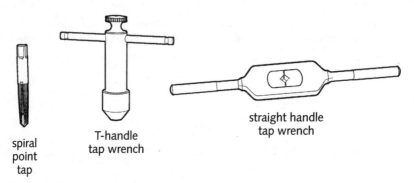

spiral
point
tap

T-handle
tap wrench

straight handle
tap wrench

FIGURE **3.18: Spiral point tap and tap wrenches.**

Cross-Reference Proper techniques for drilling and tapping holes are discussed in Chapter 7.

Miscellaneous General Tools

These general tools should be a part of every robot builder's toolbox, and are collected here since they don't fit into one of the other specialty categories.

- A *hacksaw* (see Figure 3.19) can be used to cut most materials, although power tools are generally recommended to save time and effort. Proper hacksaw technique is discussed in Chapter 5.

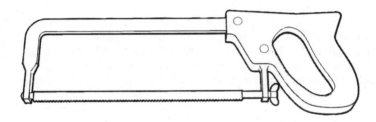

FIGURE **3.19: Hacksaw.**

- *Files* (see Figure 3.20) are used to smooth and shape surfaces. There are many different shapes and sizes, which are discussed in more detail in Chapter 6. They can also be used to remove the sharp edges left over after a metal-cutting operation. A *deburring tool* is used to remove the sharp edge. Deburring tool usage is also discussed in Chapter 6.

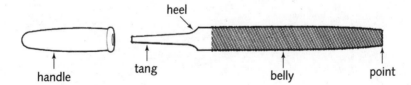

FIGURE **3.20: File.**

- The *ball-peen hammer* (see Figure 3.21) is used for most hammering tasks in the shop. It's different from a standard claw hammer in that it has one flat end and one rounded end. The rounded end is useful for shrinking oversize screw holes.

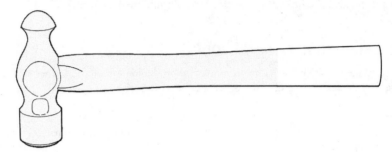

FIGURE **3.21: Ball-peen hammer.**

■ A *scale* (see Figure 3.22) is used to measure the weight of the robot or individual parts. Most bathroom scales are only accurate enough to give you a rough idea. If you're close to the limit, then you need something better. The best scale that I've found that's available to the general public is the Pelouze digital scale, model #4040, which has a 400-pound capacity and retails for around $150 (McMaster-Carr #17295T39).

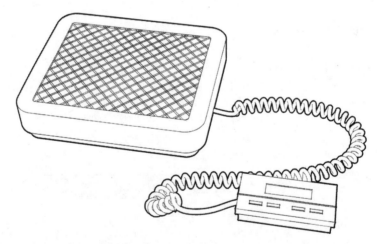

FIGURE **3.22: Pelouze digital scale.**

Tip You can beg your local shipping company to use their scales to weigh your robot. This can be inconvenient if you have a large robot, in which case it might be better to weigh it in pieces.

Note If you're really serious about getting the weight right, you can purchase test and calibration weights from McMaster-Carr. They will help you calibrate your scale so that you have a truly accurate idea of the weight.

Electrical Tools

Not every task in building a robot is mechanical. Electrical systems distribute power internally to give the robot life. The electrical tools listed here are used to prepare all of these internal connections and make sure that they're strong.

- *Flush cutters* (see Figure 3.23) are used to cut wire and cable ties. They have sharp jaws for cutting wire and plastic cable ties only. Using them on steel pins or cable will damage them.

- *Wire strippers* (also see Figure 3.23) are used to remove the insulation from a wire and expose the internal conductor. To use the manual version, you close the diamond-shaped jaws around the wire deep enough to cut the insulation and then pull it away. The automatic version grabs the wire and strips away the insulation all in one motion as you close the handles.

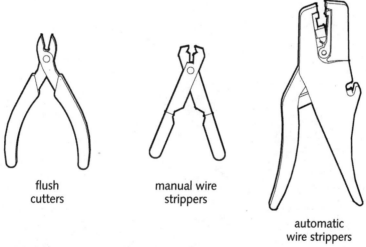

flush
cutters

manual wire
strippers

automatic
wire strippers

FIGURE 3.23: Flush cutters and wire strippers.

- *Crimpers* (as shown in Figure 3.24) are used for installing wire connectors by crushing (crimping) the barrel of the connector onto the bare end of the conductor. The crimpers for fully insulated terminals are used most often in robot building. Some crimpers have threaded bolt cutters built into the handle, so that you can easily cut down long screws. Larger crimpers for battery terminals are also used quite often in electrical system wiring, which is described in Chapter 16, "Wiring the Electrical System."

- A *soldering iron* (see Figure 3.25) is used to apply heat to an electronic component or joint. Solder is applied and melts into the joint, forming a semipermanent electrically conductive bond.

- A *desoldering tool* (see Figure 3.25) is a hand-held vacuum for removing molten solder from a joint. It works by depressing a spring-loaded plunger, and releasing it when the solder is melted enough to be sucked up.

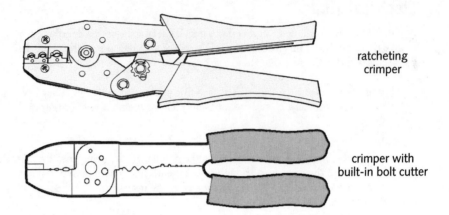

ratcheting
crimper

crimper with
built-in bolt cutter

FIGURE **3.24: Various crimpers.**

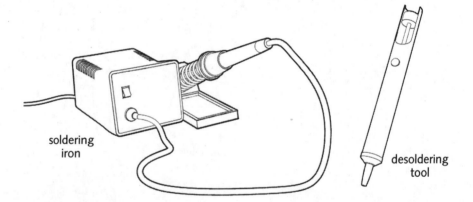

soldering
iron

desoldering
tool

FIGURE **3.25: Soldering and desoldering tools.**

- *Digital multimeters* (DMMs), as shown in Figure 3.26, are used to measure electrical characteristics such as voltage, current, and resistance. This is critical for checking the voltage of batteries, and also for testing for *continuity* (an electrical connection) between two points.

- A *heat gun* (see Figure 3.26) is used when applying heat-shrink tubing for insulation. Make sure to get a low-velocity heat gun, and not a paint stripper or a torch, which may burn the heat shrink. Master makes an excellent heat gun for these applications. If you're careful, you can use a lighter, but this can easily burn the heat shrink.

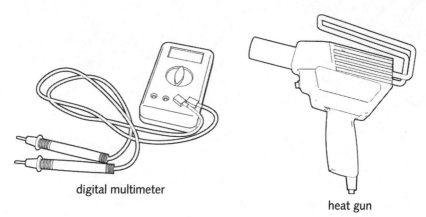

digital multimeter

heat gun

FIGURE 3.26: Digital voltmeter and heat gun.

Cleanup Tools

Cleanup tools are required to keep your shop tools clean and free from small bits of metal (chips) that can cause inaccurate cuts and holes by getting underneath the material that you're working on. Power tools rely on the flatness of their tables for accuracy. If the tables are covered with chips, you'll never maintain accuracy. Also, cleanup tools are essential for keeping the inside of the robot as clean as possible. Chips can mean instant death by short-circuit for expensive speed controls and receivers.

- You will need a small *vacuum*, as shown in Figure 3.27, for removing bits of metal from the robot. A single sliver of metal inside the robot can cause all sorts of mayhem, and could result in you losing a match. You don't need a huge vacuum for this job. Actually, a smaller unit is better, because it's portable, and you can pack it along for use in the pits. The Shop-Vac 1×1 (1 HP, 1-gallon capacity) is an excellent vacuum for robot work. It's small and light, extremely powerful for its size, and best of all, only about $25 from the hardware store.

- An *air compressor* (see Figure 3.27) is extremely handy for getting the inside of the robot really clean by blowing it out with a blast of compressed air. It's also handy in the shop for cleaning bits of metal off of tools between uses. Make sure to wear eye protection whenever you use compressed air.

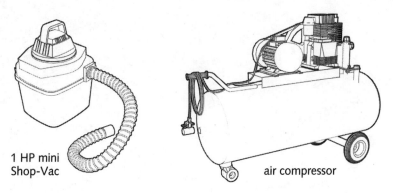

1 HP mini
Shop-Vac

air compressor

FIGURE **3.27: Shop-vac and air compressor.**

Power Tools

Power tools are used to make quick work of jobs that would require a significant amount of labor to perform by hand, especially when working with tough metals, which take effort to cut through.

■ In general, the robot builder should start out with two basic power tools: a cordless drill and a jigsaw, as shown in Figure 3.28. Almost all robot-building jobs can be done to moderate precision with these two tools.

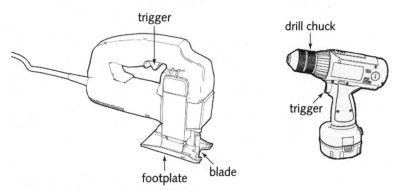

trigger

drill chuck

trigger

footplate   blade

FIGURE **3.28: Jigsaw and cordless drill.**

■ Next on the list would be a drill press (see Figure 3.29). Other tools should be purchased at your discretion following the guidelines in Chapters 5 and 6.

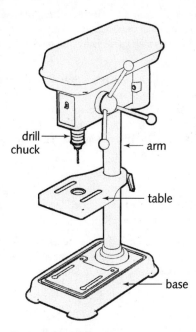

drill chuck

arm

table

base

FIGURE 3.29: Drill press.

Cross-Reference   Power tool safety is discussed in detail in Chapter 5.

Chemicals

Following are descriptions of several chemicals that are useful in robot building, including layout fluid, drilling and tapping fluids and lubricants, threadlockers, and solvents. It's important when using chemicals to have adequate ventilation, which means that they should be used outside or indoors where there is adequate airflow.

- Dykem *layout fluid* is used to paint a thin red or blue layer on the surface of metal that can be scratched away to make a mark for cutting or drilling. Since laser printers and CAD (computer-aided design) software have become so accessible, you can also simply print out layout patterns and use 3M Spray 77 *spray adhesive* to stick them onto a piece of metal.

- TapMagic, Tapmatic Gold, and Edge Liquid are all *drilling and tapping fluids* that are used to cool the metal during these operations. You can also use WD-40 as a lubricant.

- *Stick wax* is used during metal-cutting operations as a lubricant to keep the metal cool, although WD-40 can also be used. *Abrasive belt cleaner*, which usually comes in a waxy bar attached to a wooden handle, is used to clean abrasive belts and discs used in sanding and shaping operations.

- *Threadlockers* are chemicals that are used to make sure that a screw doesn't unscrew itself because of vibration or repeated shock loads. Loctite is a major manufacturer of these compounds, and there are a few different formulas, each with its own holding strength and color. Loctite 222 (Purple) is the low-strength formula, while Loctite 242 (Blue) is the medium-strength version, and Loctite 262 (Red) is the high-strength formula. Threadlockers are discussed in detail in Chapter 8.

- Other products that Loctite manufactures include Loctite 680 *retaining compound*, which is used to hold bearings (discussed in Chapter 10, "Mechanical Building Blocks") in slightly oversize holes, and Loctite 545 *pipe-thread sealant*, which is used in pneumatic and hydraulic systems to prevent leakage from threaded joints.

- Isopropyl alcohol and acetone are *general-purpose solvents* that can be used to clean drilling and tapping fluids from parts, and prep them for threadlockers. They can be used to remove felt-tip marker strokes, and acetone is good for removing layout fluid from a finished part.

Note Acetone can cloud or damage some plastics, such as polycarbonate.

- *Citrus-based solvents* like Goo-Gone are the best for removing sticky residues left by labels or spray adhesives. They are also excellent at removing the stick wax remaining after a metal-cutting operation.

- *Liquid electrical tape* is handy for insulating conductors that are irregularly shaped or would otherwise be too difficult to wrap in electrical tape. Its use is described in detail in Chapter 16, "Wiring the Electrical System."

- *Tire traction compound* is an aftermarket product from the radio-controlled car-racing industry that is applied to rubber or urethane foam tires to improve their traction in the arena. Its use is discussed in detail in Chapter 12, "Let's Get Rolling!"

Machine Tools

Machine tools are very big, heavy, specialized pieces of equipment. They are intended for industrial machine production, and their price (and physical size) are usually outside of the range of the average robot builder.

Note There are, however, miniature versions of these tools made by Sherline, Microlux, and Grizzly that are more accessible to both the budgets and the available space of most builders.

- The *lathe* (see Figure 3.30) is designed to work on round parts such as shafts and wheel hubs. The part is tightened into a chuck (holder) and spun at high speed. A cutting tool is placed in a tool holder on a precision cross-slide that allows the user to take off very little material at a time. (Wood lathes work on a similar principle but unlike metal lathes, they use hand-held tools to carve the wood.) This tool will also allow you to easily drill or bore a hole in the exact center of the part, which is difficult to do by other means.

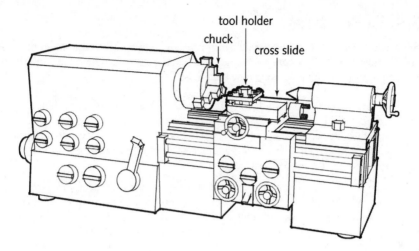

tool holder
chuck
cross slide

FIGURE **3.30: The lathe.**

■ The *vertical mill* (shown in Figure 3.31) is a machine tool that is meant for precisely cutting stock and drilling true holes. Like a drill press, the work remains firmly fastened to the table, while the cutting tool spins in the spindle. Unlike a drill press, however, the mill table can move in X, Y, and Z directions so that the part can be cut in many different directions, but in a precise and perfectly straight way.

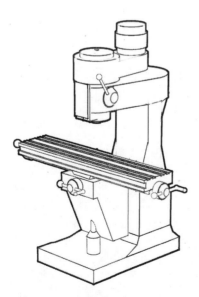

FIGURE **3.31: The vertical mill.**

Special Processes

Some jobs, such as cutting extremely hard materials, require other equipment that's not available to most people. By *jobbing out* these parts (paying someone else to perform the process), you can open yourself up to using new materials.

- *Computer numerical control* (CNC) is a process where a computer essentially controls a vertical mill in an extremely precise way, giving you parts that are essentially ready to bolt together. Usually shops work from CAD drawings, but many vendors can also design parts from hand drawings. Aluminum, steel, and just about any other material can be made into a 2-D or 3-D part.

- The *abrasive waterjet* process is used to cut materials with a computer-controlled high-pressure abrasive stream of water. This process can handle materials that are otherwise very difficult to work with such as titanium. However, it's only a 2-D process (unlike CNC) and can only produce flat parts (such as armor plates).

Wrapping Up

As you can see from this brief introduction, there are a variety of tools available to the robot builder. It's your duty to use whatever tools are available to you to get the job done. Though many tools can perform the same job, some are better suited than others to a specific task. Experience will help you develop a feeling for which tools are better for different situations. Make sure to consult each of the specific chapters for safety tips, and take care of your tools.

Selecting Materials

"**W**hat should I use to make my robot?" is one of the most frequently asked questions that I've received over the years. I always answer that it depends on what part of the robot you're talking about, what weight class you plan to compete in, and how much money you have. You could blindly start building every part of your robot out of steel. No doubt it would be a very strong robot, but you would probably be over your weight limit in a hurry. You could specify titanium armor all the way around, but that would be incredibly expensive. The goal is to find the best material for each part of your robot. This is not necessarily the *strongest* material, but the material that is strong enough to do the job *and* is as light as possible *and* won't break the bank. In this chapter, you'll learn why some materials have emerged as more popular than others for armor, internal structure/frame, and weapons, and why some are considered the "best." I'll begin with an introduction to weight budgeting and some sample weight calculations, which will also help you estimate cost. Then, I'll list the most popular materials used in robot combat in detail, comparing strength, weight, and price. Finally, I'll discuss other raw materials that may be handy in the building process.

Consider the Weight

You should think in terms of a weight "budget." In the end, all of the parts of your robot have to add up and be equal to or less than the competition weight limit for your chosen weight class. Therefore, each part of your robot "costs" you a few pounds out of your precious budget. If you decide to "spend" a large part of your weight on armor, for instance, then you may end up underpowering your drive or your weapon.

In designing my robots, I divide the robot into systems and give each system a percentage of the total weight. The major systems I use for this calculation are: armor/frame, drive, batteries/electrical, and weapons. Usually, the ratio is about 30 percent weapons, 25 percent armor/frame, 30 percent drive, and the remaining 15 percent batteries/electrical.

You should realize, however, that these values are a starting point, and can change drastically depending on the weapon design that you choose. For example, consider Hazard, a horizontal propeller blade spinner in a 120-pound weight class. The designer chose Lexan armor with a drill motor drive, which together was probably about 25 percent of the budget. That weapon's strength depended on rotational inertia, and the heavier the blade, the more damage it could cause. As a result, the weapon system was probably close to 50 percent of the weight, and the final 25 percent comprised a huge array of batteries to power the weapon and drive system. This is a risk that you have to take as a robot builder. You're gambling that slightly thinner armor can be compensated for with a deadly weapon. In the case of the above example, the gamble paid off.

How Can I Calculate the Weight of a Part?

The densities listed in Table 4.1 are in pounds per cubic inch. All you've got to do is find the volume of your part in cubic inches and multiply by the density to get the weight in pounds. Of course, finding the volume of a complicated part may be a bit tricky, but the more accurate your volume is, the more accurate your weight estimate will be. Here are a few examples:

Example: 5/8-Inch Diameter Steel Shaft 6 Feet Long

$$Volume - \pi r^2 L = \frac{\pi d^2}{4} L = \frac{(3.1416)(0.625 \ in.)^2}{4} (6 \ in.) = 1.84 \ in.^3$$

$$Weight - volume \times density = (1.84 \ in.^3)(0.283 \ lbs./in.^3) = 0.52 \ lbs.$$

Example: 1/8-Inch Plate of Aluminum 12" × 18"

$$Volume - L \times W \times H = (18 \ in.)(12 \ in.)(0.125 \ in.) = 27 \ in.^3$$

$$Weight - volume \times density = (27 \ in.^3)(0.1 \ lbs./in.^3) = 2.7 \ lbs.$$

Popular Materials Compared

The Table 4.1 lists several different materials that we use in building robots. Explanations of each of the column headings follow, as well as a detailed analysis of each material's typical physical properties. You are by no means limited to this rather short listing. Think of it as a top-ten list of materials that we use.

Table 4.1   Popular Robot-Building Materials*

Material	Alloy	TS (ksi)	YS (ksi)	E	Density (lbs/cu in)	Cost ($/lbs)
Aluminum	2024-T3	70	50	18%	0.1	$6.75
Aluminum	6061-T6	45	40	12%	0.1	$3.80
Steel	1018	70	60	15%	0.283	$2.50

Material	Alloy	TS (ksi)	YS (ksi)	E	Density (lbs/cu in)	Cost ($/lbs)
Steel (normalized)	4130	106	67	25%	0.283	$8.75
Steel (hardened)	4130	166	161	16%	0.283	$8.75†
Polycarbonate (Lexan)	9043	9.5	9	110%	0.043	$7.25
Titanium (annealed)	6AL-4V	144	134	14%	0.16	$30.00
Titanium (hardened)	6AL-4V	170	160	14%	0.16	$30.00†

*The figures are typical for each type of material. Differences in manufacturer's exact formulations and fabrication methods will yield slight differences in measured material performance. For more information on a specific material, contact the manufacturers.

† The cost listed here is for the annealed material. The hardening process will incur an additional cost set by the heat-treater, based upon the size of the part.

Sources: *Machinery's Handbook*, 26th Edition, and www.matweb.com.

Most metals can be purchased through an industrial supplier, such as McMaster-Carr, an online vendor (as listed in the materials section of Appendix D, "Online Resources"), or at a local metal supplier (check the phone book; it's there between "mental health services" and "metaphysical and occult supplies").

If a metal is provided in its "annealed" state, this means that it has been heat-treated in a specific way to reduce its strength. For many materials, this is the best (and sometimes only) way to machine them. The process called "normalizing" is very similar to annealing. Both annealing and normalizing fall under the category of "stress-relieving" the metal, rather than "hardening."

If a metal has been "hardened," this means that it has been heat-treated in a specific way to increase its strength. This process is usually done in two steps. First, the metal is heated to a very high temperature and then cooled rapidly ("quenching"). This makes the metal very hard, but also very brittle. Next, it is heated to a lower temperature and cooled very slowly to bring back some of the toughness and reduce the brittleness (tempering).

- The alloy number calls out a specific "recipe" for the material, which mixes in other elements with the base element to enhance properties such as strength. For example, you could bend pure aluminum with your bare hands, but add in some silicon and magnesium (in the appropriate amounts) and you've got robot armor.

- Tensile strength (TS) is the *ultimate* strength of the material. That is, the amount of stress required to break the part (puncture, rupture, and so on). It's listed here in thousands of pounds per square inch (ksi).

- Yield strength (YS) is essentially the amount of stress you can apply to a material before it's permanently deformed (bent, dented, and so on). It's listed here in thousands of pounds per square inch (ksi).

- The elongation (E) is a measure of how much the material can stretch before breaking. The higher the elongation, the more flexible the material. It's expressed here as how much (in percentage) of the material's original length that it can be stretched before it fails.

- The density is in pounds per cubic inch.
- The cost is an approximation based on what I've paid for these materials in U.S. dollars per pound.

Aluminum: The Workhorse

Aluminum is the basic material for many robots. It's a good choice for most of the parts for your robot. It's a lightweight material that is strong enough to be used as both structure and armor. It's easy to work with (cutting, sanding, and drilling), and it's also not that expensive on the robot builder scale of cost.

Tip

Although aluminum is great for most parts on your robot, weapon tips should not be fabricated with it. This is a job for steel.

6061-T6

6061 is probably the best material to start with if you're a beginning robot builder. It's a great material for general-purpose items such as brackets, supports, and bearing blocks and other frame components. It's readily available in plates, bars, rods, and all sorts of useful extruded shapes, such as I-beams, angles, and box channels. Note that with extruded shapes, the price will vary from the one listed in Table 4.1, which is for plate stock. I've used 6061 as armor in the past, but if you can afford something else, then you should probably upgrade.

6061 responds very well to machining (as well as cutting, drilling, tapping, and so on). I have welded 6061 with a TIG (tungsten inert gas) welder, but it takes more patience and practice than I can muster. Although I wouldn't trust my welding enough to put it into a combat robot, I have heard that there are a few teams that rely on this method rather than fasteners for joining their armor. 6061 is listed as a heat-treatable alloy, but the -T6 part of the alloy name means that the metal has already been heat-treated for maximum strength when you buy it, so you don't have to send it out for treatment.

2024-T3

2024 is a higher-performance aluminum, with almost twice the tensile strength of 6061 at virtually the same weight. The price that you pay is, well the price. It costs about three dollars more per pound, but you do get an immediate increase in performance. I have, over time, upgraded most of my 6061 parts to 2024.

Machining 2024 is essentially the same as 6061. It just takes a little longer. If you're used to working with 6061 and you happen to get a 2024 scrap, you may wonder for a second why it's taking so long to cut through it. Unfortunately, 2024 is not weldable. 2024 is listed as a heat-treatable alloy, but the -T3 part of the name means that it has already been heat-treated when you buy it.

3003-H14

3003 should not be used in combat robots. It is highly formable and is the choice if you want to sculpt something with metal, such as a car body panel. It's simply too flexible for our purposes and should be avoided.

Steel: Heavy Metal

I generally reserve steel for drive axles and weapons, because of its heavy weight and strength. Steel is the appropriate material for impact weapons like hammers or rotary blades. You need the hardness of the edge and the tensile strength to really impart some damage.

Steel can also be used for the robot's frame, although I don't recommend it for anything under a heavyweight, since you will find yourself overweight in a hurry. It's also too heavy to be used as armor. I've listed below some of the more popular alloys in combat robot building.

Caution

Steel takes a lot longer to machine than aluminum. It requires slower drill speeds and more pressure, and a lot of lubrication.

1018

This material is called "mild steel" because of its low carbon content (less than 0.2 %). This also affects the amount that the material can be hardened, and at this carbon level, it can only be case-hardened, which means that the heat treating process will produce a hard surface shell, while the core of the part remains relatively soft. This can actually be a benefit for parts that absorb a lot of shock, since the surface is hard (for inflicting or resisting damage), but the part overall is not brittle. Normally, this steel is sold cold-drawn, which means that it has been processed after being forged by pulling the material through a press. This process enhances the hardness of the material that you buy off the shelf. It is weldable and can be bent or pounded into different shapes with a hammer.

Note

By definition, 1018 is a carbon steel, not an alloy steel. The difference is that alloy steels contain elements like nickel, chromium, or molybdenum.

4130

The alloy known as 4130 or "chromoly" or "chrome moly" is popular among builders as a material for making frames and weapons. (Chromoly contains 0.80 to 1.10% chromium and 0.15 to 0.25% molybdenum, hence the nickname.) Usually, the frames are welded (by someone who already has experience welding) and then covered with Lexan or aluminum armor. Chromoly has been used in the auto racing industry for constructing race-car frames, and in high-performance bicycle frames, and is well regarded in those circles. It is heat-treatable for hardening (in the case of weapon blades), although if a harder part is your ultimate goal, 4340 is more hardenable due to its slightly higher carbon content.

As a higher-performance steel, 4130 also has a higher price tag and lower availability than mild steel. Its higher tensile strength at the same density makes this steel worth the two to three times more you'll pay.

Cross-Reference

Appendix D lists some aircraft supply houses that can supply chromoly tubing.

Stainless Steel

This alloy of steel has at least 10% chromium, which makes the material highly resistant to corrosion (and rust), hence the name "stainless." It's used in applications that would require a steel to be in contact with moisture, such as sinks in the food industry and silverware. I've seen this material (usually alloy 304) used for armor, but I still believe that it's too heavy to warrant its use in combat.

Tool Steels

For weapon systems, you can get some higher-performance steels called tool steels. These steels are used mostly in creating the manufacturing dies that stamp *lots* of metal parts. They repeatedly form other pieces of metal in big presses with a large amount of force. They are designated by a letter and number, such as M2 or H27. The letter indicates the category of tool steel (hot-work, cold-work, mold, and so on), and the number helps identify the exact chemical composition. Among the tool steels, S7 has gained popularity among builders as an extremely tough material for weapons. By the way, the "S" stands for "shock-resisting." S7 is sold in the annealed state (so that you can machine it) and has excellent hardenability.

Polycarbonate: Fantastic Plastic

It would seem that polycarbonate (GE brand name and common robot builder slang: Lexan) would be too weak for use in a combat robot. It's a plastic, right? Yes, but it's no ordinary plastic; it's what they use for bulletproof glass. I remember seeing a Lexan-armored robot in the early days of BattleBots and thinking that those designers were crazy. That was until I did the math. It turns out that they weren't crazy after all. Lexan really is a great material.

It's only half the weight of aluminum, and has about a quarter of the tensile strength. Why should it be used for armor? Because of the incredible amount of elongation. On impact, the Lexan will begin to deform, and continue to absorb energy while it pushes out of the way.

Heat-Treating Steel

Yes, you can heat-treat your own steel. There is a color chart that you can follow and a bunch of quenching solutions that you can use to bathe your steel in afterwards, but why bother? Not only am I not sure if I can generate a hot enough flame that's consistent over the entire part, but what color is "straw" exactly, and where is the best place to get "rendered animal fat"? (And what do I *do* with the rendered animal fat after I'm done heat-treating?) Luckily for us, most cities have companies that specialize in heat-treating. They deal with steel every day, and usually run nightly batches. It's pretty cheap, and the results are great. Be prepared, however, for some minor warpage depending on the alloy of steel you use and the hardness that you want to go to.

I took the S7 tool steel hammer head for Deadblow to a local bay area heat treater. I delivered it one morning and got it back the next day, perfectly hardened to Rockwell C65 for only $60. This piece had a 3/8-inch slot down the middle with a slip fit over the titanium hammer arm. It fit perfectly when I got it back. I am willing to admit that there are some things that I'm not equipped to do, and this is one of them. It was money well spent.

A Little Lexan Story

In the early days of BattleBots, my first version of Deadblow fought a robot called Son of Smashy (SOS), built by Derek Young of Complete Control fame. SOS had a powerful hammer arm, and he punched through my ridiculously thin 1/8-inch 6061-T6 top armor as if it were tin foil. The day before competition, while Derek and I were both finishing our robots in the pits (hey, we were both rookies), he gave me a 3/16-inch piece of Lexan to cover my radio control receiver with. As it turns out, this piece of Lexan withstood a direct shot from SOS's hammer, deformed to over 100 percent of its thickness, and merely dented the receiver. If that Lexan had been replaced by 1/8-inch aluminum, I would have been dead on the spot. I still have that receiver with the dent in it, and it still works!

Bear in mind that Lexan isn't perfect. It's still a plastic, and it is subject to cracking around high stress areas, such as screw holes and sharp edges. The key is that you should put a chamfer (angle) on any edges and around holes. You don't have to go very deep — just enough to remove the sharp edge from the corner of the material. Any slots or cutouts that you make in the material should have rounded corners. No sharp edges. Your armor has to be well supported. Shock loading a Lexan plate with neoprene works well.

Lexan versus Acrylic

The difference between Lexan and acrylic is astounding. They both weigh about the same, and they're both clear, but that's where the similarities end. Acrylic is more machineable than Lexan, but that's because it's brittle. Incredibly brittle. Too brittle.

Acrylic is appropriate only for the clear display case you plan to put your fighting robot in after you retire it. Acrylic has no place on a combat robot, and woe is the poor combat robot builder who confuses the two.

Cyro Industries, a leading acrylic company, has a high-impact acrylic called Acrylite Plus. Though more impact resistant than Acrylite FF, it's still nothing close to Lexan, as shown in Table 4.2.

Table 4.2   Strength Comparison between Acrylic and Polycarbonate

	Density (lbs/cu in)	Izod Impact Strength (ft-lbs/in of notch)	Elongation
Acrylic (Acrylite FF)	0.043	0.4	4.5%
Acrylic (Acrylite Plus)	0.043	0.75	4.8%
Polycarbonate (Lexan 9034)	0.043	14.0	110%

Source: Cyro Acrylite FF Data Sheet #1121D-0601-10MG

Source: Cyro Acrylite Plus Data Sheet #3059-1201-10VA

Source: GE Structured Plastics Lexan 9034 Product Data Sheet #SPD-2102B

Note The Izod Impact Strength is a measure of the material's ability to resist an impact. There are a few versions of the test, but the "notched" procedure goes like this: A notch of standard size is milled into the specimen, and then a pendulum is swung from a standard height into the sample. When the material fractures, the rebound distance is measured to calculate the amount of energy required to crack the material. These figures are given in foot-pounds per inch of notch.

Titanium: High Performance, High Price

It's high performance. It's expensive. It's difficult to obtain. It's the Ferrari of combat robot metals.

Why Is It High Performance?

Titanium is considered high performance because of its combination of light weight, strength, and shock absorption. It has emerged as the best armor against a spinning weapon. It's difficult to cut with conventional means, which makes it harder to cut through your armor.

Titanium is also great for parts that are likely to take a beating. The hammer arm of my robot Deadblow has been subjected to constant, repeated shock loads as well as attacks by multiple spinning robots, all without a scratch. It's really tough stuff.

Much of its toughness comes from its high tensile strength and incredible ability to absorb shock. It can be heat-treated to increase its strength, but is not suitable as a weapon tip on its own. That's why Deadblow has a titanium arm, but a tool steel hammer head.

Cross-Reference Chapter 5, "Cutting Metal," has some tips for cutting titanium, if you want to try machining it yourself.

What Alloy Should I Use?

Without a doubt, you should be buying Ti 6AL-4V, also known as Grade 5. The 6AL-4V designation refers to the aluminum and vanadium content in this alloy. The other alloy, 40KSI-YS (Grade 2), costs less, but it is *not* considered high performance and should be avoided. By the way, 40KSI-YS stands for 40,000 psi yield strength, which is the same as 6061-T6 aluminum. If it has the same yield strength at 1.6 times the weight, why bother? You've got better things to spend your money (and weight) on.

How Do I Get It?

You should locate a national titanium supplier. I list the contact info for both Tico and President in Appendix D.

Their prices are generally lower than McMaster-Carr, and Tico can waterjet cut to your specifications for reasonable prices. The waterjet service is great because it means that you don't have to cut it yourself. You just send them a CAD file, and they send you back a (nearly) finished part.

Tip Order early so you give your supplier enough lead time. Since a robot builder's needs for titanium are microscopic compared to most orders that come in, you'll usually find that your order will be sandwiched between large orders.

Please make sure that you're really serious about paying $30 per pound before you give these guys a call. I suggest doing all the volume, weight, and cost calculations on your parts first. And if you're thinking about asking for sponsorship, bear in mind that I'm one of Tico's biggest robot customers and after years of dealing with them, I still haven't been able to garner a sponsorship.

Tip

If you're really serious about saving money, and you're willing to cut the titanium yourself, you can find a few online suppliers that will give you a deep discount. The tradeoff is that you're limited to the surplus sizes that they have on hand. Check Appendix D for online metal suppliers.

Other Raw Materials

Not every raw material used in building your robot will be part of armor, structure, or weapons. There are a few other materials that might be handy to keep around.

Neoprene

Sometimes you need a material that's not very hard, and maybe a little springy. Why? Shock-absorbing material. You can get away with a thinner piece of armor if you can mount it with something that will move out of the way of a big impact, and that's where neoprene comes in. The neoprene fastener is strong enough to hold the armor plate in place, but springy enough to allow it to move, thus absorbing the shock.

How is neoprene measured? There are all kinds of flavors of neoprene, from super soft to hard. It's measured on the Shore-A scale, which was developed to be able to compare the hardness of soft things. For example, a rubber band is about 30 Shore-A, while a shoe heel is about 70 Shore-A. You can buy a neoprene sampler kit from McMaster-Carr so that you can get a feeling for the amount of cushioning you'd like to have.

Cross-Reference

Make sure to check out Chapter 8, "Fasteners — Holding It All Together," for some neoprene-based fasteners to mount your armor.

Wood

Wood has been used in a few robots, but never really gained popularity due to arena hazards such as saws and flame throwers. It is a great material to prototype with, especially for testing out your drive system. For example, if you weren't sure if a particular gear ratio or drive configuration was going to work, then you could quickly bolt everything to a plywood board and try it out. Of course, I would highly recommend replacing the plywood board with aluminum in the final robot.

Tip

It's good to have a few scrap pieces of 2 × 4 lumber around to put the robot up off of its wheels, or to drill into (instead of drilling through your part and into the kitchen table).

Brass

Brass is a fairly heavy metal, with a density slightly higher than steel. Unfortunately, its tensile strength is only about the same as aluminum, so that rules it out as armor or weapons material. However, brass is great for shim stock. Shim stock is very thin brass that's used to make up a gap, or correct the height of something. For example, suppose that the hole you made for a bearing is too large, and when you insert the bearing, instead of staying in place, it just falls out. You could cut a thin strip of shim stock (with a pair of scissors) and slip it in around the bearing. Then push the whole thing in, and presto! Press fit and everyone's happy. The brass is very bendable, yet has enough strength to maintain its thickness. It is sold in various thin sheets from Grainger under "shim stock."

Wrapping Up

It's true that there are easier materials you could work with to make things (wood, for example). When it comes to making an armor plate, a support block, or a weapon tip, the metals and plastics listed in this chapter are the best way to go. Remember to balance your choices based on where you need the most strength.

Cutting Metal

As you discovered in the last chapter, there are all kinds of materials that you could use to build your robot. Metal is the top choice in virtually every combat robot in existence, and this chapter will help you safely and accurately cut the metal pieces that will form the basis of your robot. We'll discuss tools and techniques for accurate marking, power-tool safety (perhaps the most important section in the book), and specific options you have in terms of tools for cutting metal (such as the jigsaw and bandsaw).

You don't need all of these tools at the beginning, just a few general-purpose tools to get started. (Some tools are very good at performing one job, while others are more general and can perform a variety of jobs.) As you have more money to dedicate to the project, you can get more specialized tools, which will make your work a lot faster and easier, as well as more precise and better looking.

Accurate Marking

Accurate marking is the first step in creating parts for the robot. It shows you where to cut the material. Careful marking will improve your precision and give you better-fitting parts. If a plate is too long, for instance, then it may not fit where you want it to. This section will introduce the necessary tools and help you develop good marking technique.

Tools

The following tools are helpful in the marking process:

➤ Steel tape measure — at least 12-foot length.

➤ Steel rule — 12" × 1/2" with either 1/64-inch or 1/100-inch markings. General makes a great line of small rules.

➤ Combination square or machinist's square.

➤ Framing square or long straightedge.

➤ Red, green, or blue (not black, which blends in with the metal color) felt-tip marker.

➤ Scribing tool or hobby craft knife and layout fluid, such as Dykem.

Tape Measure Technique

The tape measure is one of the most often used tools for making measurements. However, there is a right way and a wrong way to use it. You will notice that the tip of the tape measure is a little wiggly. (It moves back and forth to help compensate for its own thickness.) To get an accurate inside measurement, the tip is pushed up against an edge so that it is compressed in. To get an accurate outside measurement, the tip is hooked on an edge and pulled so that the tip extends out, as shown in Figure 5.1 below. Also make sure to have the tape as flat on the work as possible. You can pull the tape a little bit past the mark so that it will sit flat on the material.

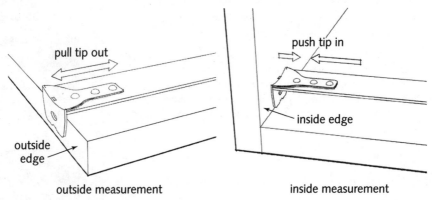

pull tip out

push tip in

inside edge

outside edge

outside measurement

inside measurement

FIGURE 5.1: Tape measure tip.

Tip

One way to get around errors due to the moveable tip on outside measurements is to extend the tape past the edge, and use the 1-inch mark as your start mark instead of the tip. Don't forget to subtract 1 inch from your measurement before making your mark.

Marking the Metal

There are two ways to mark the metal. The old-school way is to paint the metal with a layout fluid such as Dykem. This is a translucent, quick-drying paint that forms a thin red or blue layer on top of the metal. Marks are made by scratching away this layer with a scribing tool or hobby knife. This is a very precise and robust way of marking the metal that will hold up well through all your cutting and sanding operations, although it is a bit time-consuming.

A less precise (though quicker) way to mark the metal is to use a permanent felt-tip marker. The smaller the tip, the more precise your mark will be. The ultra-fine-point permanent marker made by Sharpie is the perfect choice. The marks may get smeared during cutting and sanding operations, but this is usually good enough for most robot parts.

Tip

When you've completed your cut, or if you marked something incorrectly and want to start over, you can get rid of Sharpie marks with isopropyl alcohol or acetone. Dykem requires acetone to remove completely.

Marking Sequence

Instead of a straight tick mark or a dot, use a little "V" with the point at your desired measurement as shown in Figure 5.2 below. This will give you a more precise mark because unless your tick is perfectly perpendicular to the ruler, the two ends of the mark will be at different distances from the edge.

For short lines, make a single tick mark on the work, and then use the combination square to indicate the cut line. The flat end of the square should be held firmly against the side of the stock. Long cuts require a tick mark at each end joined by a long straightedge.

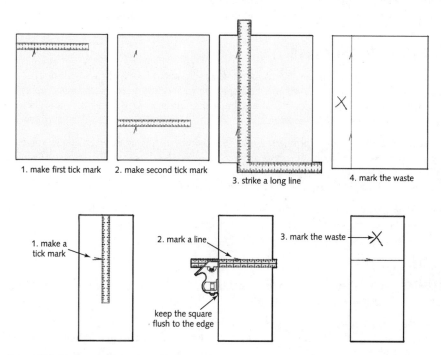

Figure 5.2: Basic marking sequence.

Note

Use a steel rule whenever possible when making marks. It will yield more accurate results than the tape. If the measurement is long, then the tape is the way to go.

X Marks the Waste

It's important to know which side is the waste side (the side that you don't want, not your part) because all saw blades have a thickness (called a *kerf*) that will result in a part that's too short if it's on the *wrong* side. In order to prevent the saw kerf from cutting into the part you want, clearly indicate the waste side with an X. Saw blades should always be lined up to the waste side of your mark. If the saw drifts a little bit, no problem. You can always remove excess material by sanding or filing. If you're wondering why your armor plate is 1/8 inch too short, go back and see if you cut on the waste side or not. Chances are, you lined up the blade on the wrong side, and the kerf cut into your part.

Power-Tool Safety

Before any discussion of using power tools can occur, safety must be addressed. Power tools are incredibly dangerous and capable of great harm. Respect your power tools.

Dressing for Success

Proper dress is required in the shop. Operators should avoid loose-fitting clothing that can catch in a spinning tool. Drawstrings that hang down should be removed. Long hair should be tied back. All jewelry, including necklaces, bracelets, and even wedding bands, should be removed prior to using a power tool.

Make it a habit to wear eye protection whenever you're cutting or sanding metal. You can keep the safety glasses on your head or buy an eyeglass strap that lets them hang around your neck. When you are wearing the glasses, make sure that the strap rests around the *back* of your neck, and does not hang in front, where it could get caught in machinery.

Some tools are also very, very loud. You should have ear protection handy in the shop. It will help your concentration to filter out ear-splitting noise coming from the tool.

Setting Up

All stationary bench-top power tools should be firmly clamped or bolted to the workbench before any power is applied to the unit.

Know your tool before you try to use it. This book provides some guidelines, but every tool has its own versions of the controls and adjustments. Read the owner's manual thoroughly before attempting to use your tool.

Make sure all guards are in place before using a tool. Defeating them does not increase efficiency and only makes it more dangerous for you. Check all adjustments and make sure that they're tight and won't vibrate loose during the operation.

When changing blades, make sure that they're installed in the correct direction according to the owner's manual. Most blades can be installed in more than one way, but will only safely cut in one direction.

Proper lubrication is required for all metal-cutting operations. Using the proper lubricant for cutting metal will prevent tool edges from heating up and losing their temper (hardness), which can lead to premature dulling. A dull blade is very dangerous to use, because it won't cut very well, and may catch on the material, causing an accident. Remember that heat is the enemy of tool edges. Each cutting procedure that follows recommends an appropriate lubricant for the job.

Powering Up

If this is the first time using your new tool, then work your way up. Build up your confidence by making test cuts in various pieces of stock.

The power tool should be off unless you are ready to use it.

Do not start any power tool in contact with the work. All power tools should be allowed to spin up before bringing the blade into contact with the work. After the operation is complete, allow the power tool to spin down completely before withdrawing it from the part.

Don't force a tool to go faster than it wants to. Your feed should be smooth and slow. Let the tool do the work, and cut at its own speed. Both thicker materials and harder materials are naturally slower to cut.

Always be aware of how to turn a tool off. Make a quick mental note when you turn the tool on. If things start to go wrong during a cut, you want to be able to turn off the tool as soon as possible.

Avoiding Distractions

Concentrate and stay focused on the work. All outside distractions should be eliminated while you work with these tools. Any friends, visitors, or observers should be told not to talk to you while you're trying to focus on using these tools.

If there is an emergency or unsafe situation of which the operator is not aware, the proper method for getting someone's attention while they're using a tool is to make yourself visible in their field of vision, and wait for them to acknowledge you.

Observers are permitted in the shop as long as they don't distract you during power-tool operation. All observers and helpers should have adequate eye protection even if they're not the ones using the power tools. Cutting and drilling operations can throw chips very far in many different directions. Observers should stand outside of the potential path of danger (that is, they should not stand directly inline with any cutting blades). Before starting an operation, you should check to make sure that all observers are aware of your intentions (sometimes operations can be startlingly loud) and have adequate protection.

Be Sure

When in doubt, don't! If there is any doubt in your mind about the operation of your machine or the sturdiness of your setup, turn off the machine and leave it off. Take a step back and look at the situation. Consult your operator's manual. If you don't have it, look for your model (or a similar one) online. Most manufacturers now have their manuals online and available for download free of charge.

Do not proceed if you feel unsure about the situation. These are dangerous tools and they can do serious harm if something goes wrong. *Respect your power tools.*

This book is a *guideline*, and you are the ultimate judge of safety in your shop. Stick to the above rules, and use your common sense. Consider the alternative of a lifetime without a finger or an eye.

Tools for Cutting

I've discussed proper marking, and you've had the safety introduction, so now it's time to get down to business. Next, I'll discuss all the major power tools for metal working that you can get from the hardware store (such as the jigsaw and bandsaw) and detailed application techniques and sample setups that you might need, such as cutting in the middle of a big plate, or making a big hole. Pay special attention to the metal blades suggested for each tool. They're the secret ingredient that makes these techniques work.

Blades for Metalworking

For robot-building purposes, having the right blade is more important than having the best machine. You can get away with using woodworking machines for metal work because their precision is good enough to generate perfectly usable parts, but not with their factory-equipped blades. They'll go dull after a handful of cuts, creating a very dangerous situation. If you're going to cut metal, then remove these blades and clearly mark them as "wood only." Put them away so they won't be confused with metal-cutting blades.

Dull blades are a safety hazard. If you find that the blade is generating a lot of heat, and isn't cutting like it used to, then it must be replaced.

Blades for the hacksaw, bandsaw, and jigsaw are specified by *tooth pitch*. Tooth pitch is the number of teeth per inch (tpi). In general, thinner pieces require a finer (higher) tpi because you want at least three teeth engaging the material at any one time to get the cleanest cut. A higher tpi means that the teeth are spaced closer together.

Harder metals like steel also require a finer tooth pitch. Softer metals like aluminum require a coarse tooth pitch, which means that there are less teeth and more space for bits of metal, helping you to avoid clogging the blade.

In general, bi-metal hacksaw, bandsaw, and jigsaw blades are the best for cutting all types of metal, and will last the longest. Of course, they're also more expensive than their high-speed steel (HSS) counterparts.

Blades for circular saws are specified by blade diameter and number of teeth. Miter saws can use the same blades as circular saws, as long as the blade diameter is correct for your model. For cutting nonferrous metal such as aluminum, carbide-tipped blades with at least 40 teeth are recommended. For steel, abrasive blades (discussed in the circular saw section) are used.

Caution Each of the individual tool sections has specific metal-cutting blades that are suggested. Make sure to purchase the right blade. Do not attempt any of these techniques with wood-cutting blades.

Cutting Reference Chart

There are a lot of different ways to make the same part. Each way has its own advantages and drawbacks. The summary below will help you get an idea of the basic tool sequence for different types of parts. Each of these tools will be discussed in more detail later in the chapter.

- Random parts (small or medium-size blocks and brackets):
    - Jigsaw ➔ disc sander ➔ deburring wheel (medium-size parts only)
    - Bandsaw ➔ disc sander ➔ deburring wheel (best results for aluminum)
    - Horizontal bandsaw ➔ disc sander ➔ deburring wheel (best results for steel)

- Armor plates (long straight cuts):
    - Jigsaw ➔ belt sander ➔ deburring wheel
    - Circular saw ➔ deburring wheel (best results, fewer steps)

- Short cuts in long bars or rods:
    - Hacksaw ➔ disc sander ➔ deburring wheel (good general results)
    - Miter saw ➔ deburring wheel (best results for aluminum)
    - Horizontal bandsaw ➔ deburring wheel (best results for steel)
    - Cutoff saw ➔ disc sander ➔ deburring wheel (steel only)
    - Reciprocating saw or jigsaw ➔ disc sander ➔ deburring wheel

The Hacksaw

The hacksaw, shown in Figure 5.3, is one of the most basic and inexpensive tools for the robot builder. It's capable of cutting almost anything that you throw at it, and the blades are cheap and easily changed. Why isn't it used more? Because it relies on *your* strength to do the cutting, which can make a big job feel slow and tedious. In this age of power tools, there are many tools out there that will do a better, faster job for a reasonable amount of cash. Still, the hacksaw deserves a place in your toolbox for emergencies, and various small jobs.

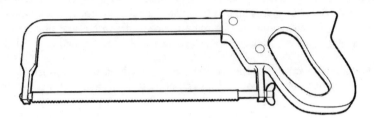

FIGURE **5.3: The hacksaw.**

When using the hacksaw, the stock should be firmly clamped into a vise and held completely immobile. If the stock is allowed to wiggle or otherwise move, the cut will take significantly longer, and will not result in a flat end.

Tip If you're concerned about marring the finish of a part in a vise or clamp, a strip of brass or scrap of aluminum between the work and the clamp will protect the part's surface and keep it looking nice, while holding it securely.

Make sure to choose the correct blade for your material. As with all metal-cutting operations, lubrication in the form of a stick wax or WD-40 is required.

One of the most important things to realize is that a hacksaw cuts only in the forward direction of the stroke. This means that downward pressure should only be exerted on the forward (cutting) stroke. Applying excessive force during the return stroke will only cause the teeth to dull, and may cause the cut to veer sideways. Also make sure to install the blade correctly with the arrow printed on the blade facing forward.

Start by holding the saw with your right hand, and put your left hand on the clamped stock. Guide the blade with your left thumb for the first few strokes, as shown in Figure 5.4. Keep guiding the blade until you've cut a deep enough groove so that the blade won't shift horizontally. Then, bring your left hand to the front of the hacksaw frame and use it to apply even pressure during the cutting stroke.

Although it can be used to cut horizontally, the hacksaw is best used when cutting straight down. The hacksaw also has a depth limitation. Eventually, the work will run into the handle. Keep this in mind when planning your parts.

The Jigsaw

The jigsaw (also called a saber saw, see Figure 5.5) can cut all types of metal, thanks to the ability to easily change blades and the variety of blades available. It's portable, yet powerful. A jigsaw can make straight and curved cuts, both inside and outside. It can cut most pieces that are too big for other tools. The maximum material thickness that you can cut is limited by the blade. For example, some Bosch jigsaw blades will allow you to cut up to 3/4-inch thick aluminum. Because the jigsaw is completely mobile, the width and length of the work is not limited.

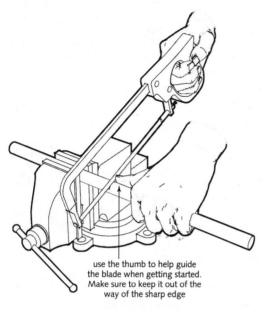

use the thumb to help guide
the blade when getting started.
Make sure to keep it out of the
way of the sharp edge

FIGURE 5.4: **Proper hacksaw technique.**

Keep in mind that all cuts will probably need to be cleaned up. Also, it's difficult to work on small parts with this tool, which is better done with a bandsaw.

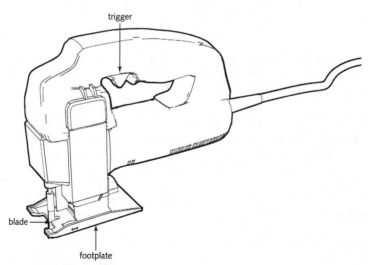

trigger

blade

footplate

FIGURE 5.5: **Parts of the jigsaw.**

What to Buy

Most jigsaws will be able to do a decent job with metal, but as a general rule, more horsepower is better. It will increase your ability to cut different materials and thicknesses. I use a Bosch model 1587AVS jigsaw. It has enough power to take care of all the robot-building jobs that I have. Changing blades is quick and easy. Best of all, it has a trigger lock that allows you to set a speed level and lock the jigsaw on, which is great for long, low-speed cuts in hard materials such as steel.

Blades are relatively cheap (only a few dollars each) and easily found at local hardware stores. I recommend the Bosch T118B blades for aluminum. This is a 14-tpi high-speed steel blade with a Bosch-style T-shank. If you have another brand of jigsaw that uses a universal shank (the other major shank style), use the model U118 blades, which are the same tooth pitch and material but with a universal shank. These blades cost about $10 for a pack of five from the hardware store. For longer blade life, you can use the T118BF, which is the bi-metal version of the blade. It costs a few dollars more, but will last longer. The T118B blades will allow you to cut at maximum speed in aluminum. Make sure to use ample lubrication (WD-40). These blades can also be used to cut steel at a slower stroke rate (about half the maximum rate). Again, WD-40 is required for lubrication during the cut.

Safety Precautions for Jigsaws

In addition to the safety precautions listed in the "Power-Tool Safety" section, the jigsaw requires a few special tips. First, make sure to choose the correct blade for your material. As with all metal-cutting operations, lubrication in the form of WD-40 or a cutting fluid is required.

The jigsaw requires a stable surface to pull against. Make sure to always start with the bottom footplate *flat* on the work surface as shown in Figure 5.6. If you're starting on the edge of a plate, you will have to keep the back end of the saw up so that the footplate is level. Line up to the waste side of the mark. Let the jigsaw reach full speed before bringing the blade into contact with the work.

Adequate support on both sides of the cut is necessary. If the work flexes down while you're cutting, this can pinch the blade (called *binding*). If the blade becomes immobile, it could cause the tool to jump up and down, giving you a bumpy ride.

Finally, apply moderate pressure, but don't force it. The saw should be doing the work. Guide the tool smoothly all through the cut. Continue to the end of the sheet, and let go of the trigger, allowing the blade to slow all the way down to a stop before retracting. Make sure to wipe off the footplate between cuts so that bits of metal don't get underneath, causing the tool to ride up slightly on one side, and giving you an unwanted angle on the edge.

Caution

Be aware at all times of where the blade will be exiting the work. Never, ever reach underneath to see if the blade has made it through the stock.

Working with Titanium

As one of the minority of robot builder's who's worked with titanium, let me share some insights into cutting it. Titanium has the ability to locally work-harden, which means that once the material reaches a certain temperature, it will harden in that spot. In our case, this unfortunately happens in the exact spot you're cutting, as a direct result of the cutting process. As the titanium begins to harden, the blade will have a tougher time with the material, which generates even more heat in a vicious cycle. It becomes harder and harder to cut, and will end up taking you ten times longer to get through, and probably dull several blades in the process.

What's the solution? Make sure it doesn't heat up. This means cutting at a very low speed with a lot of lubrication. For example, given similarly sized pieces of aluminum and titanium, what will take you 1 minute to cut in aluminum should take you 10 to 15 minutes to cut in titanium. If you attempt to cut at a faster speed, you'll dull your blade in seconds. Your patience will pay off. Don't attempt to cut hardened titanium. You should only be working on annealed material (as it should be provided from your vendor), and you should have Ti 6AL-4V (Grade 5) material, as described in Chapter 4, "Selecting Materials."

I recommend using a jigsaw for this cutting operation because replacement blades are cheap and easily installed. (You don't want to dull a $60 circular saw blade on titanium.) With the Bosch 1587AVS jigsaw, you can set the trigger lock to a slow speed, where it will stay for the duration of the cut without your having to constantly hold the trigger down, which is easier on your fingers. I use a trigger lock setting of "B" (low speed), a blade orbit setting of "O" (orbital, as recommended by the manufacturer for cutting metals), and Bosch Model T118B blades (about $10 from the hardware store for a pack of five). The slow speed coupled with the aggressive tooth pitch (for a harder material) gives good results. Also, the cutting edge should be constantly flooded with WD-40. The lubricant evaporates constantly—don't be alarmed if you see smoke. If you start to see white sparks consistently, then it's time to stop and change blades.

Also, drilling holes in titanium should be done only with solid carbide drill bits, not high-speed steel (HSS). They're significantly more expensive than standard HSS bits, and range from about $10 for a 1/8-inch bit to over $40 for a 3/8-inch bit, but you should only need one or two sizes. Set the drill to a low speed and use lots of lubrication. Do not attempt to tap titanium. You will more than likely bury the tap in the material and have to grind it off later.

Be careful of fire with any and all titanium operations. The chips are flammable and can self-ignite if they get hot enough. For example, you could take a long spiral chip (which might be the result of a drilling operation), and light it with a lighter and it would burn bright white, like a fuse, all the way to the end. It's a good idea to have a fire extinguisher nearby, and to clean up and properly dispose of your chips immediately. Bear in mind that Tico Titanium not only sells 6AL-4V plates, but can also cut any pattern you want (including holes) with an abrasive water-jet for a nominal additional charge. The waterjet produces an extremely clean edge with no cleanup required. Just something for you to consider.

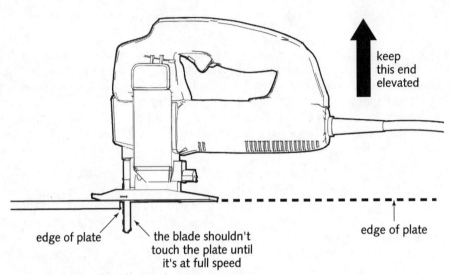

keep
this end
elevated

edge of plate   the blade shouldn't   edge of plate
                     touch the plate until
                     it's at full speed

FIGURE 5.6: Jigsaw lineup.

Setup Tips for Different Situations

There are a variety of cutting situations where you can apply the jigsaw. Described below are some common setups that you may encounter.

Normal Cuts

For most cuts, the work can be clamped to a table or supported by a sawhorse so that the waste side will fall away. If you're cutting in the middle of a large plate, make sure that the plate is supported in four places: two on one side of the blade, and two on the other, as shown in Figure 5.7. That way, the work will remain stable (that is, not close down, pinching the blade) during and after the cut.

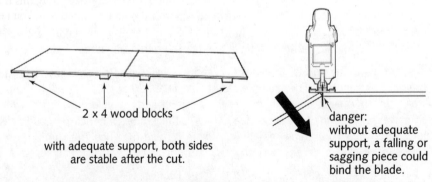

2 x 4 wood blocks

with adequate support, both sides
are stable after the cut.

danger:
without adequate
support, a falling or
sagging piece could
bind the blade.

FIGURE 5.7: Setup for cutting in the middle of a plate.

Long Straight Cuts

This setup requires a long straight piece of wood or aluminum to help guide the saw. Mark the cut line on the work, and measure the offset from the edge of the footplate to the blade. Mark another line with this offset distance for your guide. Clamp the guide in place. When cutting, make sure that the side of the footplate is held firmly against the guide. Be aware that this type of cut can drift. It's best to clamp the guide on the usable side of the work, as shown in Figure 5.8, so that if the saw drifts, it will go into the waste side, which you can sand or file flush to the correct mark.

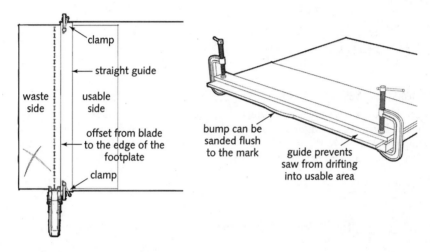

FIGURE 5.8: Straight guide jigsaw setup.

Inside Cuts, Big Holes, or Circles

In woodworking, it is common practice when making a large hole or interior cutout to *plunge* into the work with the jigsaw by tipping the saw up on the front edge of the footplate and slowly lowering the blade into the work. Because metal is much harder than wood, this is not recommended. Fortunately, the recommended practice is pretty easy to implement. You need to drill a starter hole that's big enough to fit the blade into. Insert the blade into the hole, which will allow you to rest the footplate flat on the work. Make sure that the blade is not in contact with the work when you start the saw. Open up the hole to your mark, as shown in Figure 5.9. If the circle is the part that you want, then put the starter hole on the outside of the important area.

Thin Stock

Thin stock (less than 1/16 inch thick) is a special situation because it can flex so easily. You'll probably end up with a pretty ragged edge if you don't keep the material from flapping around. The solution is to clamp some 1/4-inch plywood on top and bottom, sandwiching the material. Make sure to transfer your marks to the plywood and cut both the plywood and the metal at the same time.

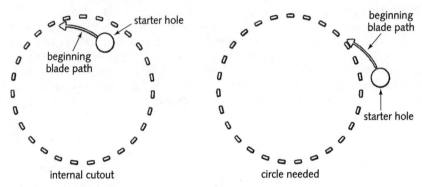

FIGURE 5.9: Starter hole placement.

Reciprocating Saws for Steel

The reciprocating saw shown in Figure 5.10 is a coarse tool. It's kind of like the big brother to the jigsaw. Usually it's used for demolition jobs such as tearing down walls. However, the reciprocating saw is useful to the robot builder because it can cut through steel really well. The only problem is that since it's hand-held, the straightness of your cut will depend on how steady you are.

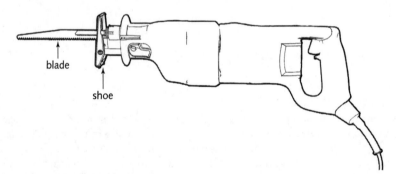

FIGURE 5.10: Reciprocating saw.

Much like the jigsaw, the reciprocating saw needs to be firmly planted. It needs something to pull against, so the shoe should be held firmly against the work. Of course, adequate lubrication should be used and the work should be clamped to a worktable or in a bench vise.

You can use many of the same setups as the jigsaw, but bear in mind that the reciprocating saw is best used with the tool held horizontally and the blade cutting straight down. Also, you'll be pulling the tool towards you, rather than pushing away, as with the jigsaw. Reciprocating saw blades tend to be much wider than jigsaw blades, so your turning radius is significantly reduced.

The Bandsaw

A bandsaw is a stationary tool with a tilting table. The blade is like a continuous loop of hacksaw blade. I consider the bandsaw to be a shop necessity for cutting medium and small pieces such as bearing blocks, brackets, and other aluminum parts. You can easily and precisely maneuver small pieces around the blade, giving you great control over your cuts.

With the bandsaw, you can cut curves as well as make fairly good straight cuts, although all cuts will probably need to be cleaned up with a sander. Unfortunately, the bandsaw can only cut along the outside of a piece, because the blade is a continuous band. For circular and rectangular holes, you will need a jigsaw. As shown in Figure 5.11, the capacity is limited by the distance from the blade to the arm.

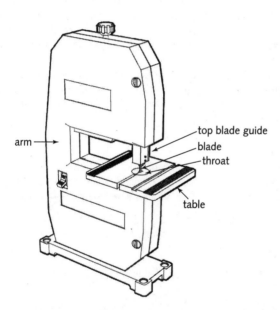

FIGURE 5.11: Parts of the bandsaw.

A standard bandsaw can easily handle aluminum and plastic, but not steel. The blade speed is too high for cutting steel. Bandsaw blade speed is measured in surface feet per minute (or sfpm). For steel cutting, the speed should be 75 to 175 sfpm for carbon steel blades, and 150 to 300 sfpm for bi-metal blades. In contrast, aluminum should be cut between 600 to 3,000 sfpm. Unfortunately, most hobby bandsaws turn too fast for steel cutting, since they run with a no-load blade speed of 1,500 to 3,000 sfpm. Although this may rule them out for steel, that's an ideal speed range for aluminum.

Note | Bandsaws that are designed to cut steel are discussed at the end of this section.

What to Buy

You can get by with an inexpensive bandsaw as long as you have the right blade. I recommend the Ryobi BS901 bench-top bandsaw, which is available for under $100 from The Home Depot and other major retailers. Note that there are other usable bandsaws in this price range from manufacturers like Delta and Craftsman.

For the Ryobi bandsaw, a 10- to 14-tooth blade is the best available choice. Because you will be cutting a variety of thicknesses, this should be a sufficient tooth pitch. And it's pretty cheap, with a cost of about $17 from McMaster-Carr (part #4179A817). If you already have a bandsaw, you should find out the correct blade length and buy a blade with similar specifications to the blade mentioned above.

 Note Unfortunately, I have become aware that some Ryobi BS901 bandsaws shipped with screws that will not allow the table to sit flat. If you have one of these bandsaws, you can simply grind away some of the plastic case with a Dremel rotary tool to allow the table to sit flat. See the Web site at www.kickinbot.com for detailed instructions and pictures if you're unsure how to proceed.

More expensive bandsaws will have greater arm distances for handling larger pieces, and higher-horsepower motors, which will allow you to cut faster without overloading the motor. Unfortunately, the price difference between the bench-top saws and the standalone saws can be significant.

Safety Precautions and Usage

In addition to the safety precautions listed in the "Power-Tool Safety" section, bandsaws require some special safety measures. First, make sure to choose the correct blade for your material. As with all metal-cutting operations, lubrication in the form of a stick wax or WD-40 is required.

With a bandsaw, the work should always be held down firmly with at least one flat edge on the table. Do not try to hold a piece at an angle. It will rip the work right out of your hands and possibly pull your hands into the blade.

Also, before you start the saw, lower the top blade guide to just above the work, leaving as little of the blade exposed as possible.

Apply even pressure on both sides of the cut as shown in Figure 5.12. Keep your fingers out of the path of the blade at all times. When pushing, consider your hand position and the direction that you're pushing in, and make sure that if your hand slips, it won't go into the blade. If the blade starts to bog down significantly during cutting, you'll be able to hear it. Don't push so hard—let the saw cut at its own pace.

If a piece gets jammed in the throat, turn off the power. Wait for the blade to slow to a complete stop before attempting to remove any material.

Finally, if you see that you're wandering off your mark during a cut, you should *turn* the part, rather than trying to push it sideways. Pushing the work sideways will only generate heat and cause greater wear on the blade and the blade guides.

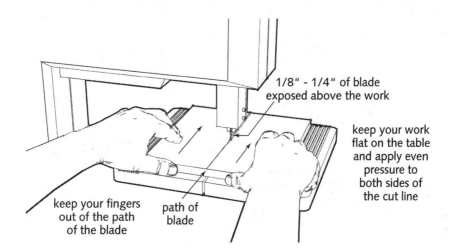

1/8" - 1/4" of blade
exposed above the work

keep your work
flat on the table
and apply even
pressure to
both sides of
the cut line

keep your fingers
out of the path
of the blade

path of
blade

FIGURE **5.12: Bandsaw guide height and pressure application.**

Note You will have to adjust your bandsaw to make sure that it tracks straight. There are several reasons why a bandsaw will want to wander away from a straight line. You should check your owner's manual to determine how to correct this behavior.

Setup Tips for Different Situations

Listed below are some of the most common setups for working with the bandsaw.

Angled and Straight Cuts with the Miter Gauge

Most bandsaws come equipped with a *miter gauge*. The miter gauge can be set to any angle and has a bar that registers in a slot in the table that's parallel to the blade. This will allow you to make clean angled cuts. When the miter gauge is set true, you can also use it to help you guide straight cuts. Stock can be clamped directly to the gauge or held against it by hand (see Figure 5.13). As you push the stock through, make sure that your fingers are out of the path of the blade.

Note Round stock should be clamped to the miter gauge because it may catch and try to rotate out of your hands. Make sure that the part is resting flat on the table.

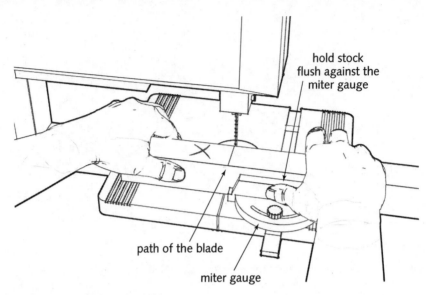

hold stock
flush against the
miter gauge

path of the blade

miter gauge

FIGURE 5.13: Using the miter gauge.

Small Pieces

Although the bandsaw excels at cutting small pieces, work that is small enough to become trapped in the throat requires a special setup. You can make a *zero-clearance* table with a thin piece of MDF or plywood. Zero-clearance refers to the fact that the gap around the blade will be very small. Make sure that the piece of wood is about the same size as the bandsaw table. Center the wood on the blade and cut to the center of the wood as shown in Figure 5.14. Turn off the saw and clamp the piece of wood to the table. With this setup, your chances of getting the work caught in the saw will be minimized. If the part is too small to hold, then you should be using some other tool.

Long Pieces

Long pieces will require extra support. Have an assistant help you. It should be made clear to the assistant that you will be controlling the cut entirely. Their only job is to help support the work, not guide it. Make sure that you guide the stock as described above, with even pressure on both sides of the blade.

Horizontal Bandsaws for Steel

The horizontal bandsaw is mounted to a base with an attached miter box, as shown in Figure 5.15. As it cuts, the saw pivots on a hinge and its own weight provides the downward pressure for the cut. Since steel takes a lot longer to cut than aluminum, the advantage with this saw is that you don't have to sit there holding the part to the blade. You just clamp it in the built-in vise and let the saw chug away until it's done.

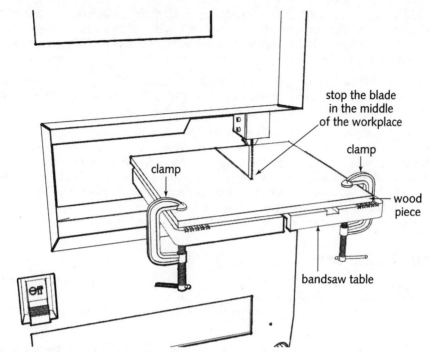

stop the blade
in the middle
of the workplace

clamp

clamp

wood
piece

bandsaw table

FIGURE 5.14: Zero-clearance table setup.

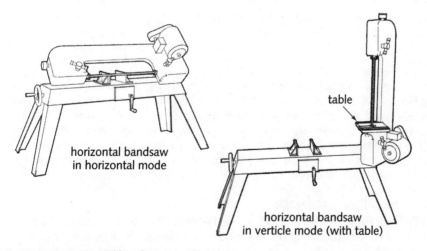

horizontal bandsaw
in horizontal mode

table

horizontal bandsaw
in verticle mode (with table)

FIGURE 5.15: Horizontal bandsaw.

These bandsaws run at a slower blade speed of around 200 sfpm, which is more beneficial for steel. They are the cleanest, most precise way to cut a steel shaft (other than using an expensive lathe), often requiring only deburring to complete the process. Unlike abrasive chop saws, there are no sparks and smoke.

Since the arm of the bandsaw is inline with the blade, there is no stock length restriction like on a standard bandsaw. The maximum thickness of the stock will be determined by the size of the vise and the amount of blade that the saw can expose.

In the $150 range, there is a line of low-cost portable bandsaws with stands available through Grizzly and Micro-Mark. (Not the Milwaukee PortaBand, which goes for around $300 not including the base.) They are lightweight and good for cutting steel shaft and bar stock, but not parts. In the $200 range, you can buy much heavier, nonportable bandsaws by Grizzly, Enco, and Rong Fu that can be used in either horizontal or vertical positions. The horizontal position is like the others, but in the vertical position (along with a removable table), you can cut steel plate just like aluminum on a regular bandsaw, but at a much slower rate, of course.

The Circular Saw

The circular saw is a portable tool that has a large circular cutting blade. The best use of this saw is for cutting sheet stock (like armor plates) of any length really straight. It can cut all types of metal with the proper blade. Unfortunately, the circular saw is only really good for straight cuts. Sorry, no curves.

You can do the same job with a jigsaw and a straight guide. The difference is that a circular saw gives a superior finish edge that requires very little or no cleanup, just deburring.

What to Buy

Most popular circular saws will handle 7¼-inch blades. Prices range from $50 to $75 at most hardware retailers. I've used the Ryobi model CSB131 circular saw, which costs $67 at The Home Depot. More horsepower will give you the ability to easily cut a wider variety of materials much faster.

Circular saw blades tend to be pretty expensive. A good saw blade for nonferrous metals such as aluminum can be anywhere from $50 to $200. Make sure to get a carbide-tipped blade for metal work, such as the 7¼-inch 40-tooth blade from McMaster-Carr (part #6910A51) used in the projects. An *abrasive disc* is used for steel cutting. This is not a metal disc, but a fiber one that has abrasive grit impregnated into the surface. Be warned that the abrasive disc will throw a shower of sparks and is not the most accurate way to cut steel. It will also generate a decent amount of smoke, so make sure that you have adequate ventilation.

Safety Precautions and Usage

In addition to the safety precautions listed in the "Power-Tool Safety" section, circular saws need special consideration. First, make sure to choose the correct blade for your material. As with all metal-cutting operations, lubrication in the form of a stick wax or WD-40 is required.

This is one of the loudest tools in the shop. Ear protection is a must. Besides, it's much easier to concentrate without all the noise.

Start on the edge of the sheet and line up the cut to the waste side of your mark, as shown in Figure 5.16. A circular saw's kerf is usually about 1/8 inch, which is larger than the other types of saws in this chapter. If you put the saw on the middle of your mark, it will cut off more than you want. When starting on the edge of a plate, the front edge of the footplate should rest on the work. You will have to keep the back end of the saw up so that the footplate is level.

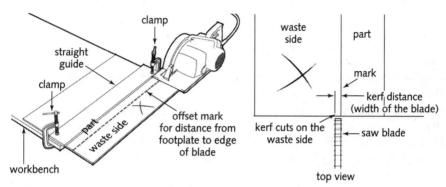

FIGURE 5.16: Circular saw lineup and straight guide.

Note

The setup tips for different situations for the circular saw are the same as with the jigsaw. Remember to pay special attention to supporting the stock so that you don't pinch the blade. Refer to that section for diagrams and detailed descriptions.

It's essential that you let the blade fully spin up before bringing it into contact with the work. Once the blade has spun up all the way, slowly and gently bring it into contact with the work. Do not jam the blade into the work, or it may catch and throw the saw.

Never back up with the saw. Always cut in the forward direction. Backing up may cause the blade to catch and cause the saw to jump, which may result in an injury. If you need to reposition or recut some part of the work, release the trigger and let the saw come to a complete stop before removing it and starting the cut over.

Although the saw is pretty precise freehand, you can increase precision by clamping a bar to the work for the saw to follow. If you see yourself veering off the line, do not try to correct by turning the saw. Instead, remove your finger from the trigger and let the saw spin down completely. Pull the saw out of the work and start again at the beginning with a slightly different angle. The path of this tool cannot and should not be corrected during the cut, or you may bind the blade and throw the saw.

Apply moderate pressure, but don't force it. The saw should be doing the work. Keep the saw moving as smoothly as possible all the through the cut. Continue to the end of the sheet, and let go of the trigger, allowing the blade to stop before retracting. Make sure to wipe off the footplate between cuts so that bits of metal don't get underneath, causing the tool to ride up slightly on one side and giving you an unwanted angle on the edge.

While it's common in woodworking to make plunge cuts with a circular saw, it's not recommended with metal, because it's a much harder material and the likelihood of kickback is significantly higher.

Finally, as with the jigsaw, you have to be aware at all times of where the blade will be exiting the work. Never, ever reach underneath to see if the blade has made it through the stock.

The Miter (Chop) Saw

The miter saw has the same type of blade as the circular saw, but instead of being portable, it's mounted to a stationary base on a spring-loaded hinge. The saw can be swiveled to various angles so that you can make straight or angled cuts up to 45 degrees easily and with great precision.

While other tools like the hacksaw and jigsaw can be used for these cuts, nothing beats the miter saw at making straight cuts in aluminum bar stock and extruded shapes. As shown in Figure 5.17, there's no limitation on the stock length — only the thickness and depth of the stock. Unfortunately, this tool is only good for straight cuts — it can't do curves.

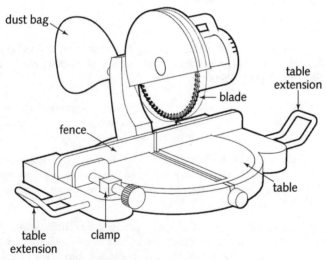

FIGURE 5.17: Parts of the miter saw.

What to Buy

Purchasing a miter saw with a 10-inch blade or larger will give you the widest variety of blade types to choose from. The basic miter saw has the ability to change the angle of the cut up to 45 degrees. Compound miter saws add the ability to tilt. Compound slide miter saws have the ability to tilt, as well as adding a slider that allows the saw to move forward and back, which increases the capacity (thickest piece you can cut). I use the Ryobi TS1302 miter saw, which is a basic 10-inch saw available from The Home Depot and other retailers for around $100. The blade I have is a 10-inch diameter 60-tooth nonferrous metal carbide-tipped blade from

McMaster-Carr (#6910A55). Like circular saw blades, the blades for miter saws are pretty expensive. This one costs a little over $50, but when used properly, cuts extremely smoothly.

Although an abrasive blade for cutting steel could fit in the miter saw, many manufacturers warn against using them in their saws, because the abrasive grit and steel particles that are produced may get into the saw hinge and motor, and shorten the saw's lifespan. If you do decide to use an abrasive blade in the miter saw, make sure to vacuum out the saw and empty the dust bag, so that there is no sawdust that might ignite from the sparks.

Note Abrasive blade cutoff saws that are designed specifically for cutting steel are described at the end of this section.

Safety Precautions and Usage

In addition to the safety precautions listed in the "Power-Tool Safety" section, miter saws have their own set of rules. First, make sure to choose the correct blade for your material. As with all metal-cutting operations, lubrication in the form of a stick wax or WD-40 is required.

And just like the circular saw, this is a really loud tool. Ear protection is highly recommended, and will help your concentration.

Line up the cut with the saw unplugged and your finger off the trigger. Bring down the blade so that it's to the waste side of your mark. Also make sure that the stock is pushed all the way up against the fence and is flat on the table. When the stock is correctly lined up, securely clamp it in place with a clamp that's designed to be used with the saw, or a C-clamp.

Caution Many woodworking books show the wood being held by hand during a cut. This is not a safe option for the robot builder. All metal pieces must be clamped and lubrication must be used.

Make sure that the miter angle knob is firmly tightened after adjustment, so that the saw will not swivel during a cut.

The tool should only be running when you are ready to cut. Do not make any adjustments to your clamps or setup while the saw is running. Power down and allow the saw to come to a complete stop before making any adjustments.

It's essential that you let the blade fully spin up before bringing it down into the work. This means that the saw should be in the fully up position with the blade cover in place before you start it. Once the blade has spun up all the way, slowly bring it down and gently bring it into contact with the work. Do not jam the blade into the work or it may catch and throw the piece.

Finally, apply moderate pressure, but don't force it. The saw should be doing the work. Guide the saw in a smooth and steady motion all the way down. When your cut is complete, let go of the trigger, but keep the blade where it is in the fully down position. Let it slow all the way down to a stop before retracting.

Caution If the piece is too small to be clamped, then it should be cut with something else.

Setup Tips for Different Situations

As with the other tools described in this chapter, the miter saw can be used in many different ways. Some of the most common setups are introduced below.

The Clamp Side and the Free Side

Every piece of metal stock to be cut with the miter saw should be clamped on one side, as shown in Figure 5.18. The clamp should securely hold the work down flat on the table and flush against the fence. The other side of the blade (called the free side) should be well supported, but not clamped. It should be allowed to move so that it won't bind (pinch) the blade and cause a kickback. Make sure that the table is level so that the free side won't roll when the cut is complete.

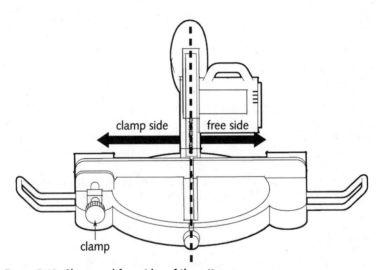

clamp side   free side

clamp

FIGURE 5.18: Clamp and free sides of the miter saw.

Cutting Long Pieces

Long stock should be supported on both sides of the cut, as shown in Figure 5.19. The free end should be supported so that it doesn't pop up at the end of a cut. The clamp side should also be adequately supported so that the work doesn't bow during the cut.

Cutting Multiple Pieces to the Same Length

Some saws come equipped with a stop block. This is a small block that can be secured at the correct length so that you don't have to measure every time. Just bring the stock up to the stop and clamp it in place.

Caution

The stop block should be on the same side as the clamp, as shown in Figure 5.20. Do not hold the free side. It may bind and kick the stock back at you.

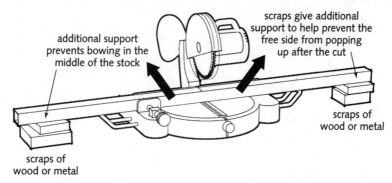

additional support prevents bowing in the middle of the stock

scraps give additional support to help prevent the free side from popping up after the cut

scraps of wood or metal

scraps of wood or metal

FIGURE 5.19: Supporting long stock.

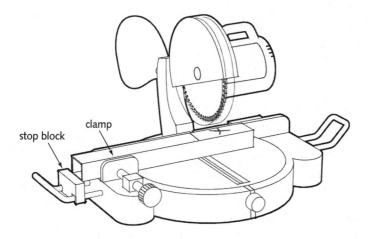

stop block

clamp

FIGURE 5.20: Stop block installed.

Abrasive Cutoff Saws for Steel

The abrasive cutoff saw shown in Figure 5.21 is very similar to the miter saw, and the same safety precautions should be observed. The difference is that the cutting blade is abrasive material, so it will throw a shower of sparks as it makes its way through the material.

These saws are generally larger and simpler than miter saws. There are no fancy compound or sliding versions.

You should make sure that all flammable liquids are put away and that you have adequate ventilation before beginning your operation.

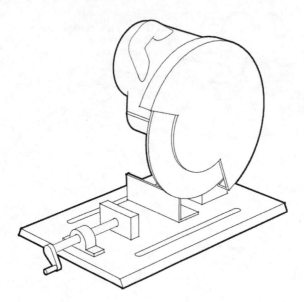

FIGURE 5.21: Abrasive cutoff saw.

Wrapping Up

This chapter introduced one of the most basic skills a robot builder can have. As you'll see in the projects, you don't need expensive machine tools to get good results, just good blades and good technique. When you apply the techniques in this chapter, remember that a tool that can cut metal has no problem with mere flesh and bone. Develop the proper respect for your tools, and they'll serve you well.

Shaping and Finishing Metal

Cutting the metal is only the first step of the process. In this chapter, we'll move on to finishing. When I talk about finishing, I mean sanding and deburring edges of parts. You may ask, "Why finish them, since no one's really going to see them?" Usually, you have to perform several operations on a piece of metal before it becomes a finished part, and rough cuts are rarely smooth and straight enough for the operations that follow. *Sanding* the part using abrasives can remove the high spots of a jagged edge, and give you a straight side to work with. Another reason is personal safety—the sharp edges left behind when cutting metal can easily cause minor cuts and scratches. You'll be reaching inside the robot, servicing things that break, and tightening loose parts. Having all metal edges finished inside the robot ensures that you won't get any cuts or scrapes from the parts you've made. *Deburring* is the process of removing the sharp parts of these edges.

This chapter will describe the various techniques for sanding and deburring using both hand and power tools. Proper file usage and care are discussed, as well as bench-top abrasive tools, like the disc sander and spindle sander. Portable abrasive tools like the belt sander and Dremel tool are introduced, and deburring techniques are addressed. Armed with the techniques in this chapter and the previous one, you'll be ready to tackle project 2, where you will be cutting metal.

Hand Files

The hand file is a basic tool for smoothing and shaping metal. When used properly, a file can safely and quickly remove moderate amounts of material. It can also be used to accurately smooth rough spots and make slightly crooked pieces straight. This section will tell you exactly how to apply the file to the work and give you tips on getting the best performance.

Types of Files

A file is a piece of tool steel with rows and rows of *teeth* that are really tiny parallel cutting edges, as shown in Figure 6.1. The small pointy tip at the end is called the *tang*, and is meant to be put into a handle. This continues up to the *heel*, which is at the base of the cutting area. At the opposite end of the file is the *point*, and the middle is called the *belly*. The length is measured from the heel to the point.

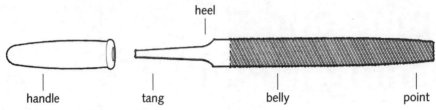

FIGURE 6.1: Parts of the hand file.

Files come in four main grades of coarseness, which correspond to the teeth per inch (tpi) on a toothed blade (such as for a hacksaw or bandsaw). These grades are coarse, bastard cut, second cut, and smooth. As the coarseness decreases, the finish becomes smoother. However, smoother files also take off less material. Generally, you start with the course file to take off a lot of material and move to the smoother ones in succession, as the surface becomes more even.

Along with the coarseness, there are also four descriptions for the way the teeth are cut into the file. These are single-cut, double-cut, rasp, and curved. Single-cut means that the file has parallel rows of teeth in a single direction. Double-cut, as shown in Figure 6.2, means that the teeth are cut in two directions in a crisscross pattern. Rasp and curved files are generally too coarse to be used with metal, and should be reserved for rapid wood removal.

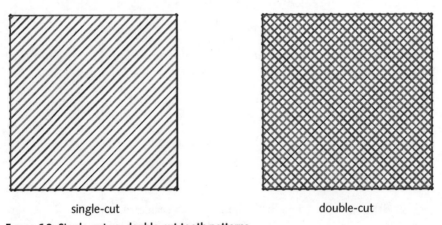

single-cut   double-cut

FIGURE 6.2: Single-cut vs. double-cut tooth patterns.

Finally, there are a variety of cross sections for files. The most common are flat (flat with straight edges), mill (flat with smooth edges), half-round, round, square, and triangular. Each profile has a different purpose. For example, the round file is intended for enlarging holes, while the flat files are intended for smoothing larger areas.

Tip

In my toolbox, I have two files that I use the most: a half-round 8-inch double-cut bastard file for rapid stock removal, and a flat 8-inch single-cut smooth file for accurate shaping where I need corners to be sharp or a surface to be perfectly flat. These two files handle most of the hand-filing chores. For all other shaping operations, I use the power tools described later in this chapter.

Seating the Handle

It is recommended that all files be used with a handle. Since these items are usually sold separately, you should buy a handle that's designed to fit your file. There are different sizes, so be sure to consult the manufacturer's information to make sure that the handle is compatible. Installing the handle on the file (*seating*) is performed by inserting the file's tang into the handle and striking the base of the handle on the table, as shown in Figure 6.3. The swinging motion should be from a moderate height, so that you have enough momentum to push the file into place. Never directly tap or push the point (tip) of the file on the table.

Caution

Never use a file as a pry bar. Because files are meant to shave metal, they have to be hardened. Unfortunately, this means that they're also very brittle. Putting prying pressure on them can cause them to shatter.

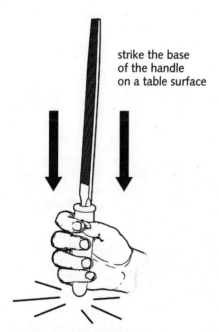

strike the base
of the handle
on a table surface

FIGURE 6.3: Seating the handle.

Proper File Technique

The work should be held immobile by clamping it down to a workbench or in a bench vise. If the work is allowed to move even a little, you may end up with an uneven surface and rounded corners. Line up the file flat to the surface. Filing is usually a two-handed operation. One hand should hold the handle and push forward, while the other hand should be at the top of the file, helping to guide it, as shown in Figure 6.4. Make sure to apply *light* pressure on the forward stroke. Excessive pressure will not result in greater stock removal.

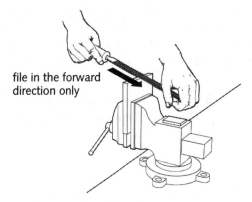

file in the forward direction only

FIGURE 6.4: Proper file technique.

At the end of the stroke, the file should be lifted up off the work and brought back to the start. Like a hacksaw, the teeth are intended to cut in one direction only — the forward direction. If you apply pressure and draw the file backward, you will cause the teeth to dull prematurely, and may also pack little bits of metal into the file, which will reduce its effectiveness.

The straight filing technique described above is used for most operations. Draw filing is another technique where the file is drawn perpendicular to the work, which can produce a high finish on the part. This technique isn't used very often, since the finished look of the part isn't as important as function. If you must have a nice shiny finish, then use the deburring wheel, described later in this chapter.

Tip If the file is merely skipping across the surface of a mystery piece of steel, chances are that it's *hardened* steel. This means that the piece won't be easily workable until it's annealed.

Cleaning a File

Over time, especially with soft metals like aluminum (compared to steel, that is), your files will become *loaded* (clogged) with little bits of metal, which reduces their cutting ability. What you need to clean them is a *file cleaner*, also called a *file card*. Basically, it's a wooden paddle with many, many rows of tiny steel bristles. The idea is that you "scrub" the bits of metal out of the file's teeth. Make sure that you brush parallel to the direction of the teeth, so that the bristles can get in there and dig the metal out. For double-cut files, you should scrub in each tooth direction. Cleaning your files should be done periodically.

Bench-Top Abrasive Tools

For most power-sanding applications, it's easiest to bring the work to the tool. You've got more control over handling small parts, and can get more precise results. This section discusses some bench-top solutions and their applications.

Stationary Sander

Hand files are great for removing small amounts of material and fixing minor imperfections. If you need to remove a lot of stock or really shape a piece of metal, then the bench-top sander is the tool to use. You can also quickly and precisely clean up the rough edges of saw cuts. You can push the part up against the wheel surface to true the edge, and sand outside curves. (Inside curves require a spindle sander, described later.)

What to Buy

You have two main choices when it comes to bench-top sanders: disc only, and combination disc and belt (see Figure 6.5). Both types have a sanding disc with an adjustable table, but the combination sander adds a stationary abrasive belt, like a belt sander.

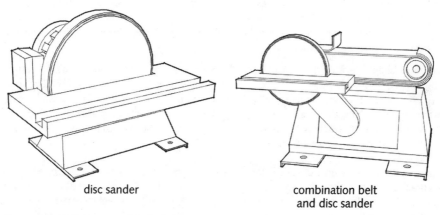

disc sander   combination belt
and disc sander

FIGURE 6.5: Stationary sander choices.

While the disc and belt may seem like a great value, combining two tools in one, I recommend sticking to the disc-only version. The disc-only models have larger discs than the combo models, and the usefulness of the stationary belt is arguable. In practice, it's very difficult to hold a plate or other part on the belt with any kind of precision.

Note The easiest position for sanding is with the abrasive surface traveling straight down. This pushes the work right into the table, where it should be. If the abrasive surface runs sideways, as with a belt sander, it's hard to maintain a true edge, since the work wants to follow the abrasive in the direction of travel, and the table that you're usually provided with on combo units is inadequate.

You should buy the largest disc you can afford, since this will determine the ultimate size of the parts that you can effectively shape and sand. Delta makes a 12-inch diameter disc sander (Model #31-120) with integral dust collection. It sells for around $175, and is the largest economy-priced disc sander you can buy.

As with all abrasives (like sandpaper), you have a choice of *grit*, which is a measure of the roughness of the surface. Lower grit numbers correspond to rougher surfaces, which take off more material at a time. Since the disc sander will be used for shaping metal quickly, you should have 60- or 80-grit discs.

Sander Safety and Technique

In addition to following the power-tool safety guidelines in Chapter 5, "Cutting Metal," you should observe some specific precautions. First, sanding belts have a direction printed on the back (inside). Make sure that the belt is properly installed so that the arrow follows the actual direction of travel.

Also, let the disc or belt spin up to full speed before bringing the part into contact with the abrasive surface. You want to keep the work moving on the abrasive. Try not to stay in one spot, or you will end up loading that spot with metal.

Like a bandsaw, the sander is equipped with a tilting table to help you put a *bevel* (angled face) on something. Make sure that you reset the table to level using a square after finishing a bevel operation. Lining it up by eye or with the built-in detents isn't enough; you've got to use a square. Keep the table as close as possible to the abrasive surface, so that you minimize the gap, where small parts might become lodged.

All work should be held flat on the table. The largest surface of the part should be resting on the table. For example, if you're sanding the edge of a long bar, the length of the bar should rest horizontally on the table, where you have more support. The bar should not be held vertically, where you only have the small cross-section edge to support it. Most tables come with a slot for a miter gauge. Just as with a bandsaw, the miter gauge can be used to help you feed material in straight or at an adjustable angle.

Sanding is done only on the side of the table where the wheel is traveling *down*, as shown in Figure 6.6. This is because the abrasive disc will tend to push the work down on the sturdy table surface. On the side where the wheel is traveling up, the abrasive is likely to catch the work and launch it upwards at you, which is very dangerous.

Removing a large amount of material can build up a lot of heat. You should have a pair of mechanic's gloves handy to protect your hands. They have a closer fit than bulky work gloves, which can get caught in the slot between the disc and the table, sucking your hand in. Hot metal parts can be cooled in water.

If you've also been using the sander for wood, make sure to vacuum out the inside before sanding steel. Steel can generate sparks that may ignite leftover sawdust inside the sander.

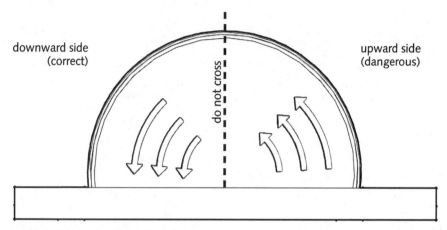

downward side
(correct)

do not cross

upward side
(dangerous)

FIGURE 6.6: Proper sanding technique.

Note

After you're done sanding, you've got to take care of the sharp edges. You should use a hand-held deburring tool or deburring wheel on a bench grinder to take care of the edges. (Deburring techniques are described in detail later in this chapter.)

Cleaning the Abrasive

Over time, the abrasive disc or belt will begin to accumulate metal bits, which impair the performance. You can clean these off and extend the life of your abrasive by using an *abrasive belt cleaner*. This looks like a handle with a wax bar on the end. Holding the handle firmly, put the tool flat on the table with the wax bar facing the abrasive surface. Start the tool and allow it to spin up. Gently bring the wax into contact with the abrasive surface, and drag it across from the outer edge of the disc to the center, as shown in Figure 6.7. Make sure that the disc is spinning down on the side that you start with, as described previously. Do not cross over to the other side, or the abrasive is likely to catch the handle and send it flying.

Tip

Buy spares of the belt or adhesive-backed disc for your sander. You will have to change them eventually, and it's better to have a spare ready and waiting rather than having to put your work aside and run out to buy another abrasive surface. Also, it's sometimes difficult to find the correct size at the local hardware store.

Spindle Sander

Disc and belt sanders can only get to outside curves. To get at inside curves, you need a *spindle sander*. A spindle sander is a bench-top tool that has a cylindrical rubber piece (the sanding drum) with a sleeve of abrasive material. The drum spins as well as oscillates up and down.

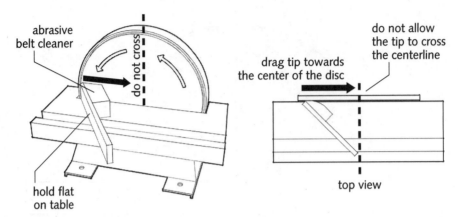

FIGURE 6.7: Cleaning the abrasive disc.

Note

A stationary belt sander can actually get inside curves, but only those with very large radiuses.

What to Buy

Stand-alone bench-top spindle sanders (also called *oscillating sanders*) can be quite expensive. If you've already got a drill press, you can buy a sanding drum that you secure in the drill chuck that does the same job as the spindle sander (without the up and down motion). Individual cylindrical drums in various sizes are made by 3M, and sold as "rubber cushion polishing wheels" for a few dollars each. They can also be found in McMaster-Carr as "expanding rubber sanding drums." The abrasive sleeves (also called spiral bands) are purchased separately. You should have a set of 50-grit sleeves to rapidly smooth sawed edges. Also, several companies make drum sanding kits that include most popular sizes of drums. When using this system, you should make an elevated base out of wood with a hole cut in it to drop the spindle down through the hole, as shown in Figure 6.8.

Spindle Sander Safety

In addition to following the power-tool safety guidelines in Chapter 5, you should observe some specific precautions. First, keep a firm grip on your part. Removing a large amount of material can build up a lot of heat. You should have a pair of gloves handy to protect your hands. Hot metal parts can be cooled in water.

Also, the work should always be fed into the sanding drum flat on the table, and *against* the rotation of the drum, as shown in Figure 6.9. If you feed the part with the rotation, the abrasive material is likely to grab the part and spin it out of your hands.

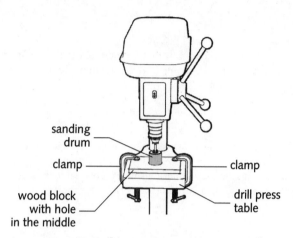

sanding
drum

clamp

wood block
with hole
in the middle

clamp

drill press
table

FIGURE **6.8: Drill press with spindle sander attachment.**

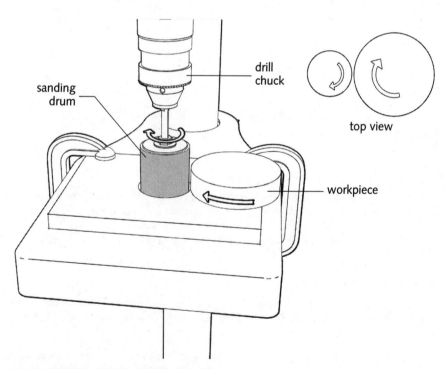

drill
chuck

sanding
drum

top view

workpiece

FIGURE **6.9: Safe spindle sander technique.**

Portable Abrasive Tools

In some situations, you can't bring the work to the tool. It may be mounted in the robot, or just too big for the stationary tool. This section describes some portable sanding tools that you can easily bring to the part.

Dremel (Rotary) Tool

The rotary tool shown in Figure 6.10 is one of the most versatile tools in the shop, and a must-have for pit repairs at a competition. Basically, it's a hand-held motor that spins at very high speed. It's an incredibly flexible tool because it has a collet that takes replaceable bits, and there are all kinds of bits for different tasks, as described later in this section.

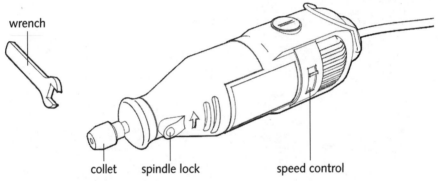

wrench

collet   spindle lock   speed control

FIGURE 6.10: Parts of a rotary tool.

What to Buy

Dremel is a major manufacturer and pioneer in rotary tools, and like Xerox or Kleenex, the name is synonymous with the product. There are several models in the line. Make sure to choose one that has variable speed (usually 5 or 6 steps), which should be available in a kit for $50 to 80. I recommend sticking to the variable speed corded units, not the cordless, battery-powered units.

Rotary Tool Safety

In addition to following the power-tool safety guidelines in Chapter 5, you should observe these specific precautions. First, eye protection is a must! For some operations, a full-face shield is also recommended.

Also make sure that the bit is secure. Depress the spindle lock button and use a wrench to tighten the collet. Be sure not to start the tool while it is in contact with the work. Make sure that it spins up completely before you begin your operation.

Finally, use both hands to guide the tool, and make sure to drag the tool opposite the rotation of the shaft.

The Drum Sander

The drum sander is a miniature version of the spindle sander described earlier. The 1/2-inch diameter sanding drum (shown in Figure 6.11) is Dremel part #407. The replaceable 60-grit abrasive drum is Dremel part #408. The drum sander can be used for cleaning up edges, opening up holes, and deburring.

The Cutoff Disc

The cutoff disc shown in Figure 6.11 consists of the disc (Dremel part #540 or #426) and the screw mandrel (Dremel part #402), where the disc mounts. Both discs fit the same mandrel. This tool can cut steel shaft just like a miniature version of the abrasive cutoff saw, and it's helpful for pit repairs.

When using the cutoff disc, make sure to keep the wheel perpendicular to the surface you're cutting. The disc is brittle, so use a light touch and do not attempt to rotate the wheel to correct the cutting path, or it will shatter. As with a circular saw, you must remove the tool and restart the cut.

Caution

When performing cutoff operations, a full-face shield is recommended. If the thin disc shatters, as it sometimes does, it can send abrasive fragments flying in all directions.

The Tungsten Carbide Cutter

This straight tungsten carbide bit (Dremel #9901) shown in Figure 6.11 is useful for opening up holes in aluminum and removing material. Chapter 8, "Fasteners — Holding It All Together," contains a step-by-step usage guide for opening up drilled holes.

The Aluminum Oxide Grinding Stone

The curved edge on the aluminum oxide grinding stone in Figure 6.11 (Dremel part #952) is helpful when deburring the insides of blind holes and around curves. This tool will eventually load up with aluminum, and should be cleaned with an abrasive belt cleaner.

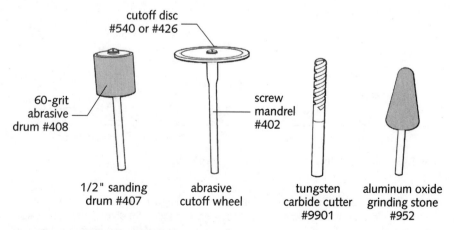

cutoff disc
#540 or #426

60-grit
abrasive
drum #408

screw
mandrel
#402

1/2" sanding
drum #407

abrasive
cutoff wheel

tungsten
carbide cutter
#9901

aluminum oxide
grinding stone
#952

FIGURE **6.11: Various rotary tool bits.**

Angle Grinder

The angle grinder is the smaller, portable version of the disc sander (see Figure 6.12) . It's like a high-power compact right-angle drill motor. You hold the body and side handle to guide the grinder around the work.

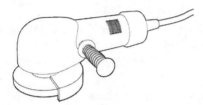

FIGURE **6.12: Angle grinder.**

There are many angle grinders on the market from 4-inch and 4½-inch sizes up to 9 inches. The size refers to the wheel diameter. For building robots, you shouldn't need much more than a 4½-inch grinder.

What to Buy

With an angle grinder, you have the choice of delivery method for the abrasive. You can either use depressed center wheels, or a rubber backing pad with pressure-sensitive adhesive (PSA) abrasive pads, as shown in Figure 6.13. Depressed center wheels usually come as one solid disc of abrasive material with a raised center. The rubber backing pad and PSA pads are usually sold separately.

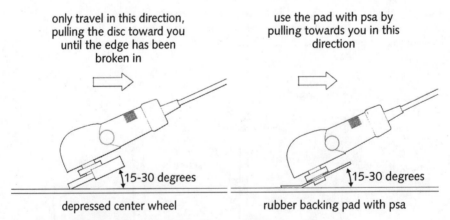

only travel in this direction, pulling the disc toward you until the edge has been broken in

use the pad with psa by pulling towards you in this direction

15-30 degrees

15-30 degrees

depressed center wheel

rubber backing pad with psa

FIGURE **6.13: Grinding wheel and rubber backing pad usage.**

Note

Aluminum is a metal, but most abrasive blades and grinding wheels that say "metal" are really intended for steel. Generally, aluminum oxide is recommended for steel, and silicon carbide is recommended for aluminum and other nonferrous metals. DeWalt and Norton have lines that are specifically designed for aluminum, which tends to load other types of discs more rapidly.

Angle Grinder Safety

In addition to following the power-tool safety guidelines in Chapter 5, you should also observe some specific precautions.

The proper angle for the grinding wheel relative to the work is 15 to 30 degrees, as shown in Figure 6.13. To smooth a surface, draw the grinder towards you. Until the edge of the disc has been rounded ("broken in"), pushing it away from you may cause the edge to dig into the work. When using the rubber backing pad and abrasive discs, maintain the same 15 to 30-degree angle, but apply downward pressure so that about a third of the disc is in contact with the work as illustrated in Figure 6.13. Keep the angle of the grinder low to minimize the risk of digging into the work with the pad.

The grinding discs can bite into the surface of a metal pretty easily, so it's important to keep the tool moving in an even, continuous motion. When about 1 inch has been worn down from the diameter of a grinding wheel, it should be removed and replaced, or it may shatter during use.

Finally, when grinding steel, you may create a shower of sparks. Make sure that all flammable materials are put away before using the grinder.

Belt Sander

The portable belt sander (see Figure 6.14) is essentially a smaller version of the belt part of the combination stationary sander. There are many models and manufacturers to choose from. Accessory holders and tables are available to convert a portable belt sander for bench-top use, but when added up, this cost is comparable to a stationary model. Some models are designed with flat tops to make it easier to mount them upside-down to a bench.

As per the discussion earlier, the belt sander has limited use for the robot builder, except for cleaning up the edges of pieces that are too long to fit on a stationary disc sander.

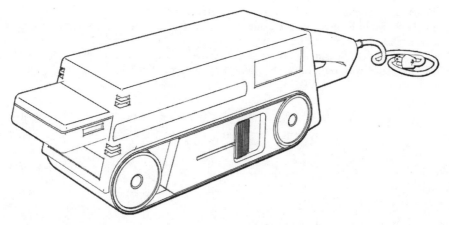

FIGURE 6.14: Belt sander.

Belt Sander Safety and Usage

In addition to following the power-tool safety guidelines in Chapter 5, you should also observe specific precautions. First, when sanding long plates, it's much safer to have the metal piece firmly clamped to a worktable, where it won't fly out of your hands, and bring the tool to the work. You'll have to concentrate to hold the tool true when sanding along the length.

Also remember to start up the tool away from the work and let it spin up to full speed. Then, bring it into contact with the work.

Try to keep light, even pressure on the work. When the belt sander is used on top of the work, the weight of the tool should be enough force. The belt sander is fairly aggressive and can take off a lot of material very quickly, so you should be careful to keep the tool moving and not dwell in any one spot, or you'll wear a low point there.

Finally, the abrasive belt cleaner mentioned earlier in the chapter should be used to clean the belts.

Deburring Tools

All metal-cutting and shaping operations should be followed by *deburring*. When you cut something out of metal, chances are it will have a sharp edge where the cut was made. Even if you use a bench-top sander to smooth the side, there will still be a sharp edge. These edges can leave you with little cuts all over your hands. It's best to smooth these edges down to make it easier to handle your parts, and prevent injury (however minor). That's where the deburring tool comes in. You have the choice of the hand-held version, or the power-tool version, which is used with a bench grinder.

Hand-Held Deburring Tool

The hand-held deburring tool consists of a plastic or metal handle with a swiveling blade at the end. As shown in Figure 6.15, the blade is curved and has a little ball hook on the end. The hook catches the sharp edge on the corners of your work and shaves it off as you draw the tool down the length of the side. These tools are relatively inexpensive and are available for $5 to $10. The blades wear out, so you should have a few extras handy.

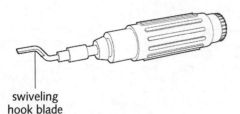

swiveling
hook blade

FIGURE 6.15: Hand-held deburring tool.

You should maintain an angle of about 45 degrees on the handle, pulling the tool towards you (see Figure 6.16). It's important to turn the work so that the edge that you're deburring is at a good angle for you to control the tool and guide it with even pressure all the way to the end. For round pieces, continuously rotate the piece so that your angle relative to the work is the same at all times. Trying to use the tool at an awkward angle is the main cause of tool slippage.

Keep in mind as you're using the tool which direction it will go if it slips. This is a pretty common mishap and requires constant attention to prevent. The greatest danger here is to the skin on your fingers. Keep your other hand out of the path of the tool. If the tool slips, the blade may take off a little skin, which is only annoying, but not fatal. Also, you could put a big gouge down the side of your part.

Note You can use a flat file to deburr the edges of most parts, but in order to deburr holes, you'll need the deburring tool.

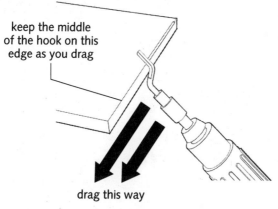

keep the middle of the hook on this edge as you drag

drag this way

FIGURE 6.16: Proper deburring tool technique.

Deburring Wheel

This wheel does the same work as the hand-held deburring tool, but in a fraction of the time, and with much less effort. Not only can it deburr, but it can also put a nice, shiny finish on your parts.

Basically, it's a wheel composed of dense abrasive fibers (like Scotch Brite pads) that are tightly wrapped and formed into a semisolid. As you use the wheel, the little bits of fiber come off along with the metal, revealing fresh abrasive material. As a result, the wheel will get smaller over time.

To use the wheel, you will need a bench grinder, which is basically a high-RPM motor with an output shaft on either side. Normally, bench grinders are supplied with two wheels intended for steel only. You should replace one of these abrasive grinding wheels with the deburring wheel.

What to Buy

You're going to need a 6-inch bench grinder as shown in Figure 6.17. The brand and model don't matter all that much. There are a lot of deburring wheels on the market. It's a delicate balance between material removal and time. You don't want to take off too much (you've already done your shaping with the other abrasive tools), but you don't want to stand there forever grinding away too little with each pass. The best compromise that I've found is the 6-inch diameter × 1-inch wide grade 9S (fine) 3M Scotch Brite EXL silicon carbide wheel (Grainger #4ZR26). Unfortunately, it's fairly expensive at $50, but when you see the results on your parts, you'll realize that it's well worth the expense. (Also, since the wheel ID is 1 inch, and the shank size of my bench grinder is 1/2 inch, I had to get an adapter flange, which was Grainger #4ZR59.)

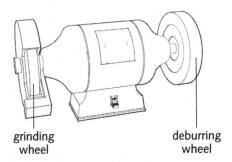

grinding   deburring
wheel   wheel

FIGURE 6.17: Bench grinder with deburring wheel.

Tip

Having a separate stand (pedestal) for the bench grinder that holds the wheel will increase your ability to maneuver a part to get all the edges.

Deburring Wheel Safety

In addition to following the power-tool safety guidelines in Chapter 5, you should also observe specific precautions. First, since they're wrapped, the Scotch Brite deburring wheels have a direction arrow printed on the side. Make sure to install the wheel so that the arrow points in the direction of turning to get the best results.

Also, to give you greater access to the wheel (and more flexibility in usage), you should remove the tool rest. This will allow you to deburr large panels more easily.

Caution

This is one of the few free-hand operations in sanding. You're not stabbing the material into the wheel, but *gently but firmly* dragging the material across the abrasive surface. If you jam the part into the wheel, it will most likely grab it out of your hands and toss it on the floor. Very light pressure is all that's required.

The work should always be brought into contact with the wheel at a 45-degree angle, on the *downward* side of the wheel just below center, as shown in Figure 6.18. This is to prevent the wheel from catching the work and throwing it.

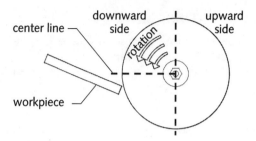

FIGURE 6.18: Proper work angle for deburring wheel.

Draw the edge all the way through to the end, and continue movement smoothly past the end. Keep the work moving on the wheel, or you'll cut a dip in the work. Make sure to get all the edges, including the corners. Gently check for spots that you've missed by feel. Be careful because the edge could still be sharp.

Tip

If you just want a nice finish on the edge, and trueness isn't critical, then move the flat side of the part on the deburring wheel until all the sanding marks go away, leaving a shiny, uniform surface.

Remember, the sanding wheel is for rough shaping. The deburring wheel should only be used for deburring and final finishing. If you're spending more than a few minutes on the wheel, then you haven't shaped enough with the other abrasive tools, and you should go back to a rougher abrasive before returning to the wheel.

Project 2: Cutting the Armor Pieces

In the last few chapters, you've learned about selecting tools and materials, marking and cutting, and shaping and finishing materials. Now it's time to put that knowledge to use. Cutting metal and plastic plates are essential skills for the robot builder, so we're starting there first. This project will get you working with the circular saw, jigsaw, and miter saw by cutting the armor and frame pieces for the project robot. While all of these cuts could be done with the jigsaw, I'll demonstrate some with the other two tools to show situations where their specialized designs allow you to make the cuts more quickly and precisely.

Caution

Eye and ear protection are mandatory for these tools. They are incredibly loud, and pieces of metal will be flying everywhere.

Caution

Review all of the general power-tool safety protocols described in Chapter 5, as well as the sections that correspond to the specific tools used below.

Cutting Metal Plates with the Circular Saw

You'll begin by cutting the 3/16" thick aluminum baseplate. The best tool for performing straight cuts in sheet stock is the circular saw. It will provide you with the cleanest and most precise cut, requiring very little cleanup. Make sure that your saw is equipped with a nonferrous metal-cutting blade. The following steps will guide you through making your first straight cut with the circular saw:

1. The bottom plate should measure 15" × 14". I started out with a 24" × 24" piece of stock. Measure and mark a 15" × 14" square on the plate.

2. In this cut, the guide will be placed on the left side of the saw. In order to set the guide correctly, you've got to flip the saw over and measure the distance from the blade to the side of the base that will be following the guide. Measuring from the desired cut line (the 15-inch line), you should draw another line with this offset, and clamp the guide in place.

3. As mentioned in Chapter 5, you must support both sides of the work, as shown in Figure 6.19, or else the blade will bind.

FIGURE 6.19: Correct support of the workpiece.

4. Make sure that the height of the circular saw is set so that it will not touch the table or floor that the wood pieces are sitting on.

5. Draw over the 15-inch cut line with the stick wax so that you have a line of lubricant all the way down the piece. You can also use WD-40 as a lubricant. I prefer the stick wax because it doesn't fly away as much when you start the saw. Regardless of which type of lubricant you use, you must use a lubricant. Do not cut dry with the blade or you will destroy it.

6. Make sure that the base of the saw is held firmly against the guide and that the blade is not touching the stock before you start. Also keep the base as level as possible.

7. Start the blade and let it spin up to full speed. Slowly and smoothly bring the blade into contact with the piece. Continue through the length of the cut. Don't rush the cut. If the saw sounds like it's starting to slow down, then lower your pushing speed, and let the saw cut at its own pace.

8. When you've made it all the way through the material, release the trigger and let the saw come to a complete stop before moving it away.

9. Take this edge to the deburring wheel and draw the ragged corner across the wheel in a smooth and steady motion, applying moderate and constant pressure. Make sure to line up the wheel below the center line as shown in Figure 6.20. Make sure to get both the top and bottom edges.

10. You can clean the remaining stick wax residue from the piece with Goo Gone.

FIGURE 6.20: Deburring the plate edge.

Cutting Plates with the Jigsaw

For the second cut in the baseplate, you'll cut on the 14-inch line. This operation can be performed with the jigsaw equipped with a nonferrous metal-cutting blade.

1. For this cut, I'll demonstrate hanging the waste piece off the edge of the table, as shown in Figure 6.21.

2. As in the previous example, you must measure the distance from the blade to the edge of the footplate and put your guide at that offset from the 14-inch cut line.

3. Use the stick wax or WD-40 for lubrication as in the previous example.

4. Begin your cut with the blade not in contact with the workpiece, and the footplate level. Let the blade get up to speed, and then slowly bring it into contact with the workpiece.

FIGURE 6.21: Jigsaw cut setup.

5. In this setup, to prevent the blade from binding, you must hold up the waste piece by hand as you near the completion of the cut, as shown in Figure 6.22. Make sure to keep your arm up and out of the way, and fingers clear of the blade at all times.

6. There will be some cleanup required with the belt sander after the cut, as shown in Figure 6.23.

7. Draw the edges across the deburring wheel as described in the previous example.

FIGURE **6.22:** Supporting the waste piece.

FIGURE **6.23:** Belt sander cleanup.

Cutting the Top Armor

The top armor will be made out of 1/4-inch thick polycarbonate. The top has the same dimensions as the base. The cutting operations for plastic are virtually the same as those for cutting metal. However, the most important thing is that you need to change lubricants. In plastic cuts, heat builds up rapidly, and you need a liquid coolant such as WD-40 or even water. To get the best results, you should keep the water or other coolant flowing into the cut.

Instead of using the deburring wheel, you should use a flat file to chamfer all edges. The deburring tool (a hand-held tool) can also be used for this operation. Its usage is discussed in detail earlier in this chapter as well as in Project 4.

Cutting the Sides

The sides will be made out of 1/2-inch polycarbonate. They should measure 14" × 3.5". Since I have a 24" × 24" sheet, I will be cutting 3.5-inch wide strips off of the sheet for the parts, and then cutting those down to the correct length of 14 inches.

One important difference with this setup is which piece the waste will be. In the previous cutting operations, we removed a smaller waste piece from a larger sheet. When cutting the side plates, you will be cutting off the part you want from the sheet, not the waste. That means that you've got to measure your offset from the edge of the base to the far side of the blade, which includes the thickness of the kerf. If you use the same offset as before for your guide, your pieces will be shorter by the thickness of the kerf.

At the end of this operation, you should have four strips that are 24" × 3.5".

Cutting with the Miter Saw

Cutting the sides to the correct length can be done with the jigsaw or circular saw, although the best tool to use in this case is the miter saw. Again, this saw should be fitted with a nonferrous metal-cutting blade (which is usually also good for plastics).

1. Mark off the correct distance, carefully indicating the waste side.

2. Clamp the piece to the saw, making sure to line up the blade with the kerf on the waste side of the cut line. The most important thing is that the work remains flat in the area around the blade. Do whatever you have to on the clamp side to make sure that the stock is flat on the table near the blade.

3. To speed up operations, you can set the stop block as shown in Figure 6.24. That way, you won't have to measure, mark, and line up all the pieces. After setting it up on the first piece, the stop block takes care of aligning all the rest of the pieces to the right length. Make sure that the stop block and clamp are on the *same* side. If they are on opposite sides, the blade may bind and throw the piece, which can cause injury.

4. Making sure that you have adequate lubrication, start the saw with the blade covered and not in contact with the workpiece.

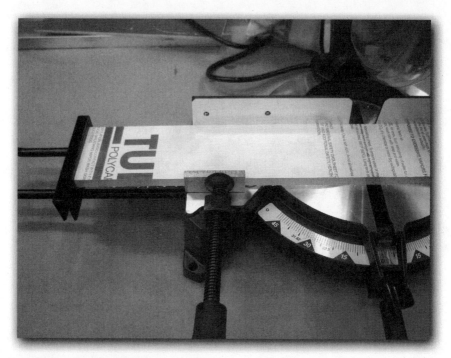

FIGURE **6.24: Stop block and clamp setup.**

5. Allow the blade to spin up to full speed, and slowly bring it into contact with the workpiece.

6. Let the saw dictate how fast it wants to cut. Keep your motion smooth and slow.

7. Continue all the way through the part and release the trigger, keep the saw held down until it comes to a stop, and then retract it.

You should now have a 3/16-inch-thick aluminum plate cut to 15" × 14", a 1/4-inch-thick polycarbonate plate cut to 15" × 14", and four 1/2-inch-thick polycarbonate plates cut to 14" × 3.5".

Wrapping Up

Hopefully, this chapter has provided you with the skills you need to finish your parts. Cutting the metal is only the first step. A part isn't finished until it's been sanded, shaped, deburred, and, well, finished. Proper sanding technique will improve the fit and accuracy of your parts, and keep you safe from rough or sharp edges. As you'll see in the projects, finishing (whether in the form of sanding or deburring, or both) follows every cutting operation.

Drilling and Tapping Holes

Holes and fasteners are the basic means for connecting robot armor and structural pieces together. While this is not a glamorous chapter, I assure you that there's more to it than just "drilling a hole." It's a basic robot-building skill that's rarely covered in detail because it seems so straightforward, but it can often be a source of frustration in assembly.

This chapter illustrates several techniques that will help you drill and thread holes safely and accurately. Numerous examples of drilling setups are illustrated to cover most of the situations that you may encounter when building your robot. Techniques for working with screws are discussed, including dealing with holes that don't line up and shortening screws that are too long. We also deal with creating threaded holes (tapping), and move onto Project 3: Drilling and Tapping Holes.

Drilling

In this section, I'll introduce you to the different types of drill bits and their applications. I'll also talk about the basic drill press safety precautions and list the basic sequence of events for successfully drilling a hole.

Types of Drill Bits

Most people are familiar with regular fractional drill bits, but the robot builder may use more than a few of the other types of bits described as follows.

Regular Drill Bits (Twist Drills)

These come in four different standards: fractional, number, letter, and metric. Although tapped (threaded) holes call for the use of number drill bits, you can get by with a set of fractional bits if you're willing to risk less strength from your screw threads. You should be prepared to buy more of your most popular sizes, since they do become dull after a while. You will be looking to buy high-speed steel (HSS) jobber-length drill bits with a 118-degree included tip angle, which should be sufficient for most jobs.

Cross-Reference

There is a complete listing of all fractional, number, and letter drill sizes in Appendix F, "Tables and Charts."

Uni-Bits (Stepped Drill Bits)

These combine many different sizes into a single bit. They have a single cutting edge, which is useful in some situations, as described in the discussion of thin stock later in the chapter. The only limitation is the depth of hole that you can drill.

Combination Center Drills

These are useful because they do not flex like regular drill bits. Their use is described in the discussion of round stock later in the chapter.

Countersinks and Counterbores

These are used to make different types of hole openings for screws, as shown in Figure 7.1. Counterbores usually come in sets matched to common screw sizes. Countersinks may be purchased individually. A 3/8-inch diameter countersink should be fine for most jobs in the shop, and an 82-degree point angle is recommended so that you can make holes that are compatible with flat-head screws. Single flute versions give a smoother finish than the four-flute models.

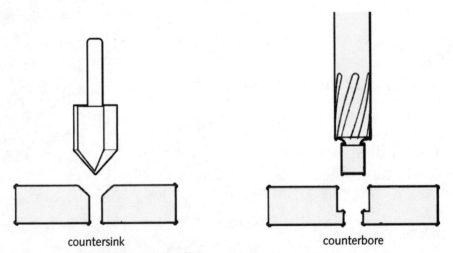

countersink   counterbore

FIGURE 7.1: Countersinks and counterbores.

Reamers

These are used to make precision-sized holes, and are discussed in detail later in this chapter.

The Importance of a Straight Hole

Almost all holes you will drill for the robot will be straight (or *true*), with the axis of the hole perpendicular to the surface of the part. A crooked hole will miss the intended location on the other side, resulting in misalignment, as illustrated in Figure 7.2. Also, drilling is the first step in creating a tapped hole. Tapping requires a true, straight hole, or else you'll end up with broken taps. (See the "Tapping Techniques" section later in this chapter.)

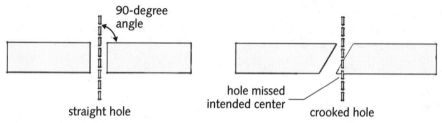

FIGURE 7.2: **True versus crooked hole.**

There are few better ways to make a true hole than using a drill press. However, there are some situations in which the part won't fit into the drill press or can't be removed from the robot. In this case, a cordless drill is what's required. Figure 7.3 shows the different parts of the drill press and cordless drill.

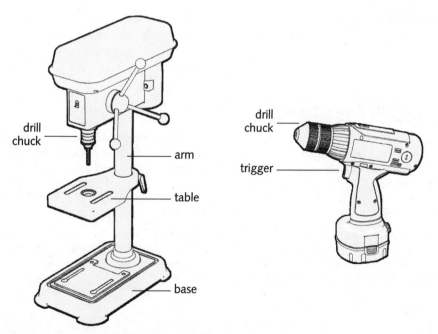

FIGURE 7.3: **Parts of a drill press and cordless drill.**

Basic Drilling Procedure

Whether you're using a drill press or a cordless drill, the basic sequence for drilling is pretty much the same. Following is a step-by-step listing. Some of these steps are covered in more detail following this section. (Before attempting to use these steps, you should review the "Power-Tool Safety" section in Chapter 5, "Cutting Metal.")

1. Mark and punch the holes to be drilled.

2. Put the drill bit in the chuck and tighten.

3. If you are holding the part with a vise, clamp it in the vise first. With the drill motor off, line up the tip of the drill bit to the punch mark.

4. Clamp the part or vise to the drill press table or worktable.

5. Apply lubricant to the part at the hole's position.

6. Put on eye protection.

7. Start the drill motor and slowly lower the bit into the part, making sure that the drill bit tip jumps into the center of the punch mark.

8. Apply firm downward pressure, but don't put all your weight on the drill bit. Let the bit do the work.

9. Retract the bit periodically from the material to clear the chips (bits of metal or plastic), which gives the bit and metal a brief chance to cool down, and lets you see if you need to add more lubricant.

10. As you get closer to the end of the hole, apply less downward pressure. Once the bit breaks through the bottom, continue down a bit more to clear the hole, and then retract the drill bit.

11. Turn off the drill motor first, then blow away the chips and remove the part.

Basic Drill Press Safety

In addition to the "Power-Tool Safety" section in Chapter 5, you should also observe extra precautions. First, as with all metal-cutting operations, adequate lubrication must be used. Heat is the enemy of your drill bits. That's why it's very important to use lubrication for all drilling operations. You should use a lubricant (such as TapMagic, TapMatic Gold, or Edge Liquid) that is specially formulated for drilling and tapping operations, although WD-40 or household oil may be used in most cases. Water can be used successfully on all plastics.

Also, the drill press should be firmly clamped or bolted to the worktable. Any adjustments to the clamping, vise, setup, table, or drill chuck should be done with the power off. The drill press should only be started when you are ready to drill. Power down and allow the spindle to come to a complete stop before making any adjustments.

Allow the drill press to spin up to full speed before bringing the drill bit into contact with the work. When you have finished drilling to your desired depth, you may retract the bit while the drill press is still running and turn it off when the spindle reaches its full spring-loaded rest height.

Note

Cordless drills and portable corded drills are only power tools that may be started in contact with the work, because they require the punch mark for accurate alignment.

Finally, the chuck key, which is used to tighten and loosen the chuck, should be kept in its holder, which is usually somewhere on the drill press's body. Always keep your hand on it and never, ever leave it in the chuck. It should be returned to the storage area immediately after any tightening or loosening operation is complete. If you forget it and turn on the drill press, the chuck key will become a projectile capable of great harm. Always store it in the holder, and never in the chuck.

Successful Drilling Techniques

Here, I'll describe in more detail some parts of the basic technique outlined previously, and provide a variety of setup and drilling procedures for different situations and types of stock.

Marking and Punching

Marking is the process of indicating where on the part you want the holes to be drilled. It used to be common practice for a machinist to paint a thin layer of a marking solution (such as Dykem) on a part to make the surface a uniform color. Then, lines would be scribed into the part at the intersection of two measurements, forming an X where a hole should be drilled. With the advent of computers and CAD (computer-aided design) software, this process has gotten a lot easier. Nowadays, you can print out a 1:1 scale drawing of your part, coat the back side with a spray adhesive such as 3M Spray 77, stick the drawing to your part, and you're done. You can even skip a step by printing the drawing on adhesive-backed label paper (although it's a bit more expensive). You've still got to make sure to punch at the intersections, but it saves a large amount of time in the marking department.

Cross-Reference

Other techniques for marking are described in Chapter 5.

Tip

Citrus-based adhesive solvents such as GooGone work well to remove the sticky residue left over from the Spray 77 or label paper.

To complete the process, you've got to punch the marks for your holes. *Punching* means putting a small divot in the surface of the part to indicate the exact center of the hole. Without a punch mark, the drill will tend to wander, as shown in Figure 7.4. The punch mark helps to center the drill, because the center of the divot catches the tip of the drill bit, and holds it in the right location.

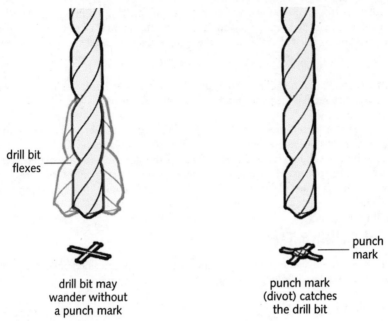

drill bit
flexes

drill bit may
wander without
a punch mark

punch
mark

punch mark
(divot) catches
the drill bit

FIGURE 7.4: Drill bit and punch mark.

Punching is usually done with a manual center punch (a long cylindrical tool with a sharp hardened tip) and a ball-peen hammer. The tip of the punch is lined up to the mark, and the head of the punch is struck with the hammer. A better and faster way to accomplish this is with an automatic center punch. This is a spring-loaded tool that does not require a hammer. You line up the tip to the mark, and then push straight down, which compresses a spring. You continue pressing until the spring fires at a preset (and adjustable) pressure, making a mark on the part just like a manual punch. Automatic center punches cost about $25, and I highly recommend them as a way of saving time in the shop. Both punches can be seen in Figure 7.5.

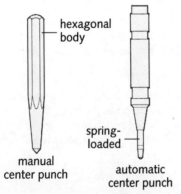

hexagonal
body

spring-
loaded

manual
center punch

automatic
center punch

FIGURE 7.5: Manual and automatic center punches.

Drill Press Setup

The first thing to check with your drill press is that the table is level. Put a square on the table and bring the spindle down to see if the angle matches. Many small bench-top units have an angle adjustment on the table. You should make sure this is perfectly true to the spindle and leave it. If you need to drill an angled hole, then use the means described later in this chapter. Do not adjust the angle of the table. If it's set back incorrectly, all your holes will be crooked, and you'll be wondering why your parts aren't lining up.

It's also a very good idea to get a 3/4-inch-thick piece of wood (plywood or MDF) that's a little bit larger than the dimensions of the drill press table. This will give you a disposable surface that you can drill into to complete your holes, and will protect the metal surface of the drill press table as well as your drill bits.

 Note Cleanliness counts! Make sure that you keep the table surface clear of chips (bits of aluminum or plastic) at all times. If you get a chip underneath a part that's clamped to the table, the small angle could cause you to drill a crooked hole. The same goes for the vise. Make sure to blow everything off and wipe away any chips before clamping your part in place.

Drilling True Holes with a Cordless Drill

The drill press automatically lines you up to drill a true hole. What if you've got to drill a hole in a part that won't fit into the drill press or that is already mounted to the robot? No problem. There are a few easy ways to ensure that you drill a true hole with a cordless drill.

- You can set up two squares next to the hole at 90 degrees to each other as shown in Figure 7.6. This will help you visually line up your drill to the true angle.

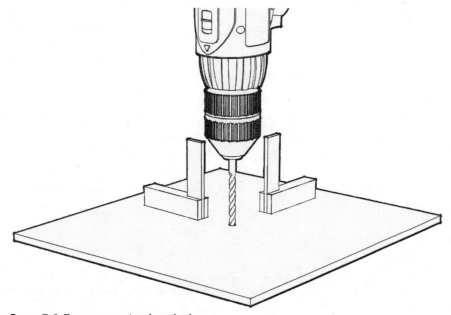

FIGURE 7.6: Two-square visual method.

- If you've got a friend handy, have them help you sight and correct the drill angle.

- If you have access to a drill press, you can take a scrap of aluminum and make a block. The block should be wide enough to help you determine whether it's flat on the surface. A 1" × 1" × 1/2" thick block should suffice. Drill a true hole in the block that's the correct size using a drill press. Chuck the bit in the cordless drill and slide the block on. Use the tip of the drill bit to line it up to the punch mark, as shown in Figure 7.7. Slide the block back down to the surface of the part, and adjust the drill angle until the block sits flat.

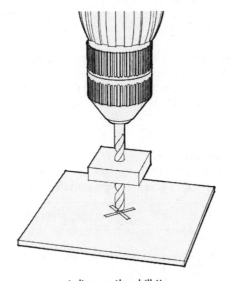
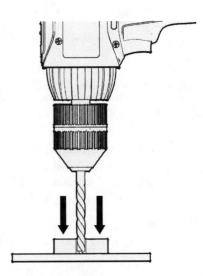

1. line up the drill tip
to the punch mark

2. slide the block down
so that it sits flat and
hold it there

FIGURE 7.7: **Block method.**

Caution

It's important that the part does not move while you're drilling. It must be clamped into a bench vise or worktable, or the robot must be up on blocks to prevent rolling.

Tip

It's easier to maintain a true angle when drilling straight down. If you can, rotate the part (or the robot) to your advantage.

Clamping Stock

All raw metal stock and parts to be drilled must be clamped down to the drill press table, clamped to a worktable, or held in a drill press vise or bench vise. There are three important reasons for this: First, despite the use of lubricants, the drilling process will produce some heat. This can cause a piece of stock to become too hot to hold with your hand. Second, the motion of the drill bit can cause the stock to move slightly, leaving you with a hole that's slightly off the mark. Finally (and most importantly), the drill bit can catch in the stock and cause it to spin out of your hands, which may lead to an injury. Following is a discussion of some common stock shapes and how to clamp them effectively.

Caution Don't underestimate the amount of torque and speed that a drill press can transmit through a bit caught in a piece of metal.

Tip It's handy to have three or four big C-clamps around the shop for drill press setups and other jobs.

Using a Drill Press Vise

The drill press vise (shown in Figure 7.8) is an essential tool in drill press setups. It securely holds parts that are too small or whose shape is not easily clamped directly to the drill press table. For effective use of the vise, the following guidelines should be followed:

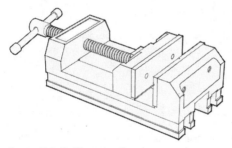

FIGURE 7.8: Drill press vise.

- The vise should always be clamped to the drill press table. Simply securing a part in the vise does not guarantee your safety if the part catches on the drill bit. Clamping the vise to the drill press table is the best insurance against harm.

- The jaws of the vise should be blown off and wiped clean with a rag before clamping. No chips should be present on or underneath the vise. Stray chips can cause you to drill crooked holes.

- Make sure that you don't drill into the vise by keeping in mind where the drill bit will exit the part. You can also use a piece of scrap wood underneath to help prevent this.

- Small parts should be propped up to a height that's just above the jaws of the vise, so that the drill bit doesn't have to dive down between the jaws. This is best done with a set of parallels, as described later.

What Are Parallels?

Parallels are usually 1/8-inch-thick, 6-inch-long steel bars in a set of various precision heights. Their precision guarantees that a part is parallel to the base of the vise. Sets of parallels can be pretty expensive; a more cost-effective alternative (albeit lower precision) is to use 0.100-inch brass stock. K&S Brass makes an assortment of sizes of brass bars.

If you find that the brass pieces are falling down (collapsing like a house of cards) when you try to stack them, put a small piece of double-stick tape on each piece of brass, and then stick it to the sides of the vise jaws. Start your stack at the bottom, so that you make sure all the pieces above it are straight, and make sure that the tape doesn't fold over the edge of the brass, which would cause a high spot, leading to inaccuracy. You can also bend a thin piece of brass into a hump. As you close the jaws, the bending strip acts like a spring, keeping outward pressure on the parallels, to keep them from falling down.

Plate Stock

This is the most common configuration for armor. Clamping a plate is pretty straightforward. You'll bolt it on top of the wood plate (as described earlier) directly to the drill press table. Try to have as much of the plate supported by the table as possible. If the plate is very large, have a friend help support it on the side so that it rests flat on the table. You can swing the table horizontally to get better access to the edges of the plate while still allowing most of it to be supported by the table.

Thin Plate Stock

Thin plates (less than 1/16-inch thick) should be treated as a different type of stock from thicker plates. This is because of their tendency to flex when clamped down at the edges. Also, a stuck bit will pull the plate upwards. By adding long strips of wood or metal and clamping them as shown in Figure 7.9, you can prevent this flexing from happening. The idea is that if the part can easily flex, you should get the clamping as close to the hole as possible.

Tip A *Uni-bit* or stepped drill bit (shown in Figure 7.10) works very well in thin stock. Unlike regular twist drills, a Uni-bit has a single flute (cutting edge), so it is much less likely to grab the thin plate and pull it up as it breaks through the bottom. Make sure to indicate the step you want to go to with a marker before you start drilling.

Bar Stock

This is one of the most dangerous types of stock to drill because the bar can easily catch and swing around, causing a serious injury. Bar stock should be clamped directly to the drill press table like plate stock, or secured in a vise that's clamped to the table. If the bar stock is fairly long, you can take an extra precautionary step by clamping it so that the excess length is lined up on the left side of the drill press arm. That way, if it breaks free, it will immediately stop at the arm, instead of swinging all the way around past you.

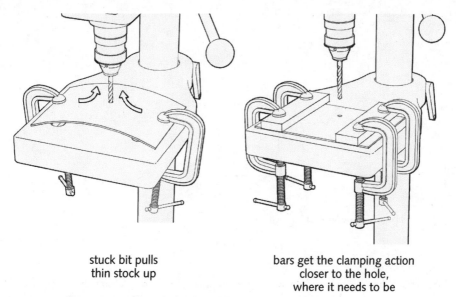

stuck bit pulls
thin stock up

bars get the clamping action
closer to the hole,
where it needs to be

FIGURE 7.9: Thin plate stock clamping technique.

Round Stock and Tubing

Round stock and tubing should be held in a drill press vise that's clamped to the table. Don't try to clamp the stock directly to the table.

Drilling round stock also requires special attention, because the drill bit is more likely to skip around, even with a punch mark. Before using a regular drill, make a punch mark as usual, and then use a combination center drill to make a wide dimple, as shown in Figure 7.11. Because the center drill will not flex around the side of the tube like a standard drill would, it's better suited to starting the hole.

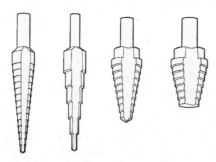

FIGURE 7.10: Various Uni-bit styles.

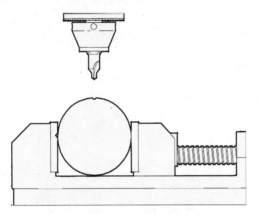

FIGURE 7.11: **Round stock drilling procedure.**

Irregular Shapes

You should use your own judgment in clamping irregular shapes. Think about where you're applying pressure, and try to design your clamping scheme to handle the force. Consider what might happen if the drill bit caught in the stock.

Precision Holes

For most operations, a regular twist drill is more than enough to do the job. However, some items, such as bearings, need a really precise hole diameter. For large precision holes, you need a *boring head*, as discussed in Chapter 10, "Mechanical Building Blocks." Small precision holes can be made with a *reamer*. Regular twist drills tend to flex and wander a bit, even under ideal conditions. A reamer is specially designed not to flex at all, and should be used at a lower speed with a lower feed rate, and plenty of lubrication.

You have to drill a hole to start with, since the reamer cannot drill a hole itself. Unlike a drill bit, it has straight flutes that are designed to cut away only a small amount of material at a time. For the best results, your hole should be the next smaller fractional size from the reamer size that you're using.

Large-Diameter Holes

Usually, you'll have to mail order any drill bits that are more than 1/2 inch in diameter. You should make sure that the shank of the drill bit is small enough to fit in your chuck, which is usually 3/8 inch on cordless drills and 1/2 inch on drill presses. This is called a *reduced-shank drill bit*.

Tip

Often it's the case that if you're using a really big drill bit, it's easier to start by drilling a little hole, called a pilot hole. Because larger-diameter drills have a large point, sometimes it's hard for them to catch the small punch mark that you've made to indicate the hole's location. Also, you may have a lot of chatter because the bit is trying to take off a lot of material in each rotation, which can lead to poor hole quality.

Another way to drill large-diameter holes is by using a hole saw, as shown in Figure 7.12. You've got to drill a pilot hole at the center of the large hole to help you maintain alignment. Hole saws should be used at the slowest speed you can set the drill press to, with a lot of lubrication. Very large sizes (greater than 2 inches in diameter) should be cut out with a jigsaw.

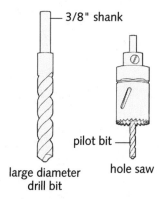

FIGURE 7.12: Reduced-shank drill bit and hole saw.

Drilling to a Specific Depth with a Drill Press

Most drill presses are equipped with a depth gauge that will limit how far down you can move the spindle. This works great in theory, but it's likely that you'll have to make several adjustments before hitting the desired depth. This would mean tightening and loosening the nuts several times. I find that it's easier to tighten the nuts in a position that's close to the correct depth, and then use the table height crank to make fine adjustments.

Drilling to a Specific Depth with a Cordless Drill

Since you don't have a fancy depth gauge system, you're going to have to be a little more creative. You can mark the depth you want with Dykem or a marker. There's also the low-tech approach of putting a piece of tape on the drill bit, which works pretty well. After a few holes, however, the tape tends to get pushed up, and you'll have to replace it. A better approach is to use a block of wood (as shown in Figure 7.13) that's long enough to touch the chuck when the drill is at the correct depth, and prevent you from going any deeper. If you have a standard fractional size drill bit, you can also use a one-piece shaft collar. (Shaft collars are described in more detail in Chapter 10.)

Multiple Sets of the Same Holes

By clamping straight bars or blocks to the drill press table, you can create a setup that will help you quickly and accurately line up multiple parts so that you can drill the same hole in them. You can use wood or metal, or even a ruler. It doesn't matter as long as you create a consistent edge to line up to. Clamping the bars into an L-shape usually works best for aligning parts.

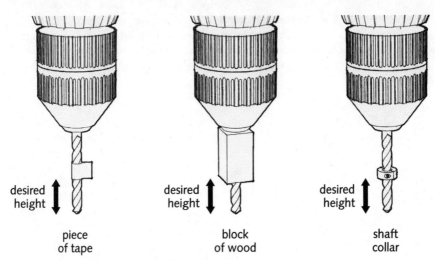

desired
height

piece
of tape

desired
height

block
of wood

desired
height

shaft
collar

FIGURE 7.13: Cordless drill depth stop techniques.

You can also tape several similar-sized plates together with double-stick tape and drill the holes all at once. Double-coated paper tape #410 made by 3M is very thin, but strong, and works well for this procedure. It is available from Grainger and various art supply stores. Make sure that the parts are clean (no lubricant) before you stick them together.

Angled Holes with a Drill Press

In this situation, instead of tilting the table, the best way is to tilt the part and clamp it into a vise. First, use an angle finder to mark a parallel line to the intended drilling angle, as shown in Figure 7.14. Then you can use this line to clamp the part in a vise. Much like round stock, angled holes don't provide the drill bit with a good surface to dig into, so you should start the hole with a combination center drill, giving yourself a wide chamfer (angle on the hole's opening) to guide the drill bit into place.

Angled Holes with a Cordless Drill

The best way to make an angled hole using a cordless drill is to make an angled block. Begin by using a bandsaw or jigsaw to cut a small block of aluminum. Grind or sand the surfaces as flush as possible. Use the drill press to make sure that the guide hole is true. Use an angle finder to indicate the correct angle on a piece. Rough cut the angle into the block with a bandsaw or jigsaw and sand or grind to the mark.

Slide the angle block onto the drill bit and line up the tip to the punch mark as shown in Figure 7.15. Slide the block down to the surface, and adjust the angle of the drill until the angled edge sits flat on the surface. Be careful not to let the block slip.

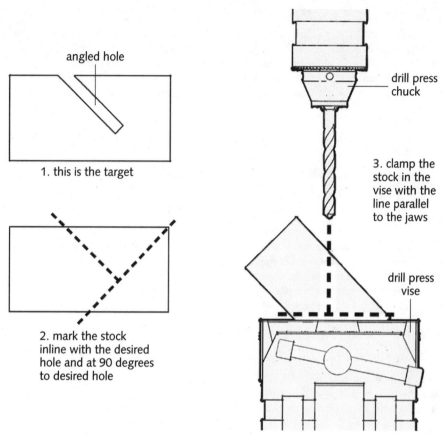

angled hole

1. this is the target

2. mark the stock inline with the desired hole and at 90 degrees to desired hole

drill press chuck

3. clamp the stock in the vise with the line parallel to the jaws

drill press vise

FIGURE 7.14: Angled hole drilling setup on a drill press.

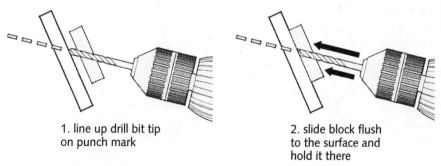

1. line up drill bit tip on punch mark

2. slide block flush to the surface and hold it there

FIGURE 7.15: Angled hole block for cordless drill.

Drilling into Hardened Materials

As described in Chapter 4, "Selecting Materials," certain metals can be hardened by the heat-treating process. While this makes them ideal for weapons, it makes it very difficult to modify them after the fact. In order to successfully drill a hole in a hardened material, it has to be *annealed* (heated) first to make it softer. If you don't do this, you can have some frustrating results destroying drill bit after drill bit with little progress. The process of annealing goes like this:

1. Cut or grind the head off of a nail and chuck it into the drill press. You will be sacrificing this nail for the sake of your project.

2. Use the tip of the nail to line up to the punch mark.

3. Set the drill press to a very high speed. Unlike all other drilling operations, the point here is to generate a large amount of heat. Also note that no lubricant will be used for this step.

4. Lower the spinning nail to the part and apply pressure. Continue pressure until the nail heats up the area around the punch mark to a bluish color. This will indicate that the area has been sufficiently annealed. Raise the spindle and turn off the drill motor. Be careful to avoid touching the part, since it will be very hot.

5. Allow the part to cool down.

6. Insert the correct-sized drill bit and apply lubricant to the part.

7. Start the drill motor, and drill the hole normally.

8. Raise the spindle, and turn off the drill motor.

To make holes in exotic materials like titanium, I like to use solid carbide drill bits. They are significantly more expensive than high-speed steel bits, but I usually only need one or two sizes, and they work really well. Note that carbide bits will also work with hardened materials. They are available for purchase in individual sizes from McMaster-Carr.

Fixing Drilling Mistakes

Yes, sometimes things don't turn out exactly as you expect, even for a veteran robot builder. No problem. This section deals with fixing some common problems with drilling holes.

Fixing a Punch Mark in the Wrong Place

You may occasionally put a punch mark in the wrong place, just shy of the actual mark. To correct this, you will be "pushing" the mark over in the right direction, as shown in Figure 7.16. This is accomplished by punching at the inside edge of the wrong mark a few times towards the correct direction. Then, finish up by punching straight down on the real mark. It might be necessary to repeat this process a few times.

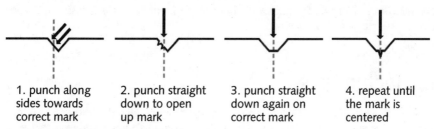

1. punch along sides towards correct mark

2. punch straight down to open up mark

3. punch straight down again on correct mark

4. repeat until the mark is centered

FIGURE **7.16: Fixing an incorrect punch mark.**

Fixing Holes That Are Too Big

For some holes, you may want a tight tolerance, such as a slip fit for a pin. Twist drills may flex, or the drill chuck may not be perfectly straight, giving you an oblong hole that's slightly oversize. You can make the hole diameter essentially smaller by closing down the opening. There are two ways to do this, as shown in Figure 7.17.

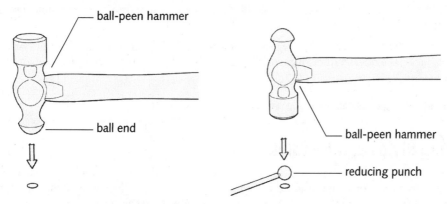

ball-peen hammer

ball end

ball-peen hammer

reducing punch

FIGURE **7.17: Hole-reducing techniques.**

First, using the ball end of the ball-peen hammer, strike the hole squarely. The rounded face will deform the edge of the hole's opening, closing it down. You must make sure to strike the hole squarely when you do this, or you may cause a slight misalignment.

Second, Micro-Mark sells an inexpensive set of peening handles, or reducing punches, that are essentially a hardened steel ball on a little handle. These allow you to accurately hold the ball in place while striking, and do the same job as the ball end of the hammer, but with much less hammering skill required.

Note

You can use an automatic center punch to make a series of divots in a ring around the oversize hole. This will push some of the material inward a little bit, shrinking the effective size of the hole.

Several techniques for fixing holes that are drilled in the wrong place are discussed in Chapter 8, "Fasteners—Holding It All Together."

Tapping Techniques

Making threaded holes is another basic skill for the robot builder. It's pretty easy (as long as you follow the guidelines) and very useful. It offers speed in assembly and requires less space than a nut. This section will introduce the builder to this skill and describe the step-by-step sequence of tapping. Also, you will find out about precautions to avert disaster and how to correct some mistakes.

Screw Sizes and Drills

Since the ultimate goal of tapping is to fasten parts with screws, the tap sizes are specified by machine screw sizes, as shown in the table in Appendix F. Before tapping, you must create the correct-sized hole, since a tap cannot drill a hole for itself. For the taps to work correctly, these holes have to be specific sizes. Number drills are called out for all tap sizes, but some fractional drills can be substituted if lower performance is acceptable.

The *tap drill* is the size drill that must be used on a hole that will be threaded. The *clearance hole drill* (also called a through-hole drill) is used for holes that the screws pass through (no threads).

 Screw sizes and fastener types are discussed in detail in Chapter 8.

Tap Styles and Handles

Spiral-point taper taps are the easiest to get started, and will do the job for most of your robot work. Bottoming taps are intended for blind holes that don't go all the way through the part, although you can easily grind the point off of a spiral tap to get a similar effect.

There are a few different styles of tap handles as shown in Figure 7.18. Generally, smaller taps are intended for use with the T-handle wrench, since it's smaller and offers excellent control, while large taps use the straight wrench because of the leverage.

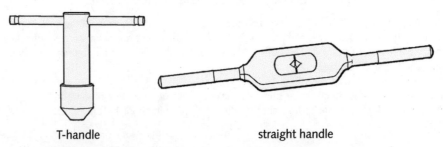

T-handle   straight handle

FIGURE 7.18: Tap wrenches.

Tapping Procedure

All tapping operations are generally done by hand with a tap wrench, which yields the highest precision and control. If you're willing to risk breaking a tap for faster operation, you can chuck the tap into a reversible cordless hand drill. A tapping block should be used to ensure a straight entry. The following basic technique should be used for hand tapping:

1. Just as with the drilling procedure, lubrication must be used at all times. In most cases, WD-40 or some type of household oil will suffice, although you can also purchase chemicals that are specially formulated for tapping, such as TapMagic. For plastics, water will do just fine.

2. The tap must enter the hole perpendicular to the surface of the part. In most cases, a tapping block created from a scrap of aluminum will help you get the tap started straight. The hole that's drilled in the block to guide the tap should be the through-hole size.

3. As you turn the tap trying to get it started, it will tend to pull to one side or the other. Taps are by nature self-centering, but you should try to keep them as straight as possible while starting.

4. If the hole is straight and the tap goes in correctly, then you should encounter very little resistance. If it becomes very difficult to turn the tap, then stop immediately, unscrew the tap from the hole, and check things out.

5. You must break the chips that are created by the thread-cutting process. For every two turns forward, go back a quarter turn. This will break up the chips and keep them from getting in your way.

6. For holes that go all the way through the material, you've got to continue past the bottom of the hole a little way to make sure that the threads go all the way through. If the hole doesn't go all the way through the material, make sure to stop tapping before you hit the bottom of the hole, or the tap may become stuck in the material.

7. When you're certain that the threads are fully formed, unscrew the tap from the hole.

8. Wearing your eye protection, make sure to blow out the hole to remove any chips that are trapped in the hole. For blind holes (where you don't go all the way through the part), you can use a pair of thin needle-nose pliers or a needle to dig the chips out.

9. For metal parts, make sure to rinse the hole in acetone or isopropyl alcohol to remove any lubricants, leaving a clean surface for bonding with the threadlocker, as described in Chapter 8. Acetone and threadlockers should be avoided in polycarbonate parts.

Caution

Taps should never be chucked into a drill press. Not only are drill presses not reversible for tap removal, but also the speeds are too fast for this operation.

Tip

If the hole is threaded all the way through, make sure that you have clearance on the bottom for the tap to exit the part. If you're tapping a blind hole, make sure that you drill a little bit deeper than your thread needs to be. You need to allow space at the bottom for broken chips to compress during the tapping process. Also, you can grind the tip off of a spiral point tap to form a bottoming tap, which will allow you to extend the thread a little deeper.

Making Sure Holes Line Up

Sometimes, you'll mark two pieces and drill them exactly on the marks, but they still don't line up. (It actually only takes a few thousandths of an inch to prevent two pieces from lining up.) The following procedure will help you create aligned holes for pieces that are intended to be fastened together.

1. Clamp the two pieces together in their final orientation relative to each other.

2. Mark and punch the locations of the holes that join the two parts.

3. Drill through both parts with the tap drill. If you accidentally use the through-hole drill, then you're out of luck. You'll have to use a larger size screw or a nut on the bottom side.

4. Clearly mark the correct orientation for the two pieces and unclamp them.

5. Tap the piece that gets the threads as described earlier.

6. Now go back to the piece that will get the through hole. Use a countersink to put a chamfer (a small angle) on the inside rim of the hole. This will help guide and center the larger through-hole drill.

7. Drill the through hole.

8. Now you should be able to fasten the two parts together using the correct size screws.

Dealing with Broken Taps

Unfortunately, it's actually pretty easy to break a tap. They're not very wide, and you're generally applying a lot of force to them. Taps are threaded all the way around, so that they're susceptible to breakage due to excessive side loads. What's worse is that they're made of tool steel and don't usually break off flat, which means that you've got to grind them down. A broken tap also blocks a hole that you usually need.

This section will describe some common causes for tap breakage, so you can avoid them. Also included are some solutions, just in case you find yourself in that regrettable circumstance.

Common Causes for Binding and Breakage

Binding, or encountering a force that causes the tap to stop, often leads to breakage. Most breakages occur because the applied force (by you) causes the tap to snap on a screw thread. Following are a few common situations that may cause a tap to bind and break:

- The hole is the wrong size, requiring excessive force to tap.

- The hole isn't deep enough, causing the tip of the tap to crash into the bottom of the hole, or the chips that have been compacted together at the bottom.

- The builder did not use lubrication. Without lubrication, heat builds up and chips tend to compress together in the threads, causing the tap to bind.

- The builder did not break the chips. The long chips resulting from the thread-cutting process jam together and get stuck in the threads, causing the tap to bind.

- The hole is crooked or has a bend in it caused by changing the drill angle during the drilling process. The tap goes in easily, but then binds when it reaches the angle change.

- The tap did not enter the hole straight. The tap will work okay for a while, but will bind as it goes farther into the hole at an angle and tries to correct itself. In addition, the downward force that you apply to the turning motion will start to bend the tap, and may easily result in breakage.

Solutions

Usually, dealing with a tap that's wedged itself in a part isn't that hard. You can use an adjustable (crescent) wrench on the square part of the shank to unscrew it. However, you've got to be very careful no to put any side load on the tap or it will break off.

Basically, it's best to avoid breaking taps at all cost. However, if you've already broken the tap, then the following techniques describe how to deal with it:

- You can try to grind a slot in the broken tap with a thin abrasive disc in a Dremel rotary tool. Then you can use a flathead screwdriver to try and twist the tap out.

Cross-Reference For a review of Dremel rotary tool safety, check out Chapter 6, "Shaping and Finishing Metal."

- Tap removers are expensive and only work well under perfect conditions. I've actually had very little success with these tools.

- Grind the broken tap flat with an angle grinder and move on.

Project 3: Drilling and Tapping Holes

In this project, you'll use the armor/structure plates you cut out in Project 2 (Chapter 6) and apply the drilling and tapping techniques you learned in this chapter. I'll cover techniques for using the drill press and cordless drill, as well as methods for tapping holes. You will be drilling holes into the aluminum base plate you cut in Project 2, as well as the Lexan side panels. You will also create threaded mounting holes in these pieces, so that you can assemble them in the next project.

Caution Eye protection is required for this operation. Pieces of metal may fly around. Review all of the general power-tool safety protocols described in Chapter 5, as well as the sections that correspond to the specific tools used below.

Note This project makes use of the patterns available on the Kickin' Bot Web site at www. kickinbot.com. You should download them and print them out at 1:1 size. Tape the sheets together to form a large pattern.

Drilling Holes in the Base Plate

You will be attaching the drill pattern and marking holes using the automatic center punch. While a cordless drill can perform all of these operations, the drill press will give you the best results and allow you to work the fastest, because you won't have to concentrate on holding the tool straight to create straight holes. Some of the holes are inaccessible with the drill press, however, and that's where you will use the cordless drill. You'll finish up the base plate by tapping some of the holes for screw threads.

1. Attach the pattern (available online at www.kickinbot.com) to the base plate using 3M Spray 77 spray adhesive. It's best to tape the pattern together first, making sure that your alignment marks match up. Then, apply the spray adhesive to the whole pattern.

2. Using an automatic center punch, mark all the holes indicated. To accurately line up the punch, hold it at a 45-degree angle and center the tip over the X mark, as shown in Figure 7.19. Then raise the body upright and push straight down.

3. Drill the holes using the drill press. Make sure to line up the hole and then clamp the plate to the drill press table. Holes should be drilled with the number drill indicated. The unmarked holes should be drilled with a 1/4 -inch drill bit. Make sure to use a cutting fluid or WD-40 for lubrication.

4. After you've drilled all the holes, remove the pattern and clean the sticky residue from the plate's surface with Goo Gone or another solvent.

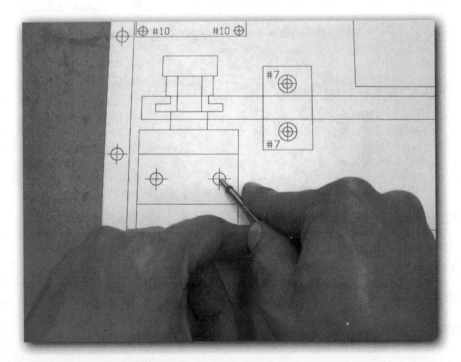

FIGURE 7.19: Correct punch lineup.

5. Some of the holes indicated may not be accessible with the drill press. Use a cordless hand drill to drill these holes. You can use the double-square method described in Chapter 7, "Drilling and Tapping Holes," to make sure that the hole is straight.

6. Chamfer all holes with a countersink. You're only removing the sharp edge from the hole, so apply light pressure (usually the weight of the drill will do) and let the countersink slowly turn three to four times.

7. Tap the holes according to the procedures and precautions described in Chapter 7. Create #10-24 holes for the master power switch, #6-32 holes for the speed controls, and ¼"-20 holes for the battery plate and brackets, chain tensioners, secondary power switch bracket, and receiver and receiver battery brackets. Make sure to use plenty of lubrication.

Drilling Holes in the Side Panels

Again, the drill press is the tool of choice when you've got a lot of straight holes to make. For this series of steps, you will be working on the Lexan side panels. You'll start by using the hole saw (or Uni-bit) to make large-diameter holes for the axles, and then drill the holes in the front and back armor pieces. Finally, you'll drill all of the mounting holes in the top and bottom of each panel, and tap them to form a screw thread.

1. Peel off the protective coating on one side of each of the side panels. Leave the other side covered. Attach the face, top, and bottom patterns to each of the side panels using 3M Spray 77 spray adhesive, as shown in Figure 7.20.

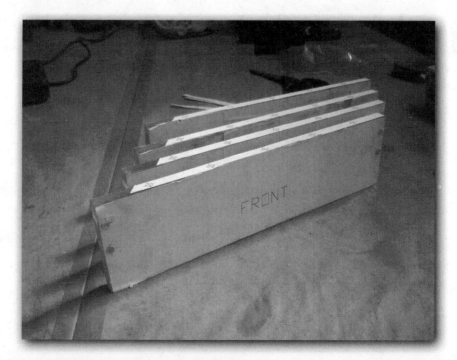

FIGURE **7.20: Patterns applied to the side panels.**

2. Using an automatic center punch, mark all the holes indicated.

3. Lower the spindle speed on the drill press to the slowest setting possible. Consult your owner's manual for instructions on how to do this for your machine.

4. Start by drilling the axle holes in the panels marked "side" (as opposed to those marked "front") with a 1-inch hole saw or large Uni-bit. Add a scrap piece of wood underneath the side panel as shown in Figure 7.21 before clamping the workpiece into place. This will allow you to drill the large diameter axle hole all the way through the piece without damaging the drill press table. Make sure to use a generous amount of lubrication to keep the plastic cool. You can use water or WD-40.

 Note If you use a Uni-bit, make sure to drill a 1/4-inch pilot hole first to help guide the tip into place. Also, you should use the large version with a 3/4-inch maximum diameter to make the initial hole. After you're done, clamp the piece to the table and use the Uni-bit with a 1-inch maximum diameter to open up the hole to the desired 1 inch. You can't start with the 1-inch maximum diameter Uni-bit because it has a flat tip and a minimum diameter of 1/2 inch.

FIGURE 7.21: Scrap wood to protect the drill press table.

5. Next, drill the holes in the "front" pieces using a 1/4-inch drill bit.

6. Drill the top and bottom holes in each of the panels by clamping them into a drill press vise and securing the vise to the drill press table. After lining up and drilling the first hole, you can quickly set up for drilling the next hole simply by loosening the vise and

moving the panel over. You can leave the vise clamped to the drill press table, as shown in Figure 7.22.

7. Tap the top and bottom holes with a ¼"-20 spiral-point tap. Although you can save a lot of time by using a cordless drill instead of the tap wrench to tap holes, I can't officially recommend it as a standard technique because it increases the likelihood of breaking the tap. (This is a personal choice for you.) With a large, strong ¼"-20 tap like this, you probably won't break it, but it's much safer to use the tap wrench.

8. Remove the patterns with Goo Gone and chamfer all holes with a countersink, using light pressure as mentioned before.

FIGURE 7.22: Lineup for top and bottom hole drilling.

Wrapping Up

As you can see from the length of this chapter, it's not always the same old thing when drilling holes. You can be faced with many different situations, requiring the use of different techniques. Carefully following the techniques in this chapter will help improve the accuracy (and fit) of your parts, making it easier to assemble your robot's frame and armor pieces. As mentioned earlier, don't worry if all the screw holes don't line up. Adjustments are a part of any robot-building experience.

Fasteners—Holding It All Together

You will rely on fasteners to hold your robot's frame and armor pieces together. They keep drive components in alignment and prevent heavy items such as batteries from banging around inside the robot. This chapter describes the nuts and bolts of, well, nuts and bolts. You'll learn about the different fastener types and their functions, and why there are so many screw sizes and threads. You'll also learn how to correctly use your fasteners, including detailed procedures for dealing with misdrilled holes, making a screw shorter, and using threadlockers. I'll introduce you to flexible fasteners for shock protection and a listing of fasteners that should be avoided in combat robots.

Screw Types

I use almost exclusively alloy steel and stainless steel cap screws. (The stainless screws are more for appearance — shiny metal versus the dark black alloy screws.) The screws found in the following illustrations require an Allen key to tighten, which provides you with more mating surface than standard slotted or Phillips machine screws, making them less likely to strip out when you turn them. They are more compact than hex cap screws and are widely available with a variety of different head types for different applications.

Socket-head cap screws, as seen in Figure 8.1, are used as general-purpose screws and the thicker heads provide great strength. I use them for almost everything.

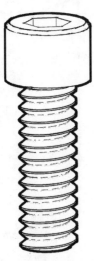

FIGURE 8.1: Socket-head cap screw.

Flat-head cap screws (see Figure 8.2) are used when you need a fastener that's flush to the surface. With an 82-degree countersink, you can sink the screw into the piece you're fastening.

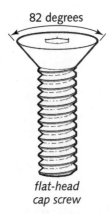

82 degrees

flat-head
cap screw

FIGURE 8.2: Flat-head cap screw.

Button-head cap screws (shown in Figure 8.3) are used where you need a lower profile than a socket head, but can't countersink the screw because the piece that you're attaching is too thin.

Shoulder screws (as seen in Figure 8.4) are a special type of screw that's used for assemblies that are meant to slide or rotate, like mounting an idler sprocket. The smooth shoulder acts like a bearing surface and allows movement in a situation where a normal thread might dig in.

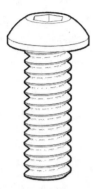

FIGURE 8.3: Button-head cap screw.

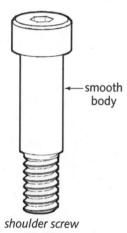

← smooth body

shoulder screw

FIGURE 8.4: Shoulder screw.

When to Tap and When to Bolt

In the last chapter, you learned all about creating threads with a tap. Having tapped holes is pretty convenient when you need to assemble and disassemble something quickly, since you don't need to hold a nut on the bottom with a wrench. You just tighten your screw, and you're ready to go. But there are some situations where it is preferable to drill a through hole and use a nut instead of cutting threads. You must consider if the thing you're fastening is likely to tear the screw out of the threads. Remember, you'll be relying on the strength of the threads in the material you've tapped, which is most likely aluminum. In a high-stress application, you might pull the screw right out of the tapped material, stripping the threads in the process. In this case, you should drill a though hole and put a nut on the other side.

Screw Sizes and Threads

Most U.S. builders use the American Standard for Unified Screw Threads for screw sizes. Why? Because it's the standard that was agreed upon by the United States, Canada, and England. It ensures that the screws you buy will be the same size and thread even though they may come from different manufacturers.

These standard screw sizes are broken down into two categories: Unified Coarse (UNC) and Unified Fine (UNF). The coarse screws have fewer threads per inch, while the fine screws have more. Each screw has a two-part designation. The first number indicates the standard size (diameter) and the second number is the number of threads per inch. They are commonly available in 1/8-inch length increments. You can buy these screws right out of the McMaster-Carr or Grainger catalogs. Table 8.1 lists the most commonly used screw sizes and threads. A more complete listing is provided in Appendix F, "Tables and Charts."

Table 8.1   Common Screw Sizes and Threads

Screw	Standard Screw Size	Actual Diameter (Inches)	Threads per Inch	Series
#2-56	#2	0.086	56	UNC
#4-40	#4	0.112	40	UNC
#6-32	#6	0.138	32	UNC
#8-32	#8	0.164	32	UNC
#10-24	#10	0.190	24	UNC
#10-32	#10	0.190	32	UNF
1/4"-20	1/4"	0.250	20	UNC
5/16"-18	5/16"	0.3125	18	UNC
3/8"-16	3/8"	0.375	16	UNC

Note: Usually, you just drop the number sign (#) and refer to the screw as a "10-32," for example.

Tips for Working with Screws

Really, there's more to it than just blindly turning an Allen wrench. Listed next are some important tips about working with your fasteners, including how to get all the screws in, opening up stubborn holes, and locking them in place with a chemical threadlocker.

Starting All the Screws

You were really careful in drilling the holes. You kept the tap perfectly straight while threading the holes. You put the first screw in, tightened it down, and none of the other screws lined up. Don't worry—it's normal. Unless you've used a vertical mill or CNC machine, this will almost always be the case. Here's what you do: Loosen that first screw almost all the way out; get all of the other screws started by hand; when you've got them all in, start to tighten them down with a wrench.

What If There's Still One (or Some) That Won't Go In?

It's still no disaster, and it happens all the time. You can go one of two ways. The first way is faster, but you end up with a larger hole. The second method is more precise, but it takes a lot longer, and produces the best results on #8 size and larger holes.

Drill the Hole out Larger

Usually, a Uni-bit is the fastest way to do this because it self-centers, meaning that it guides itself to the true center of the hole. If you don't have a Uni-bit, then you can countersink the hole a little so that your drill has a centering guide, and drill the hole out with the next larger bit (the next 1/32-inch) as shown in Figure 8.5. Repeat this until all the screws fit. If the hole is *really* far off and you drill it out too big, you may need to slip a washer under the screw head so that you've got more surface area attaching the material.

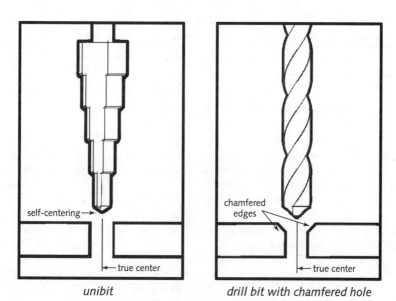

FIGURE 8.5: Uni-bit method versus chamfered-hole method.

See Chapter 7, "Drilling and Tapping Holes," for a discussion of the Uni-bit.

Use a Dremel Rotary Tool and Grind It Out

First, mark the true location where the hole should be. Get out the Dremel tool and your eye protection (and probably some ear protection, too). Using a straight 1/8-inch carbide bit, grind out the hole towards your desired direction, as described in Figure 8.6.

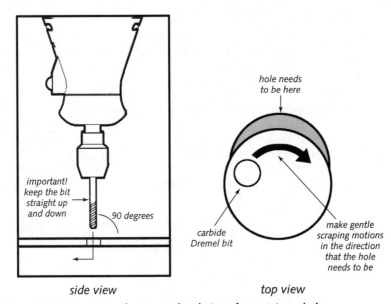

FIGURE **8.6: Proper Dremel rotary tool technique for opening a hole.**

Turn off the tool and clean the hole with a vacuum so that you can keep track of where you're going. The vacuum is a good way of cleaning the hole because this process produces chips that are needle thin and easy to get stuck in your fingertips.

Keep grinding the hole until all the screws fit. It's not the fastest way, but you'll eventually get there. Unfortunately, if the hole is smaller than the carbide bit, then you're out of luck, and you've got to drill it out as described earlier.

Use eye and ear protection, and make sure to review the "Power-Tool Safety" section in Chapter 5, "Cutting Metal," and the Dremel tool usage and safety guidelines in Chapter 6, "Shaping and Finishing Metal," for proper Dremel tool use.

How to Make a Screw Shorter

Often it's the case when you're building a robot that you won't have exactly the right length screw. In fact, it's usually the rule. If you don't have a long enough screw, then it's a trip to the hardware store or a call to McMaster-Carr. Fine. What if you have a screw that's a little bit too long? Then you're in luck. Here are two methods you can use to shorten a screw.

Abrasive Cutoff Wheel in the Dremel Tool

Are you getting the idea that a Dremel tool is useful in building robots? You bet. For this operation, you need to get an abrasive cutoff wheel, as shown in Figure 8.7 It's a standard Dremel attachment that's basically a little disc on a shaft, called a mandrel. The disc is composed of abrasive material that cuts the screw by grinding down whatever gets in its way.

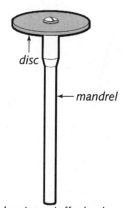

disc

← mandrel

abrasive cutoff wheel

FIGURE 8.7: Abrasive cutoff wheel.

You install and remove the disc by tightening the mandrel in the Dremel tool and preventing the shaft from turning (usually by pressing a button on the tool). Next, you unscrew the tiny slotted screw on top of the mandrel. The abrasive discs come with little holes in the center that are threaded onto the screw and then tightened down.

 Caution

This is a case for full face protection, since the disc may shatter, and usually does. This is why they make the discs easy to replace. Make sure to review the "Power-Tool Safety" section in Chapter 5 and the Dremel tool usage and safety guidelines in Chapter 6 for proper Dremel tool use.

Approach the screw gently. Don't attack it with the wheel, or you'll be showered with bits of broken wheel and you'll have to stop and insert another one. Keep the blade parallel to the screw's head or you'll cut a diagonal through the threads, and end up with an angled bottom on your screw that may be difficult to get started in the hole. Be patient. It will take a few minutes to make it all the way through, depending on the size of the screw. Make sure to clean up the end of the screw on the deburring wheel afterward.

 Caution

Cutting the screw will create a shower of sparks that are harmless to your skin, but may ignite something flammable nearby. Make sure all open containers of flammable liquids are closed and moved away from the work area.

Bolt Cutters Built into a Crimper

Some crimpers have threaded bolt cutters built into them, as shown in Figure 8.8. All you do is open up the handles and turn your screw into the hole that matches your screw size and thread. One side of the handle has the screw threads and holds the screw in place. The other side acts as the cutter. As you close the handles, the bolt is cut off flush to the length that you've screwed into the handle.

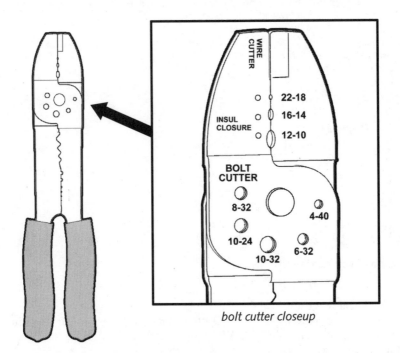

bolt cutter closeup

FIGURE 8.8: Crimper with a built-in bolt cutter.

This method is pretty easy, but it has some drawbacks. You can't cut any screws smaller than #4-40 or larger than #10-32. Also, it can sometimes be quite difficult to cut the screw, requiring a lot of effort. Make sure to clean the end of the screw on the deburring wheel afterward.

 You should wear eye protection when attempting to cut bolts. The cutoff part could fly out and strike you in the eye. Ouch!

The Good Old Hacksaw

The hacksaw can be used to cut most metals, including steel screws. Once you determine the correct length, mark the screw and clamp it in a bench vise. Make sure to spray a little WD-40 on the screw to lubricate the blade and keep it from heating up too much during the cut. Don't touch the screw immediately after the cut. It will most likely be hot. Wait until it cools down, or use a pair of gloves to handle the screw, and drop it into a cup of water. Make sure to clean up the end of the screw on the deburring wheel afterward.

Review the hacksaw safety guidelines in Chapter 5.

Using Loctite

If you haven't already heard about Loctite, then this is what it is: a chemical threadlocker. You put it on a screw and thread it into a tapped hole (or nut) and it cures to form a strong bond that "locks" the screw in place and prevents it from backing out.

Why You Should Use Loctite

I know you will tighten your screws to the best of your abilities. But even the perfectly tightened screw will back out eventually when subjected to enough vibration. And combat robots are, by nature, subjected to a huge amount of vibration, as well as huge shock loads. All of this means that you should use Loctite on all screws in your robot. You say, "But that's a big pain to have to take everything apart, clean it, apply the magic fluid, and put it back together." All right, apply Loctite *only* to the screws on your robot that you don't want to shake loose at the worst possible time, costing you a match.

Avoid using Loctite and other anaerobic threadlockers with polycarbonates (such as Lexan). They do not react well together, and the threadlocker will weaken the plastic.

What Flavors of Loctite Are There?

There are several different flavors of Loctite. Each has its own color and strength. Loctite 222 (Purple) is called the "low-strength" formula. This is the formula that I use the most in my robots. Loctite 242 (Blue) is the medium-strength version. Loctite 262 (Red) is the high-strength (You didn't need that to come off again, did you?) formula.

If you really, really need to get something loosened that has cured 262, I've heard that heat is the trick. To be more precise, a lot of heat, followed by a sharp blow with a hammer. Good luck.

Clean Parts Make a Good Bond

In order to make a strong bond, all of the parts must be clean. That is, there should be no cutting fluid or WD-40 on the parts *or* screws. Why? Because the lubricants form a barrier between the Loctite and the metal. If the Loctite can't get to the metal, then it can't seep in and form a strong bond. So, the moral is: Clean off your parts with brake cleaner, acetone, or isopropyl alcohol. Flush all tapped holes with any of the above solvents. Give them a second or two to dry, and then apply the Loctite.

How Much Should I Use?

You must overcome the overwhelming urge to flood the hole and soak the screw in Loctite. Because it is an anaerobic compound, it cures in the *absence* of air, and the less you use, the better. The instructions indicate that you should only use a drop per screw. Put it right on the thread. If you do end up soaking the screw, just wipe off the excess with a clean rag.

How Long Does It Take?

I usually leave it overnight. Besides, if you're putting Loctite on something, you probably don't want to take it off immediately, right? Just let it sit a good couple of hours, and it will be fine.

Breaking Bolts

Be nice to your bolts — they will break if used improperly, and broken bolts mean armor plates flying off, or drive components losing alignment and becoming inoperative. When using screws to connect two pieces of material (whether it be wood, metal, plastic, or something else), you can subject them to two kinds of loading: tension and shear.

What are tension and shear loads? Imagine that you are attaching the top armor plate to your robot. As you tighten down the screws, you're squeezing the top armor down onto the frame of your robot. This squeezing together is a tension load. A two-piece shaft collar also squeezes together, clamping onto a shaft. The direction of the load is in the direction of the screw body. Now if you can imagine a sliding or slicing force, that's shear. That force is applied perpendicular to the direction of the screw body, as shown in Figure 8.9.

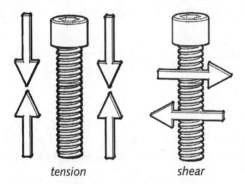

tension   *shear*

FIGURE 8.9: Tension and shear.

Carroll Smith, engineer and racecar constructor, says in his *Nuts, Bolts, Fasteners and Plumbing Handbook*, "Always bear in mind that clamping is the function of bolts and that location is the function of dowel [pins]."[1] I agree with Mr. Smith and provide the following example to support that doctrine.

Consider the rotary weapon blade shown in Figure 8.10. A robot builder wants to attach the weapon blade to the hub, which will rotate. That's fine, but on impact, all of the screws are likely to be sliced right off because of the huge amount of shear (sliding, slicing) force that they will be subjected to.

Now, the usual course of action is to get bigger, stronger bolts to hold the blade in place. This will probably work, but over time, you will see that the repeated impact will result in wear on the bolts and threads.

What would be a better way to overcome the shear force? This would be the proper job of something other than the bolts, whether it is the structure (redesign the part) or hardened dowel pins. As illustrated in Figure 8.11, you can use the screws in tension to hold the bar in place, and the dowel pins will absorb the shear load.

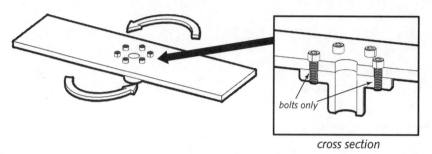

FIGURE **8.10: Rotary weapon example with bolts only.**

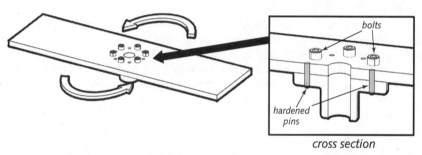

FIGURE **8.11: Rotary weapon example with bolts and pins.**

Flexible Fasteners

Some components of your robot are much more fragile than others. In fact, some are so fragile that an impact could cause them to shake an essential part loose (think speed controls). In these cases, it's handy to have something to absorb the impact energy, and that's where Neoprene- (and rubber-) based fasteners come in. Also, a few builders use these fasteners to shock mount their armor pieces against impact, allowing them to drive around the arena and smash into walls at full speed without blinking.

Tip   These should absolutely be used in compression only, since a shear load could cause the neoprene to rip apart.

Cylindrical Sandwich Mounts

These fasteners are called "sandwich" mounts because you have two mechanical fasteners (male or female) bonded to the top and bottom of a neoprene or natural rubber disc, forming a little sandwich as shown in Figure 8.12. They are available from McMaster-Carr in all standard screw sizes.

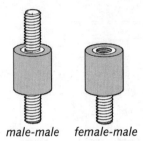

male-male   female-male

Cylindrical sandwich mounts

FIGURE 8.12: Cylindrical sandwich mounts.

Rubber-Insulated Rivet Nuts

I know the name says "rivet nuts," but you don't actually rivet these shock isolators. Instead, you drill a hole large enough to press them into, and then put a screw through the item you're mounting into the thread in the middle of the rivet nut. As illustrated in Figure 8.13, when you tighten down, the thread sucks the rivet nut upwards, and expands the neoprene just like a standard rivet. The cool thing is that you can remove it by loosening the screw. They are available from McMaster-Carr in all standard screw sizes from #6-32 to 3/8"-16.

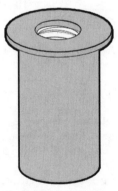

FIGURE 8.13: Rubber-insulated rivet nut.

Velcro

I wish I could devote more space to Velcro. It's saved me in the past, but what more can I say? I carry adhesive-backed Velcro in the toolbox wherever I go. It does two really useful things: First, it sticks an item down wherever I want it, and second, it gives me a little shock protection, because it allows the part to move just a little bit. I keep the items that I secure with Velcro limited to small things such as radio control receivers and their batteries. In the past, I've used Velcro in combination with cable ties to hold down larger items.

Other Fasteners (and Why We Don't Use Them)

Not every fastener under the sun is up to the rigorous punishment that we combat robot builders will be putting them through. In fact, most other fasteners, though widely available and handy in some situations, are simply not appropriate. Next I've listed a few alternatives that are fine for non-combat situations, but should be avoided in the arena.

Rivets

Rivets are quite easy to install. You drill a hole that's the right size through both pieces of material that you wish to fasten together. You put the rivet in the tool, and place the rivet body in the hole. Squeeze the tool and the rivet expands and breaks off the remaining stem in one smooth "pop" motion, leaving a strong connection. The problem is that in order to remove the connection, you've got to drill the rivet head out, which can be time consuming. Also, you're limited to relatively thin sheet materials. And finally, you're relying on that tiny flange, which you were able to deform by hand (with a tool) to maintain your connection. Skip the rivets and drill and tap. You'll be happier later on.

Rivet Nuts (or "RivNuts")

The best application for rivet nuts is to quickly put a threaded hole into a material that's too thin to tap, or that is likely to tear its threads. Similar to the procedure for a rivet, you drill a through hole of the appropriate size. You then screw the rivet nut onto the tool's mandrel, put the rivet nut in the hole, and squeeze. Simple installation, but I don't trust any formed flange technology in a combat robot.

Nut Inserts (or "NutSerts")

Nut inserts are very similar to rivet nuts except that they aren't riveted. When you turn the screw handle on the insertion tool, it draws the nut insert up and compresses it, forming a flange on the top and bottom just like a rivet nut. A cool solution, but it falls into the same category as the previous two.

Sheet Metal Screws and Self-Drilling Screws

Sheet metal screws are great for, well, sheet metal. You can get them in pretty long lengths, but really, they are limited by the thickness of metal that they can accommodate.

The difference between sheet metal screws and self-drilling screws is that the self-drilling screws have a little drill-bit-like tip that works in theory, but in actual practice, it sucks, and it's best to drill a pilot hole for both types of screws.

Tip

The key to properly installing sheet metal screws and self-drilling screws is to use a cordless drill with a clutch. Set the clutch to trip at a pretty low torque (in the bottom 25 percent of the range) and put the screw in place in your pilot hole and tighten it down. If you don't use the clutch, you'll probably end up with a screw spinning in a stripped out and useless hole.

What about Welding?

Welded frames are certainly not rare in the fighting robot world. However, most of these frames have been welded by competitors who already have experience welding, or by someone they know who is a welder, or by a professional welding shop. It's not a skill that you can pick up immediately. It develops over time, with lots of practice. While it's true that you could probably make a joint your first time with a welder, the strength of that joint will certainly not be optimum. Creating strong welded joints takes practice to master, which is time that you generally don't have when preparing for a competition. You will be relying on these joints in battle, and you want them to be as strong as possible.

Also, you won't always have access to welding equipment at an event (and it's usually not allowed in the pit area), making it difficult for you to perform repairs. Some larger competitions have on-site welding support from a sponsoring welding supply vendor, but this is a rare case, and you can't count on it. Rookie builders should rely on fasteners for frame and armor components. You'll be better off when it comes to repairs and upgrades.

Sounds pretty easy, right? The problem is that this is a one-way trip. These types of screws are best left alone. If you try to remove and reinstall them, you will end up stripping the threads, and then you're out of luck. Experience says, don't bother with these. They're more trouble than they're worth for combat situations.

Wrapping Up

This chapter introduced you to the different types of screws that you have to choose from, and how to correctly implement them in your design. It also presented some important techniques for dealing with misdrilled holes, which are a fact of life for the robot builder.

Correctly using your fasteners will allow your armor to do its job protecting the robot. Fasteners are what hold your robot together. If enough fasteners fail, you could be left with your insides exposed to attack, or worse, incapacitated in the arena with an inoperative drive system.

---

1 Smith, Carroll. *Carroll Smith's Nuts, Bolts, Fasteners and Plumbing Handbook*. St. Paul, Minnesota: Motorbooks International, 1990.

Selecting Drive Motors

Having a good drive system is critical to your robot's success in competition. Remember, if you can't move, then you're out. If you can't maneuver very well, then you won't be able to deliver your weapon, and you'll probably get the snot kicked out of you by a robot that's faster and more nimble. At the heart of the drive system are the motors, and this section will help you find the right motors for your robot by first introducing what a motor is and what some of the more important characteristics are. Then, the top ten most popular motors in robot combat are listed, so you can see what other builders are using and why these motors are so popular. Included is a discussion of gearing and gear ratios to help you make the most of your motor's power. Finally, if you don't find something that works for you in that list, I'll discuss what to look for if you want to find a motor of your own through surplus suppliers.

What Makes a Good Drive Motor?

There are lots of good drive motors to choose from. Each one has different strengths and weaknesses that make it better or worse for a particular application. The trick is to match the motor to your robot. Do you want your robot to be methodical and precise? Perhaps you want to fly around the arena and use your speed as a weapon. Whatever your strategy, you can examine the different motor properties such as RPM, torque, horsepower, and weight to decide what's most important to you.

DC Motor Basics

Almost all combat robots use DC (direct current) motors for their drive systems. It's a well-developed locomotion technology, and good parts are readily available, and in some cases, quite affordable.

The basic idea is that electrical energy (in the form of a DC voltage) is applied to the motor's power input leads, causing the output shaft to spin. The turning force of the output shaft is called *torque* and is usually measured in inch-pounds. The speed at which the motor turns is measured in *RPM*, or revolutions per minute. Applying a lower voltage will cause the motor to spin more slowly, and reversing the polarity of the voltage will change the direction of the spinning.

While the shaft is spinning, the motor is also drawing *current* from the battery. Current is the flow of electrons, and if your wire is not large enough to handle this flow, or if your battery isn't large enough to supply the amount of current requested by the motor, you will end up starving the motor, never reaching full power.

Cross-Reference Appropriate wire size is discussed in Chapter 16, "Wiring the Electrical System," and battery technologies are described in Chapter 15, "Choosing Batteries."

The parts of a permanent-magnet DC motor are shown in Figure 9.1.

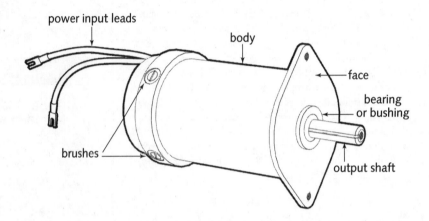

FIGURE **9.1: Parts of the permanent-magnet DC motor.**

Note Although this discussion will be focused on permanent-magnet DC motors, brushless motors (by manufacturers such as Aveox) are starting to gain some popularity in the combat community, and may be a consideration for you. They generally have higher efficiencies than brushed motors, and good power-to-weight ratios. However, you will have to use a special controller and the cost will probably be a bit higher than your average brushed DC motor.

Combat Robot Motor Comparison

Table 9.1 compares some of the most popular motors used by the top competitors in our sport. More detailed descriptions of the different categories follow, along with an explanation of their importance in your selection.

Table 9.1   Basic Characteristics of Popular Motors

MOTOR	Running Voltage (volts)	Stall Torque (in-lbs)	No-load Speed (RPM)	Peak Power (W)	Weight (lbs)	Price
AME D-Pack	12	173.1	6960	3561	7.75	$46
Andrus 5015-1	24	60.8	4107	738	7	$100
Andrus 5017-1	24	40.8	4292	518	4.6	$75
Andrus 5019-1	24	147.7	4434	1935	11.9	$140
AstroFlight Cobalt 15	12	2.6	38568	299	0.47	$130
AstroFlight Cobalt 40	24	12.5	16368	605	0.78	$160
AstroFlight Cobalt 60	36	22.5	12492	831	1.375	$250
AstroFlight Cobalt 90	48	33.1	12288	1203	2	$300
Bosch GPA	24	97	4100	1175	8.4	$172
DeWalt 18V (gearbox in high)	24	150	2133	946	1.6	$40
Dustin Motor (36V hi speed)	36	185	3000	1640	3.375	$260
EV Warrior	24	48	5000	709	3.5	$15
MagMotor mini (S28-150)	24	123.1	6000	2183	3.8	$299
MagMotor 3" (S28-400)	24	232.5	4900	3367	6.9	$349
MagMotor 4" (C40-300)	24	240	4000	2837	11.9	$299
NPC 1200	24	160	3445	1629	9.6	$215
NPC 2212 (12V only)	12	180	285	152	5.1	$155
NPC 2423	24	22.1	3700	242	4	$62
NPC Black Max (Scott)	24	281	3400	2823	15.7	$265

Continued

Table 9.1 (continued)

MOTOR	Running Voltage (volts)	Stall Torque (in-lbs)	No-load Speed (RPM)	Peak Power (W)	Weight (lbs)	Price
NPC R81/R82	24	896	235	622	15	$285
NPC T64	24	825	230	561	13	$286
NPC T74	24	1480	245	1072	14.4	$291
NPC 02446	24	42.6	4000	504	4.8	$72
NPC 41250	24	375	174	193	7.5	$155
Sullivan Dynatron	12	21.3	4800	301	3.69	$35
Sullivan Hi-Tork	12	13.1	5500	213	2.64	$25
Sullivan Model 4	12	37.5	4000	443	6.42	$50
ThinGap TG3200-35	12	10	22000	600	0.625	$239
ThinGap TG3200-42	12	6.9	17700	350	0.625	$239

Sources: NPC Robotics, Team Delta, Robot Marketplace

Note: Prices can change over time. Use these figures as a guideline for budgeting.

Cross-Reference The stall current and torque constant (K_t) are other motor characteristics that are useful in estimating the battery requirements of your robot, as described in Chapter 15. These numbers for the motors listed in Table 9.1 are listed separately in Appendix F, "Tables and Charts."

Voltage

Every motor has a rated voltage, but most motors can be run higher than their rated voltage without damage. The thing that you have to watch out for is the heat buildup in the motor. By doubling the voltage, you can actually quadruple the power output, since power is proportional to the square of the voltage. Note, however, that your current consumption will also increase along with the voltage. Also, higher voltage means more batteries in series, which is more weight and expense.

Stall Torque

This is the maximum amount of torque that the motor puts out. A stall condition is where the shaft is stopped and no longer rotating.

No-Load RPM

This is the number of complete rotations the shaft makes per minute with nothing (no load) attached to it. It is also called the free-running speed.

Peak Power

The first impulse of the robot builder is to compare raw torque ratings and select the one with the most muscle. This isn't a bad technique, but in selecting a drive motor, the true measure of a motor's muscle is power, which takes into account both the torque and speed of the motor.

Cross-Reference If you're dying to know exactly how to calculate power, skip to the end of the chapter to the "Discovering Your Own Motor" section.

Weight

Weight plays a big part in motor selection for lightweight and middleweight robots. Since they have less weight to dedicate towards drive, bigger horsepower motors are usually out of the question. Larger robots have more options when it comes to motors because they can dedicate more of their weight to the drive system.

Expense

In general, good motors cost more. There are higher production costs for tighter precision construction, better magnets, and quality brush holders. However, there are a few *great* values to be found among the most popular motors, and cost is one of the reasons that might make a motor popular in the first place.

Mounting Options

Some motors have a convenient faceplate with a known bolthole pattern (see Figure 9.2). Some motors have a mounting bracket attached to the motor itself. Some motors are just plain hard to mount. This should be taken into account when selecting your drive motor because of the time it may take to make a bracket. Note that some of the more popular motors have custom aftermarket brackets.

Gearboxes

Does the motor have a built-in *gearbox*? A gearbox reduces the output speed of the motor to something that could drive a wheel. If a motor has a built-in gearbox (a *gearmotor*) then in most cases, you can simply attach a wheel and you're ready to go. If not, you'll have to buy or make your own. There are a few good gear sets emerging. NPC Robotics is currently offering complete packages that even include the bearing blocks and wheels. You could also go with sprockets and chain, but that might weigh more, and the higher the geardown, the larger the sprocket, limiting you to a maximum usable size of 8:1 for most robots. If you need more geardown than that, you'll have to add additional stages.

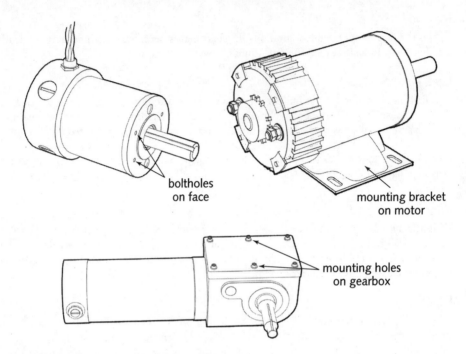

boltholes
on face

mounting bracket
on motor

mounting holes
on gearbox

FIGURE 9.2: Some different mounting configurations.

The Top Ten

The motors listed in Table 9.1 are among the most popular in robot combat. In this section, I'll describe in detail the top ten (or so) motors chosen by builders today and some of their strengths and weaknesses.

AstroFlight Cobalt

If weight savings is your main concern, and price is not too big a factor, then the AstroFlight motors (see Figure 9.3) are the ticket. Of all the top motors, they pack the most power into the smallest packages. With very small diameters, they can fit into ultracompact spaces for a really low profile. Since they are so light, they can be used in multiples for high performance. The Cobalt 15 and 40 are *two-pole* motors, meaning that they have two brushes. The Cobalt 60 and 90 are *four-pole* motors, which allows them to carry more current in four brushes with less loss due to internal resistance.

Active cooling is generally recommended for these motors, since their small size leaves them little heat-sink area to get rid of all the heat. These motors will require geardown even if they are purchased with the optional (expensive) AstroFlight gearboxes.

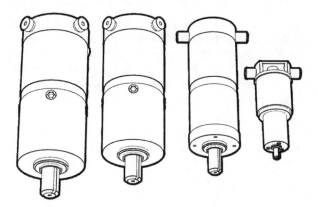

FIGURE **9.3: AstroFlight motors.**

Bosch GPA

Early on, the Bosch GPA (see Figure 9.4) held the distinction as the motor that everyone had, or wanted to have, because of its combination of light weight and horsepower. Although many U.S. competitors have moved on, the Bosch GPA is still popular among European competitors on the Robot Wars circuit. They do require some modifications to handle large shock loads. This motor has a few aftermarket accessories, such as a handy little PC board made by Team Delta with capacitors to reduce motor noise that fit inside the case. This motor has a no-load speed of 4,100 RPM and will require some geardown.

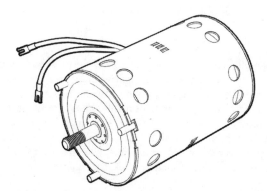

FIGURE **9.4: Bosch GPA motor.**

Cross-Reference   See Appendix D, "Online Resources," for a listing of Web sites that deal with motor modifications.

Briggs & Stratton Etek

This motor doesn't really belong in a discussion of drive motors, but you can't mention any motors for robotic combat without including the Etek (see Figure 9.5). It is not meant to be a drive motor. It is meant to drive the biggest, baddest weapon you can muster. Very few speed controls can handle the amount of current it requires to drive this motor, and you would be better off using a relay to switch this motor on and off. It's the most horsepower you can get in a relatively compact and lightweight package.

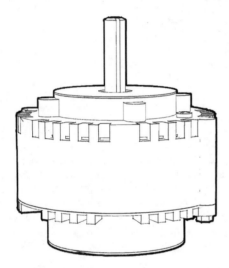

FIGURE 9.5: Etek motor.

DeWalt Drill Motor

The DeWalt drill motor shown in Figure 9.6 emerged a few years ago as the most widely used motor in robot combat. It's no mystery when you consider the application for which it was developed: a cordless drill that was intended to be lightweight and compact, powerful yet efficient on current, sturdy, geared to a relatively low RPM, and affordable. These are all ideal qualities in a combat motor. No other motor can claim to have been used in all weight classes of combat robots, from lightweights up to superheavyweights (using many motors per side). Looking at the numbers in Table 9.1, you can see that the drill motors have truly outstanding power-to-weight ratios. In addition, because of the attached gearbox, the motors could be used to drive a wheel directly.

This is not to say that these motors are ideal right out of the package. They have inherent weaknesses, which include a plastic gearbox (which is prone to shattering) and a small shaft coupling area (which makes it difficult to connect a shaft to the gearbox). Fortunately, over time (just as with the Bosch GPA), builders have developed solutions for these weaknesses, which are available online as aftermarket accessories from the suppliers listed in Appendix D.

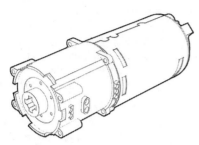

FIGURE 9.6: DeWalt drill motor.

Tip

Check out the Colson caster wheels for use with drill motors. They have a proven track record as a good match to these motors.

EV Warrior

The next great motor that everyone had to have is the EV Warrior (see Figure 9.7). It even sounds like a motor intended for combat. In reality, this little brother to the Bosch GPA was used in a failed electric bicycle venture (hence the name EV — electric vehicle). The EV Warrior is quite popular among lightweight and middleweight competitors.

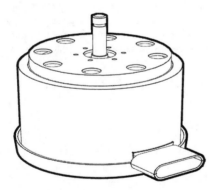

FIGURE 9.7: EV Warrior motor.

Its proliferation stems from the fact that it is one of the cheapest (if not the cheapest) motor solutions for a combat robot, while still giving a surprisingly high level of performance in a lightweight package. As with many of the other motors, the EV Warrior requires some tweaking, which includes removing the tabs that short the case to one of the power leads to prevent the conducting current and frying your speed controls. Several excellent Web sites document the processes required to make this a competition-ready motor.

Cross-Reference See Appendix D for a listing of Web sites that deal with motor modifications.

In addition, these motors were sold in left/right pairs that were *timed* according to their side. Motor timing is the process of adjusting the brush position so that you achieve more horsepower in one direction than another. With this in mind, you should seek out these motors in pairs, usually designated as a CW (clockwise) motor and a CCW (counterclockwise) motor.

The pancake design makes for a larger diameter motor, which translates into a taller robot. In the past, it was a challenge to mount this motor to the robot's frame, since it has no usable bolt pattern on the face, like most other motors. Nowadays there are lots of custom aftermarket aluminum mounts for sale, and some interesting cast urethane ones as well. Geardown will be essential to achieve the maximum benefit and prevent overheating.

Now, here's the bad news. Like the Jensen motors described later in this section, at press time, the EV Warrior is in short supply. Since its only known availability is in the surplus market (no new production), supplies were limited to begin with. That supply is dwindling due to the meteoric rise in popularity of these motors. Hopefully a new source can be located to continue the line. Otherwise, this motor may become extinct.

MagMotors

At the top of the food chain is the MagMotor, shown in Figure 9.8. Designed from the ground up as a motor specifically for robot combat (with the help of veteran Carlo Bertocchini), this is the Rolls-Royce (or more accurately, the Ferrari) of motors. The MagMotors are built for maximum strength-to-weight ratio and boast high-energy neodymium magnets for high torque in a small and lightweight package, and high efficiency through a large power range. These are robust motors with four large brushes and heavy gauge wire. To reduce radio noise, the motors even have built-in capacitors. The only drawback is the price. Then again, would you expect any less from the finest motor you can buy?

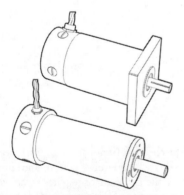

FIGURE 9.8: MagMotor motor.

Since the motor comes with no mounting system, you'll need to take care of mounting through an aftermarket mounting block/heat sink, or by using the bolt hole pattern on the motor's face. Also, to make the best use of this power in a drive system, you'll have to gear it down, since the no-load RPM is in the thousands.

Tip If you're building a lightweight robot with a spinning weapon, the mini MagMotor (S28-150) has the highest horsepower in a single small motor on the market.

NPC 2212

The NPC 2212 shown in Figure 9.9 is a good motor choice for lightweight robots, and is gaining popularity among builders. The right angle drive is handy for mounting and keeping the inside of the robot uncluttered. It has dual output shafts, so you use it as a right- or left-side motor (hacking off the shaft you don't use). It has an output speed appropriate for direct drive and good torque ratings for a smaller robot. Although it's not as powerful as a DeWalt motor, it's an all-in-one package, and you could be up and running quickly and with minimal effort, which counts for a lot if you're running behind schedule.

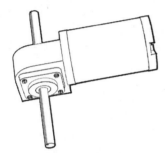

FIGURE 9.9: NPC 2212 motor.

NPC R81/R82

The R81 and R82 motors (see Figure 9.10) are the replacements for the 60522 motor that was popular among middleweights and heavyweights. One of the features that made this motor a favorite was the built-in right-angle gearbox. This is a great solution for robots that require a lot of interior space. In addition, the gearbox has convenient mounting holes and the output RPM is great for directly mounting a wheel. At this weight and power level, this motor is appropriate for use in a pair on a middleweight robot, or in multiple pairs for larger robots. The R81/R82 indicates whether the output shaft comes out of the right or left side of the gearbox.

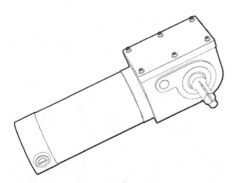

FIGURE 9.10: R81 and R82 motors.

NPC T64 and T74

Formerly known as the NPC 64038 and NPC 74038, the NPC T64 and T74 motors (see Figure 9.11) have made their way into some of the most famous combat robots on the U.S. circuit. From middleweights to superheavyweights, these motors are favorites among the builders because of their ease of mounting in the robot, ease of connecting a wheel, built-in gearbox, and reliability. They have excellent speed and torque characteristics, although it should be noted that if you want a lot of speed, this is not the motor to choose. The relatively low RPM makes this a precise motor for positioning your robot. You can get more speed by using larger wheels (10 or 12 inches in diameter), but these make large targets for spinner robots, and should be protected.

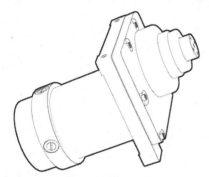

FIGURE 9.11: T64 and T74 motors.

Lots of upgrades to these motors are now available, such as custom billet gearboxes, different (faster) gear sets, and a variety of wheel-mounting hubs.

Tip These motors are often used with Carefree Battletreads or NPC Flat-Proof tires. NPC Robotics carries a full line of custom hubs to match their wheels to these motors.

Sullivan Starter (Jensen Equivalent)

In recent years, a popular choice among robot builders for performance and price was the Jensen starter motor. Many lightweights and even some middleweights used these motors in their drive systems. Since then, these motors have become somewhat scarce. (For a while, they became a hot commodity and even went *up* in price.) Fortunately for the robot builder, the Sullivan starter motors (see Figure 9.12) are equivalent in size and performance to these motors, and the prices are really cheap.

Since these are starter motors (meant for starting model airplane engines), they are intended to produce a *huge* amount of torque for a *short* amount of time (that is, intermittent duty). While robot combat might qualify as intermittent, there are times where you will call upon a motor to push in a stalled condition. In this situation, a starter motor can quickly overheat and destroy itself. This is why active fan cooling and geardown is necessary for these motors.

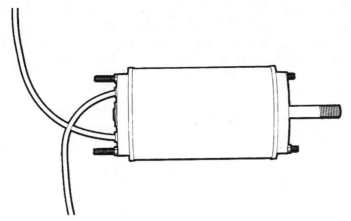

FIGURE 9.12: Sullivan starter.

Runner-up Choices

There are other great motors that aren't in quite as widespread use as the top ten, but are good choices nonetheless.

AME D-Pack

I remember looking at this motor in a surplus catalog some years back and thinking how it was amazing that they got all that horsepower into a regular-size package (3-inch diameter × 7 inches long). The efficiency peaks at 80 percent at 1.5 HP, and surprisingly, it remains above 70 percent up to 3.5 HP, which is pretty good for a motor in this price range. Note that the AME D-Pack (see Figure 9.13) is more suited to running a spinning weapon than drive.

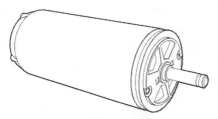

FIGURE 9.13: AME D-Pack motor.

NPC Black Max (Scott Motor)

If horsepower is the goal, then the Black Max shown in Figure 9.14 is one of the biggest. This should be considered for heavyweight and superheavyweight robots only, since the price you pay for all that horsepower is the weight, which is about 15 pounds. In addition, you'll need big batteries and powerful speed controls to handle the amount of current required to drive this monster.

Note The true rating of the Black Max (Scott motor) is 1 HP. The reason for this is that the efficiency is not that good at higher power levels, and at the peak of power (4 HP), the motor is running at only about 50 percent efficiency, which means that only half of the current going into the motor is going towards turning the shaft, and the rest is being burned as heat. What this means for you is bigger batteries and speed controls to make up for the extra current you're going to need.

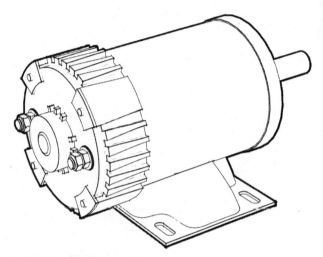

FIGURE 9.14: Black Max motor.

NPC 1200

A good motor can get lost among all the other choices out there, especially when there are motors with built-in gearboxes that have equivalent performance. Don't pass up the NPC 1200 (see Figure 9.15) simply because it doesn't have a gearbox. It maintains over 70 percent efficiency up to about 1 HP. With the proper gearing, this could make an excellent drive system.

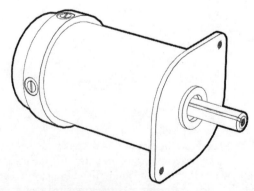

FIGURE 9.15: NPC 1200 motor.

The New Breed

New companies and technologies emerge all the time in this sport. Listed below are some newer systems that I haven't seen in battle, but deserve a mention all the same.

Andrus Engineering

Andrus Engineering is a relatively new company to the combat robot scene. They have an ambitious product line that includes not only motors, but custom molded wheels, robot base plates, and pulley systems for drive. Their motors are pretty standard, although the GX4-13 gearbox (shown in Figure 9.16) was custom designed for combat and has a lot of attention paid to the little details.

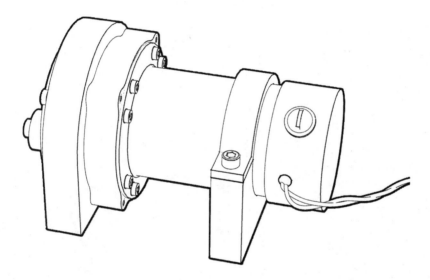

FIGURE 9.16: GX4-13 motor.

Dustin Motor

This is a relatively new custom mod for the DeWalt motor, which uses a 24-volt hammerdrill motor instead of the old 18-volt motor. The main drawbacks of the DeWalt motors are melted brush housings, a bad joint between the gearbox and motor, difficult mounting (no front face to bolt onto), and that darn plastic gear in the transmission. All of these issues are addressed by the Dustin motor (see Figure 9.17), and the system looks very sturdy.

Team Whyachi T-Box

The Team Whyachi T-Box motor and gearbox combo (see Figure 9.18) is based on the Titan 550, an R/C hobby motor in the 100–150 watt range. This gearbox is strictly for lightweights, but provides an ultracompact and weight-efficient package. At only 1.6 pounds (including the wheel!) this motor should perform pretty well when used in multiples for all-wheel drive.

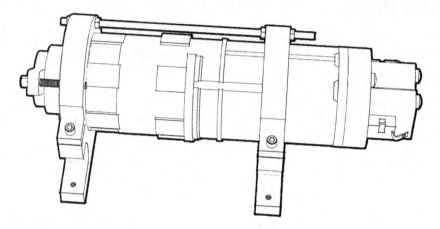

FIGURE 9.17: Dustin motor.

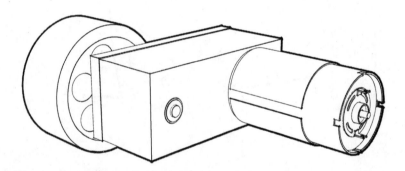

FIGURE 9.18: T-Box motor and gearbox combo.

ThinGap Gearbox

At press time, this motor and gearbox combo (shown in Figure 9.19) is just being introduced to the public by NPC Robotics. The gearbox can be configured on the fly (if you've got the gear set) to give a reduction from 11:1 to 37:1. NPC is selling the gearbox and motor combo with a midrange reduction (probably 15:1) for about $395, with additional gear sets available at around $50 each.

The interesting thing is that the motor used with this low-profile gearbox (made by a company called ThinGap) employs a unique core design that does not use windings like a conventional DC motor. Instead, it is built with copper sheet metal to produce an "ironless core armature." ThinGap claims higher efficiencies with less heat buildup in a compact package. According to the specifications, these motors should be equivalent to or better than the AstroFlight motors in power-to-weight ratio. Time will tell if they become a technology of champions.

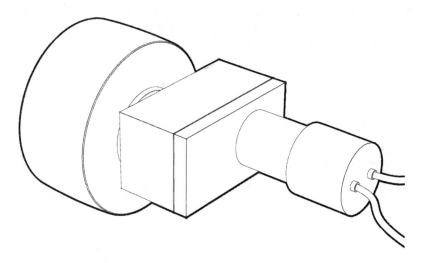

FIGURE 9.19: ThinGap gearbox.

What Can Gearing Do for You?

Not all of the top choices above have an output speed that's appropriate to connect to a wheel. You can add a gear reduction that will help you get the speed right while increasing torque.

Speed versus Torque

Off-the-shelf motors usually turn at about 1,000 to 3,000 RPM. This is way too fast to connect directly to a wheel. Besides, the amount of torque that you have to begin with is pretty small. Using a gear reduction, you can trade all that excess RPM for more torque. This has several benefits: It gives your robot more starting power, it lowers your top speed to something more manageable (controllable), and it increases your pushing power and maneuverability.

The price that you pay for these benefits is a more complicated system. The gears have to be properly aligned, and there are extra shafts, bearings, and brackets that need to be added. That's why the T64 and T74 gearmotors are so popular. The gearbox is built in, so all you have to do is bolt on a wheel, and you're ready to go.

Note The tradeoff between power and torque is one that you'll always have to make. The only way to get around it is to get a motor with more horsepower. As mentioned before, the horsepower spec takes into account both the RPM and torque of the motor.

Smaller motors usually have very low torque, but compensate by turning *very* fast (around 20,000 RPM or more), thus giving them a respectable horsepower rating. They're counting on a gear reduction to make them usable.

Note Skid-steer tank drive systems (usually four-wheel drive) require more torque than two-wheel (wheelchair) drive systems. The extra torque is needed for turning, since all four wheels have to slide to their sides during a turn.

Finding the Right Speed

The ideal output shaft speed that you connect to a wheel should be around 150 to 450 RPM. The popular gearmotors are all in this range, and wheel choices (discussed later in this chapter and in Chapter 12, "Let's Get Rolling!") can be used to tweak your speed up or down.

Gear Ratio Basics

Here's the quick and dirty version of how gears work. Figure 9.20 shows a simplified gear system.

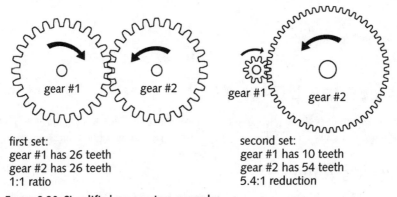

first set:
gear #1 has 26 teeth
gear #2 has 26 teeth
1:1 ratio

second set:
gear #1 has 10 teeth
gear #2 has 54 teeth
5.4:1 reduction

FIGURE **9.20: Simplified gear system examples.**

First, assume a few things:

- The gears have teeth, and the number of teeth is proportional to the size of the gear. A gear with a lot of teeth will tend to be very large.

- The gears are kept in contact so that their teeth always mesh together. If this is true, then gear #2 will always turn when gear #1 is turned.

In the first set, both gears are the same size. When gear #1 turns, gear #2 also turns with the same speed and torque. That's easy enough to understand. Now, take a look at the next case, where gear #1 is made much smaller, and gear #2 is made much larger. In this situation (called a gear reduction), gear #1 turns with torque T1 and speed S1. Gear #2 turns with a speed S2 that's much slower than S1. Because of its size, gear #2 takes longer to complete a full revolution than gear #1. The difference in size between gear #1 and gear #2 forms a *gear ratio*, which you can use to calculate difference in speed. Because gear #1 has fewer teeth than gear #2, the speed S2 will be slower than S1.

$$S2 = \frac{Gear1\#teeth}{Gear2\#teeth} \times S1$$

As I mentioned before, speed and torque are a tradeoff, and the torque of gear #2 will increase based on the gear ratio. Because gear #2 has more teeth than gear #1, the torque T2 will be larger than T1.

$$T2 = \frac{Gear2\#teeth}{Gear1\#teeth} \times T1$$

That's about it. If gear #1 has 10 teeth, and gear #2 has 50 teeth, then you can say that this will give you a 5:1 reduction in speed.

Note Unfortunately, making your own gearboxes is not recommended if you don't have access to a vertical mill. To work properly, gears require tighter tolerances than you can achieve with a drill press and bandsaw. Fortunately, you can buy a complete kit from NPC or use a roller chain and sprockets, as described in Chapter 11, "Working with Roller Chain and Sprockets."

Adding Stages

If you need a big ratio, one approach would be to have a very small gear and a very large gear. This is not always practical and you could end up with a ridiculously large gear that might not fit inside your frame. Instead, you can split the ratio into multiple smaller *stages*. Each stage is a complete gear reduction, and you can connect the output of the first stage to the input of the second, *multiplying* the two ratios. For example, if you need a 24:1 reduction, you can have a 6:1 stage and a 4:1 stage connected together. They can be connected by having the output gear of the first stage share the same shaft as the input gear of the second stage.

Mixing Types

You don't necessarily need gears to achieve a gear ratio. You can use a roller chain and sprockets, V-belts and sheaves, or timing belts and pulleys. You can also mix types in different stages. For example, you could use a small gearbox as your first stage, and then use sprockets and a chain to get power from the gearboxes to the wheels in a second stage.

Wheel Diameter

The last part of the equation for the drive system is wheel size. In effect, this is also a way to gear your motor up or down.

You can get more speed by using larger wheels. For every revolution of the output shaft, a larger diameter wheel will cover more ground. This puts a greater load on the motor, so you should be sure that your shafts and bearings are up to the test. For example, the T64 and T74 series gearmotors by NPC have become popular with large, 10-to 12-inch, tires. With an output speed of 230 RPM and a 10-inch tire, your robot will have a maximum speed of about 10 feet per second, which is about average in the arena. Incidentally, thanks to the large torque on these gearmotors, this combination handles very precisely as well. As I mentioned before, you should be careful to protect your wheels, since they will be large targets sitting out there for spinners to attack.

Conversely, you can lower your speed (and get more torque) by switching to smaller wheels. To illustrate this, consider the 18 V DeWalt gearmotor in low gear running at 24 V. This motor should have an output speed of about 650 RPM. These have become popular with the 3-and 4-inch Colson caster wheels. With a 4-inch wheel, your top speed in the arena should be an above average 11.3 feet per second.

Why not go all the way? Well, there is a limit. For example, you would never want to put a 10-inch wheel on a DeWalt drill motor. While the motor would probably be able to move the robot, the back force on the motor would probably destroy it on impact.

Look at it from the other side. In an impact, the wheel transmits a back force to the rest of the drive system. As shown in Figure 9.21, the larger the wheel's diameter, the larger the back force. In this situation, the wheel is trying to turn the motor, and a large wheel may be able to damage your drive system.

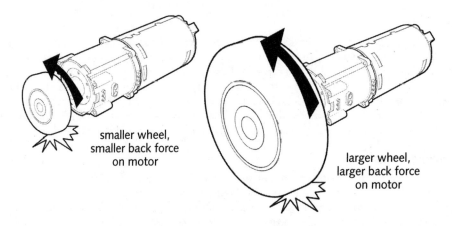

smaller wheel,
smaller back force
on motor

larger wheel,
larger back force
on motor

FIGURE 9.21: Back force on a motor by a wheel during impact.

Cross-Reference   Many different wheel choices and sizes will be discussed in Chapter 12.

Discovering Your Own Motor

At this time in robot combat, many people have put a lot of time and effort into finding and developing good motors for drives and weapons. Although many of these motors are very expensive, all have been battle-proven, which counts for quite a bit when it comes to your selection. While my personal feeling is that your time would be better spent researching your weapons system, maybe you're dying to try out a surplus motor that you found in a catalog to see if it might be appropriate for combat. The best combat motors have come from many different sources, including cordless drills, starter motors, truck radiator fan motors, and wheelchair motors. Who knows? You might discover the next "hot" motor. In this section, you'll see some of the important features to consider in evaluating the battle-worthiness of your motor candidate.

Look at the Power

As I mentioned before, the true measure of a motor's muscle is the horsepower. Listed below are a few ways to look at the horsepower, including examining the motor's characteristic curve and extrapolating from the motor's specifications.

Check the Curves

The best way to evaluate a motor is to check out the motor characteristic curves. As in Figure 9.22, the X-axis should be torque. Horsepower, efficiency, RPM, and motor current should all be represented on the same graph. The peak horsepower is the top of the hump on the horsepower curve, and current and efficiency should be noted at this point. Also pay attention to the efficiency over the whole torque range, which will tell you how much of the torque range is actually usable.

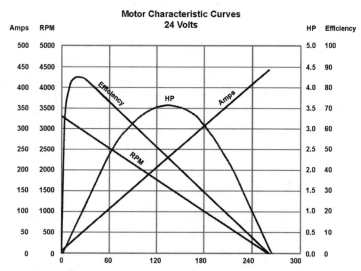

FIGURE 9.22: Example characteristic motor curve.

The sample motor in Figure 9.22 has a lot of power, with a peak HP of 3.6. However, the efficiency curve peaks sharply at 1 HP and then tapers off to 50 percent at peak horsepower. This is not a bad motor, but you'll need really big batteries and speed controls to take advantage of the power. If the motor were more efficient, you could get away with smaller batteries (less weight) and smaller speed controls (less money).

Do the Math

If you don't have access to the motor curves, then you can still evaluate many of the motor's characteristics based on a few of the specifications. First, you'll need the free-running (no load) speed in RPM. As mentioned previously, this is the speed the motor shaft turns when nothing is connected to it. Next, you'll need the stall torque in inch-pounds. (You can convert from other units using the tables in Appendix F.) Given these two items, you can calculate the peak power in watts according to the following formula:

$$Peak\ Power = \frac{Stall\ Torque \times No\ Load\ Speed}{338.4}$$

Where peak power is in watts, stall torque is in inch-pounds, and no load speed is in RPM.

Converting peak power in watts to peak horsepower:

$$Peak\ HP = \frac{Peak\ Power}{745}$$

Note that the above equations use the extremes of *stall* torque and *no-load* RPM to figure out *peak* power. In fact, what you're doing is calculating the power at half of the maximum (stall) torque and half of the maximum (no load) speed. Below is the equation for instantaneous power, which you can use for any point where the torque and RPM are known values:

$$Power = \frac{Torque \times Speed}{84.6}$$

Where power is in watts, torque is in inch-pounds, and speed is in RPM.

Converting instantaneous power in watts to horsepower:

$$HP = \frac{Power}{745}$$

Sometimes you will get a *dynamometer* performance chart. A dynamometer is a device that measures the motor's torque output, current consumed, and RPM. These readings can be used to calculate the efficiency at a few points by comparing the mechanical power out (instantaneous power) to the electrical power in, as shown in the following equations:

$$Power\ Out = \frac{Torque \times Speed}{84.6}$$

Where power is in watts, torque is in inch-pounds, and speed is in RPM.

$$Power\ In = Volts \times Amps$$

Where power in is in watts, volts is the voltage that the motor was run at, and amps is the current that was drawn by the motor at that torque.

$$Efficiency = \frac{Power\ Out}{Power\ In} \times 100$$

Where efficiency is in percent, power out and power in are in watts.

Note   I've focused on the equations for power in watts because not all countries use the term horsepower. If you're more comfortable working in horsepower, then you can easily convert back and forth as shown above by multiplying or dividing by 745.

Timing

Since the drive motors in a combat robot will be subject to constant reversals as you steer and maneuver around the arena, brush timing becomes an important evaluation factor. Motors that are timed in a specific direction will turn faster and have more torque in that direction. When they are run in the opposite direction, they will have less torque and run at a slower speed. Usually, a robot will have one motor powering one side, and a similar motor powering the other. The trick is that the motor on one of the sides will have to turn backwards for the robot to drive forward, as shown in Figure 9.23. If these motors aren't turning with the same speed, the robot will tend to drift off course. What you want is neutral timing, which is what most of the popular motors have. Or, if you can get a motor timed in each direction (as in the case of the EV Warrior), then this will counteract the drifting.

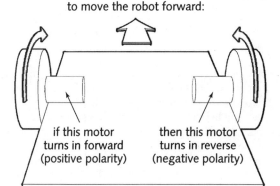

to move the robot forward:

if this motor turns in forward (positive polarity)

then this motor turns in reverse (negative polarity)

FIGURE **9.23: Motor rotation while driving forward.**

Weight

As with any part that you consider for your robot, weight should be something you pay attention to. Sure, the motor may have an incredible amount of torque, but if it weighs half the limit of your desired weight class, then you might have to rethink your choice. Bear in mind that even the motors used by 340-pound superheavyweight robots are usually no more than 15 to 20 pounds.

Voltage

You may discover a motor with excellent power and efficiency, but find out that it runs on 100 volts DC. This may be a drawback because some competitions limit the maximum DC voltage for safety. Also, you will need more batteries in series to get a voltage that high, which means more weight and expense. Finally, most speed controls won't run that high, so you'll be limiting your selection of off-the-shelf controls. Moreover, you're better off finding a motor that has a 12 to 24 V DC nominal rating.

Construction

In your search for a motor, you should take a look at how the motor is constructed. Are the brushes securely mounted? For example, if not properly cooled, some drill motors have plastic brush housings that will begin to melt and deform, leading to failure.

Does the motor shaft have ball bearings or bushings? Higher-quality motors have ball bearings that reduce the turning friction on the shaft, and allow them to be more efficient and run cooler. Cheaper motors use bushings to do the same job.

Price

The final part of the equation is price. In fact, this may be the reason that you're considering using an unproven motor in the first place. Although some motors may look like good deals in the surplus catalogs, make sure that the time and effort of testing and proving the motor will be worth the monetary savings.

Testing

Remember that by choosing your own motor instead of a battle-proven favorite, you will be ultimately responsible for its performance in the arena. As mentioned above, most motors that have become favorites also require extensive modifications to make them ready for battle. Without that previous experience, you will have to allow *extra* time to put your new, unknown motors through their paces. Only through testing and beating them up will you be able to identify any weaknesses that they may possess and be able to fix them in time for competition. You should also be prepared to scrap a motor if it looks like it's going to fail in combat.

Wrapping Up

This chapter has presented a lot of information about motors and their various characteristics. This is an important choice for the builder, because it will help determine what your robot's personality will be like in the arena. Will it be fast and zippy, running all around the arena, or will it be methodical and precise? That's part of the fun of the process—weighing the advantages and drawbacks and then making choices for your design.

Mechanical Building Blocks

Mechanical building blocks are basic parts that are used in just about every robot out there. Each of these "blocks" is a mechanical component that has a specific useful function or conveniently solves a mechanical problem, such as a bearing that reduces friction, or a shaft collar that prevents sliding. In this chapter, I introduce each one and give some details, such as bearing mounting options, axle types, and materials. You'll then be able to apply this information to "Project 4: Assembling the Parts."

Bearings

Without bearings, combat robots would waste a lot of power just trying to turn their wheels or weapons. A ball bearing is a mechanical item that reduces rotational friction by using a circular array of tiny balls that roll inside of a pair of rings. By lowering the friction, a ball bearing allows a shaft to turn more easily, which improves efficiency and reduces the amount of power wasted.

All bearings have a *bore* size. The bore is the diameter of the hole in the middle. The bore size can be specified in inches or millimeters and should be matched to the shaft that you intend to use.

 Note If the shaft won't fit through the bearing's hole, it's probably not the bearing's fault. The bore sizes are held to very close tolerances, and it's more likely that your shaft is slightly oversize, or that the end that you're trying to push through the bearing has a little burr (bit of metal) on it, which you may be able to feel by hand. It doesn't take much to interfere with the fit because the tolerances are relatively tight. You'll have to sand or file down the end. You can tighten the shaft in a power drill or drill press and use sandpaper to reduce the diameter of the shaft. I also put a small 45-degree angle (called a chamfer) all around the edge of the shaft with a disc or belt sander, and then smooth out the shaft surface with a fine flat file. Never force a bearing onto a shaft by pounding it.

Bearings can be *open*, *shielded*, or *sealed*. Open bearings have the balls exposed, while shielded bearings have metal covers. Sealed bearings have a liquid-tight neoprene seal. For combat robots, I suggest using shielded bearings. They don't weigh much more than open bearings, and you never know what bits of metal or plastic may find their way into your robot.

Unless the gearboxes are specifically designed to be direct drive (as is the case with the NPC gearmotors listed in Chapter 9, "Selecting Drive Motors"), the motor's output axle should *not* be used to support the robot. Instead, the motor should drive an axle supported by two bearings. The axle then connects to the wheel and supports the robot. The motor itself should not support any of the load, and should provide turning force (torque) only. Bearings should be used in pairs to support an axle, since a single bearing will allow the shaft to wobble, while using three or more will probably cause the shaft to bind from misalignment.

Bearing Mounting Options

There are a dozen or so different ways to mount ball bearings. Listed below are the most popular types that are used in combat robots.

Standard Ball Bearings (Radial Bearings)

When mounting standard bearings, you need a press-fit hole (see sidebar). It's also recommended that you capture the bearings. This means that you should have a step at the bottom of the hole to prevent the bearing from being pushed out of the other side. You can capture the bearing on both sides with a step on one side and a cover plate (that you manufacture) on the other, or a plain hole with a cover plate on each side. Usually, however, bearings are used in pairs to support a shaft, and in this case it's not necessary to capture the bearings on both sides as long as the steps are facing each other, as shown in Figure 10.1.

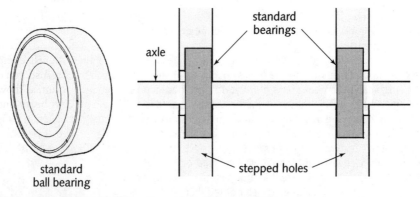

FIGURE 10.1: Standard ball bearing and mounting example.

Flanged Ball Bearings

Flanged ball bearings are just like standard ball bearings, except that they have a little lip on one edge, as shown in Figure 10.2. This lip helps with capturing the bearing, since it prevents the bearing from being pushed through a plain hole. They're also easier to implement because you don't need to make a stepped hole. These bearings are usually mounted with the flanges facing each other.

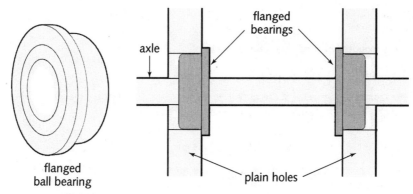

FIGURE **10.2: Flanged ball bearing and mounting example.**

Tip

Team Delta carries an excellent line of premade aluminum bearing blocks, which are convenient and lightweight. You won't need to make any press-fit holes for the bearings. They're already mounted.

Mounted Bearings and Pillow Blocks

Mounted bearings are more convenient for the robot builder who doesn't have access to a vertical mill to make custom press-fit holes. They are often self-aligning, which can help compensate for mounting holes drilled in the wrong place or frames tweaked in battle. Also, you don't have to worry about capturing the bearings. They come in square- or oval-mount, as shown in Figure 10.3. Although they're more convenient, mounted bearings can be fairly heavy since the mounts may be made of cast iron or steel and are much more expensive than plain or flanged bearings.

Pillow blocks are another type of mounted bearing shown in Figure 10.3. While the square- and oval-mounted bearings are more suited for shafts that are perpendicular to the mounting plane, pillow blocks are meant for shafts that are parallel, such as drive axles. Cast aluminum rather than cast iron pillow blocks are recommended because of weight considerations.

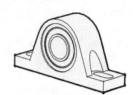

oval-mounted   square-mounted   pillow block
   bearing   bearing

FIGURE 10.3: Mounted bearings and pillow block.

Unlike the radial bearings in all of the examples above, *linear bearings* are meant to slide along the length of a shaft. These have limited use in a combat situation.

Other types of bearings that are more appropriate for weapons are discussed in Chapter 18, "Choose Your Weapon."

How Do I Get a Press-Fit?

As mentioned above, the most desirable mounting situation for a bearing is to have a light press-fit. Achieving that situation is the hard part and requires the precision of a vertical mill. A press-fit usually means a hole that is 0.001 inch larger than the outside diameter of the bearing. This 0.001-inch tolerance must be approached carefully and gradually, since making the hole too large may mean that you have to scrap the part. You can compensate for a hole that's *slightly* too big (as described later in this chapter), but you can't go too far.

First, the part is secured in the vertical mill's vise and a hole is drilled that is several thousandths undersize. (Usually, a drill that is the next smaller fractional size than the target is used.) Next, a device called a *boring head* is installed in the mill and adjusted to about the diameter of the hole, as shown in Figure 10.4.

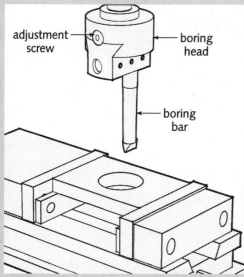

FIGURE 10.4: Boring head setup in a vertical mill.

A boring head is a block that is horizontally adjustable in very fine increments via an adjustment screw. The head has a tool called a *boring bar* sticking out of the bottom. As you adjust the head outwards, the bar cuts an increasingly larger hole, as shown in Figure 10.5.

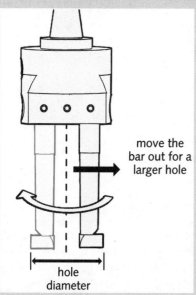

FIGURE 10.5: Boring bar cutting path.

Continued

Continued

The mill is started and the boring bar is lowered into the predrilled hole. As the bar turns, the cutting edge shaves off a little bit of the inside wall at a time. You keep adjusting the bar outwards until you reach your target hole size. This is a relatively slow process the first time because it requires several adjustment cycles and hole measurements to "sneak up" on the correct size. Once the correct size is found, the setting should remain very close for the rest of the parts. The bearings are pressed into the hole by evenly pushing on the outer race with an arbor press, which helps make sure that the bearing is pushed straight in. This operation may require a ring or other fixture to make sure that you do not press on the inner race, which can damage the bearing.

If you make an error and the hole is oversize, you can compensate by slipping a very thin piece of brass (called shim stock) around the bearing edge and pressing it in. You can also use Loctite 680 retaining compound (not threadlocker) to make up the difference. Make sure to thoroughly clean all surfaces first with alcohol or acetone.

Axles

The main job of the axle is to support the weight of the robot through the wheels without bending. The axles give the wheels a center revolution point, and in some cases, transmit power from sprockets or gears to the drive wheels. In this section, you'll find out some important design considerations for axles, such as axle type, materials, mounting situations, and diameter.

Note Technically, in strict mechanical engineering terms, an *axle* does not rotate, while a *shaft* rotates and transmits power. However, since robot building absorbs parts and technology from so many places, it's much easier to use the terms axle and shaft interchangeably.

Live Axle versus Fixed Axle

This refers to whether or not the axle turns along with the wheel. In some cases, the axle is mechanically connected to the wheel and spins along with it. This is called a *live axle*. In other cases, the axle is firmly attached to the frame, and the wheel just spins around it. This is called a *fixed axle*. The choice of which type of axle to use is up to the builder. To help you decide, the advantages and drawbacks of each type are listed.

The live axle is most popular among builders because most power transmission components such as gears and sprockets (see Chapter 11, "Working with Roller Chain and Sprockets") as well as hub and wheel solutions (see Chapter 12, "Let's Get Rolling") are set up for use with a live axle, which gives you a wide selection of choices. They have keyways (see the sidebar "Keys and Keyways") that allow them to lock to the shaft, giving you a very strong mechanical connection. These components turn the axle, and the axle turns the wheel. You must make sure that the bearings that support the shaft are lined up properly, or the system will bind, causing an increase in friction, which will waste power and slow you down.

Although most of the solutions on the market are focused on live-axle applications, fixed axles have their own benefits (as well as weaknesses). With a fixed axle, you won't have to worry about precisely lining up bearings. In this type of system, the bearings are built into the wheels (which will affect your wheel choices; see Chapter 12). In addition, you don't use any keys, and you don't have to cut keyways in any components, which will also allow you to use hollow steel shafts to save weight. I recommend alloy 4130 Chromoly steel, as described in Chapter 4, "Selecting Materials," for the maximum strength-to-weight ratio.

Almost invariably, fixed-axle systems use chain and sprockets to get power to the wheels. This means that you need to mount your sprockets directly to the wheel, instead of coupling them to the shaft, as in the live-axle case. Since most sprockets aren't configured for this, you need to drill your own bolt holes in both the wheels and sprockets and bolt them together. (This may be a little tricky to get perfectly on center. If you miss, it may cause the chain to alternately go tight and slack as the off-center sprocket rotates.) Fortunately for the builder, this system is popular among go-karts, and Chapter 12 lists some manufacturers that have predrilled sprockets that you can mount to their wheel rims.

Keys and Keyways

Most drive components (gears and sprockets) have a *keyway* or notch cut into their center holes. The size of the notch is determined by the diameter (bore) of the hole according to an industry standard that promotes part interchangeability between different manufacturers. This notch is half of a square. The shaft has a similar notch cut into it that is the same size as the one on the gear or sprocket, also a half of a square. A *key* is a piece of steel that is square and long, and slides into the square-shaped hole created by the sprocket and shaft. The key and keyway solution is easily installed, easily removed, and repositionable, yet incredibly strong.

Keyways are cut into gears, sprockets, and hubs with a tool called a broach. It's a long skinny cutter that slides into a collet. In Figure 10.6 you will see that the round collet goes into the hole (bore) where you want to cut the keyway.

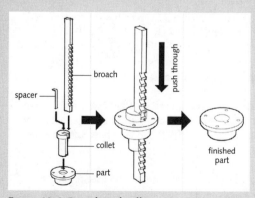

FIGURE **10.6: Broach and collet.**

Continued

Continued

You push the broach through with an arbor press, which is basically a vertical gear rack that pushes straight down as you turn the handle, illustrated in Figure 10.7.

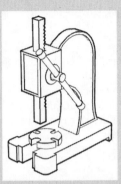

FIGURE 10.7: Arbor press.

The broach is tapered, so that it cuts the slot a little bigger the further down it goes. You push the broach all the way through the piece, and it drops out of the bottom. Next, you put a little spacer behind the broach and run it through again, cutting a little deeper still. After two or three passes (depending upon the specification for the key size), you should have a notch cut into the part.

Live-Axle Materials

For basic shaft material, you can use general-purpose chrome-plated 1045 steel shafts, or precision-ground drill rod. The important thing is that the positive tolerance (the amount that the shaft might be larger than specified) should be less than +0.001 or the shaft won't easily slide into the bearings, and you'll have to (painfully) file or sand it down.

By far, some of the best parts that have become available since this sport began are prekeyed shafts. McMaster-Carr and Team Delta both carry prekeyed shafts made out of 1045 steel. This is the stuff that you want. You'll have to use a hacksaw or reciprocating saw (with a steel cutting blade) to cut the shaft to the length you need, but the keyway is already cut into the shaft at the correct depth, making it easy to mount a gear, sprocket, or wheel.

Don't make the mistake of ordering *prehardened* precision ground shafts from McMaster-Carr. It may sound like a good idea, but you'll find out that if you try to cut these shafts, almost every tool will go dull. Instead, you should buy the general-purpose shafts and then have them professionally hardened after you machine them.

Usually, live axles are solid, because keyways need to be cut into the axle to accommodate mounting sprockets and wheel hubs. Also, the ends may be tapped for a safety screw to make sure that the wheel stays on the shaft. Builders who have access to a lathe can put grooves into the axles to accept retaining rings (described later in this chapter) to keep the sprockets and wheel from moving along the shaft.

Bending Force and Mounting Situations

Some robots have the drive wheels protected, while others have them exposed. Exposed drive wheels offer the benefit of being able to get away from a lifter because you are less likely to be propped up on the edge of your frame, since the wheels are exposed at the outermost edges. However, fully exposed drive wheels place specific needs on your axles that protected wheels do not. Mainly, they are large targets for arena hazards and spinning weapons. This means that your axles should be made thicker to accommodate the extra abuse that you're expecting them to endure. Technically, an exposed drive wheel is an *overhung* load. It's only supported on one side (overhanging the side of the robot). This type of mounting allows the wheel to act as a long lever with mechanical advantage to deflect (bend) the axle.

When the axle is mounted on both sides of the wheel with bearings, then it is double-supported and the overhung load situation no longer exists. In this case, the wheel does not have the same mechanical advantage as in the overhung case, and is less likely to deflect. This allows you to use a smaller diameter axle, which means less weight.

Axle Size

The thickness of your axle should be scaled to the weight of the robot. Also, if you have overhung shafts (as mentioned above), you may want to move up to a slightly thicker axle. Table 10.1 lists some guidelines for minimum recommended axles sizes by weight.

Table 10.1   Axle Size

Weight	Shaft Diameter
0-60 lbs	1/2"
60-120 lbs	1/2" to 5/8"
120-220 lbs	3/4"to 1"
220-340 lbs	1" to 1¼"

Casters

Casters solve the problem of providing an extra contact point with the ground (for stability) without dragging something along that would slow you down. They come in a few different varieties.

Swivel Casters

The swivel caster consists of a free-turning wheel on a bracket that can swivel around in any direction like an office chair, as illustrated in Figure 10.8. This provides a low friction contact point that is omni-directional. This is a valuable component in a drive system that must be able to swing the robot around and maneuver all over the arena.

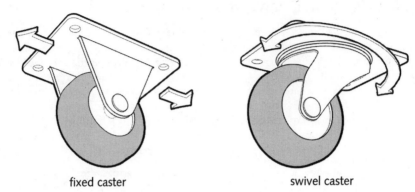

fixed caster   swivel caster

FIGURE **10.8: Fixed and swivel casters.**

Fixed Casters

A fixed caster has a free-turning wheel similar to a swivel caster, but the bracket does not swivel, as shown in Figure 10.8. It is fixed in one direction. Although it may not seem as useful as a swivel caster, the fixed caster can be used to help correct drive problems, as detailed in Chapter 19, "Troubleshooting." Since it does not swivel, the fixed caster provides resistance to turning that will help some robots drive straighter inside the arena.

Ball Transfers

Many robots have size and space restrictions, or very low ground clearance. In these situations, a ball transfer does the job of a swivel caster, but in a much smaller package. However, one of the big things that you have to worry about is getting dirt and grit inside. Foreign substances will reduce the rolling effectiveness of the ball transfer and eventually lead to failure. In combat, the arena is usually littered with bits of metal and plastic, so ball transfers should be checked often and changed when they accumulate too much drag. They can also get dented and deformed, which can cause them to seize up. Fortunately, they're relatively inexpensive, so changing out isn't a financial burden.

Ball transfers come in a few different styles, but the two most popular are flange-mount and stud-mount, as shown in Figure 10.9.

Note Ball casters make for an extremely loud robot, because they tend to rattle. Just something to keep in mind when you're test-driving your robot around the neighborhood.

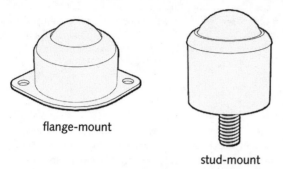

flange-mount

stud-mount

FIGURE 10.9: Flange-mount and stud-mount ball transfers.

Hose Clamps

Hose clamps consist of a steel band perforated with diagonal slots, as shown in Figure 10.10. The slots are used to tighten the band by means of a worm screw attached to the end of the band. They solve the problem of firmly mounting a cylindrical object. They come in many different band lengths, and the excess length can be trimmed off with a rotary tool and a cutoff wheel, or by repeatedly bending them back and forth until metal fatigue causes them to snap. In both cases, you should clean up the sharp edge left behind with a deburring wheel.

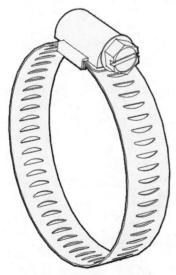

FIGURE 10.10: Basic hose clamp.

Hose clamps are more robust than you might imagine, and several competitors in the past have used hose clamps to mount air tanks, batteries, and even motors. Although you could simply drill two holes in the base and slip the band through them for mounting, it's best to make a bracket that cradles the cylindrical object and allows you to tighten the hose clamp around it for a solid connection.

Shaft Collars

Shaft collars are more or less a staple of this sport. They are rings that slide over a shaft that can be tightened down to prevent horizontal movement along a shaft. They come in a variety of *bores* and styles. The bore is the size of the shaft that it's designed to clamp around. The most common shaft collar styles (shown in Figure 10.11) are single piece collars with setscrews, one-piece clamp-on collars, and two-piece clamp-on collars. The two-piece collars are the most useful, since they do not require you to slip them over the end of the shaft. Instead, they can be split and clamped anywhere on a shaft that's already installed.

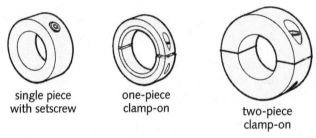

single piece
with setscrew

one-piece
clamp-on

two-piece
clamp-on

FIGURE 10.11: Typical shaft collars.

Shaft collars are typically used to hold the horizontal position of a sprocket or wheel on a shaft, as shown in Figure 10.12.

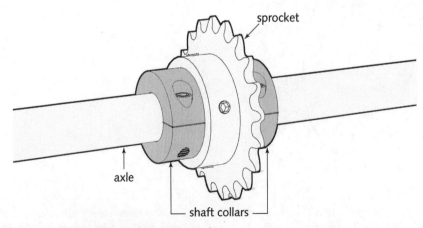

sprocket

axle

shaft collars

FIGURE 10.12: Shaft collar application example.

Retaining Rings (Clips)

Retaining rings are a much lighter and significantly smaller alternative to shaft collars. When used with the correct size groove, they can be as strong as or stronger than a shaft collar. The difficulty for the average robot builder is that they're more time-consuming to use, and you need access to a lathe to make a special groove in the shaft. This groove has specific tolerances that must be met to get the best performance.

There are three types of retaining rings shown in Figure 10.13. For the purposes of combat robots, I'll focus on external retaining rings and E-clips.

external   internal   E-clip

FIGURE 10.13: External, internal, and E-type retaining rings.

External Retaining Rings

External retaining rings need a special pair of pliers for installation and removal, called retaining ring pliers. The pliers have little pins on the end to match the holes in the tips of the clip. They spread the ring a little so that it can be slipped over the end of a shaft and slid into place. When the ring is in the correct position on the shaft, the pliers are released and the clip snaps into the groove. (Retaining rings are also called *snap rings* for this reason.) Removal is performed by reversing the order of the steps described above. Sounds pretty easy, right? Wrong. These are very difficult to work with, especially in tight places. If you do have the inclination to use these clips, you should purchase a set of retaining ring pliers with removable tips, so you can insert straight tips, 45-degree tips, and 90-degree tips. You'll need them, along with some patience. The only reason that I endure the frustration is that they are one of the strongest ways to keep something from sliding sideways on a shaft, and will withstand a significant amount of side force. (My robot Deadblow has fully exposed wheels, which means that the snap rings sustain a lot more punishment than with other drive-wheel configurations.)

E-Type Retaining Rings (E-Clips)

A far easier, but not as robust solution is the E-clip. E-clips are easier to work with because they can be snapped onto the shaft directly from the side. They don't have to be stretched and slipped over the end of the shaft and slid into place. You don't need any special tools for installation or removal. A pair of needle nose pliers can be used to snap them in place, and they can be pried off the shaft with a small flathead screwdriver. The price you pay for this convenience is that E-clips are not as positive a locking device as external rings. It's possible that when it encounters a high shock load from the side, an E-clip can pop off, causing all sorts of havoc with your drive system.

Shaft Couplers

Shaft couplers are used to connect two inline shafts and transfer turning force between them. They also help compensate for minor misalignments. Although rigid couplers definitely have their applications, for combat purposes, I'll focus on the *flexible spider coupler*.

In addition to the above benefits, a flexible spider coupler can also help absorb impacts in systems that are subject to high shock loads. This type of coupler consists of two halves that are made out of steel or aluminum. Each half is keyed so that it can be locked into place on the shaft, and you have a choice of many different bore sizes. The two halves sandwich a *spider*, which is a multispoked insert made out of a flexible material, as shown in Figure 10.14. The spider is what forms the connection between the two halves, and also absorbs the shocks. The thing to keep in mind is that the two shafts that you're going to join must be very close to (if not exactly) inline. The shafts should also be independently held in place by shaft collars or retaining rings, since the spider is only sandwiched by the coupler halves, and does not fasten them together.

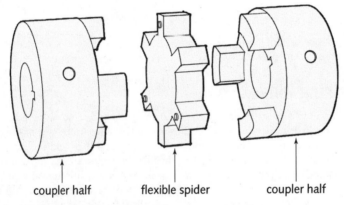

coupler half   flexible spider   coupler half

FIGURE 10.14: Flexible spider coupler.

Universal Joints

The universal joint solves the problem of "severe misalignment," which means having a drive shaft (such as a motor shaft) not inline, or even parallel with a shaft that you want to transmit power to. Due to a unique design, the universal joint flexes while it transmits rotational energy at an angle. As illustrated in Figure 10.15, one U-joint can help with angular misalignments, but you will need two joints to correct for parallel misalignment.

Unfortunately, the universal joint has some major disadvantages to be aware of. First, the power transfer is pretty inefficient. Power is lost in trying to turn the U-joint, and the loss increases as the two shafts are moved away from being parallel (that is, the U-joint is flexed more). Also, it's not as robust a solution as you might think, since the U sections tend to be relatively thin. You should avoid using this type of linkage in favor of gears or chain, if you can.

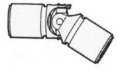
single universal joint

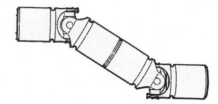
double universal joint

FIGURE 10.15: Single and double universal joints.

Other Common Blocks

This section will briefly mention some other common mechanical items that most people already know about.

Hinges

Hinges allow for a pivoting motion, just like on a door. In combat robots, the most common type of hinge used is the *piano hinge*. These are smaller and much longer than door hinges and are useful in making flaps that hang down to the floor. They are usually made of steel or brass, although some competitors have used aluminum versions. My advice is unless you're really dying for weight, use the steel hinges.

Springs

Three major types of springs are used in combat robots: extension, compression, and torsion springs. Extension springs pull back when they are extended. Compression springs push back when they are compressed. Torsion springs try to hold a specific angle, and depending on their winding, push back when they are curled or uncurled.

Note

Torsion springs require a core or shaft that passes through the center and supports the spring as it flexes.

Project 4: Assembling the Parts

So far, you've cut out all the armor plates and drilled all the holes for your project robot (Chapters 6 and 7, Projects 2 and 3). And now that you have a basic understanding of fasteners, drive motors, and mechanical blocks building, you're almost ready to bolt everything together. In this project, you will prepare the polycarbonate plates, axles, motors, and other drive system components for assembly and fasten them to the base. You'll get practice in preparing Lexan (no sharp edges, as mentioned in Chapter 4), as well as your first taste of soldering with the motor leads, and cutting steel for the axles.

Eye protection is required for most of these operations.

Review all of the general power-tool safety protocols described in Chapter 5, "Cutting Metal," as well as the sections that correspond to the specific tools used below.

Deburring the Side Plates

One of the rules for working with polycarbonate is *no sharp edges*. Sharp edges that are the result of cutting and drilling operations can occur on every side and hole (including tapped holes). These are the first places to fail by cracking under extreme loads. You can relieve these by chamfering them, which puts a slight angle on the edge. Holes are chamfered with a countersink, while the edges of the part can be done with a deburring tool.

1. Clamp the part to the table so that the piece does not move while you use the deburring tool.

2. Start close to one end and draw the tool along the edge in a nice, even stroke. Keep the angle of the tool low so you don't scratch the polycarbonate with the ball end.

3. Continue all the way down the length, keeping constant pressure towards the center of the plate, and past the corner in one smooth stroke. You should see some plastic curling up into a little spiral, as shown in Figure 10.16. You can run the deburring tool along the edge a couple of times if you want, but you don't need to take a huge amount off. All you're doing is getting rid of the sharp edge.

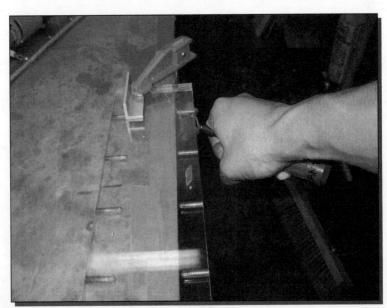

FIGURE **10.16:** Deburring the edge of the plate.

4. Next, you'll flip the plate around and take care of the remaining part (the very tip) that you didn't get at the start of the first pass.

5. Repeat this process for all edges of the plate. Also, you should use the deburring tool to take care of the large diameter axle holes.

6. Don't forget the corners. It will be difficult to use the deburring tool, so use a flat file and swing your arm up in an arc as you push the file forward, so that you can round over the edge. Remember to only file in the forward direction, since that's where the file's cutting edges are pointing.

Installing the Side Panels

Assembling the side panels requires one more step, where you will have to mark a hole location using an existing hole, which is more convenient in some cases, and necessary if you don't happen to know all of the hole locations in advance. You'll put the panels together, mark the holes, and then pull them apart to do the actual drilling and tapping.

1. Install the side panels on the base with 1/4-20 × 3/4-inch long button-head cap screws. As with any assembly operation, try to get all of the screws started before tightening any of them down all the way. Do not use Loctite threadlocker on the polycarbonate. Polycarbonate and anaerobic threadlockers are incompatible.

Note

Some of the screws may not go in. That's okay. Mark the holes on the base that miss. You'll need to use one of the techniques described in Chapter 8," Fasteners—Holding It All Together," to open up the holes.

2. Once all of the screws are installed in the side panels, drill through the 1/4-inch holes in the front and rear panels to mark the hole positions. You're not drilling a new hole, just making a small divot to mark the hole center, like a center punch mark.

3. Remove the front and rear panels so you can drill and tap the remaining holes.

4. Make a block to help you line up the drill bit correctly. You can rough-cut a piece of scrap. All that's important is that the holes are straight.

5. Use the drill press to drill both a #7 hole for the tap drill, and a 1/4-inch hole for the tap itself.

6. Set the depth of the tap drill using a shaft collar, a piece of tape, or one of the other techniques described in Chapter 7, "Drilling and Tapping Holes." It's important to hold your block in place while you set the depth, or the hole will be too shallow by the thickness of the block.

7. Hold the block in place as shown in Figure 10.17 and drill until you bottom out on the shaft collar or other method of depth setting. Make sure to use water or WD-40 to keep the plastic cool while you drill.

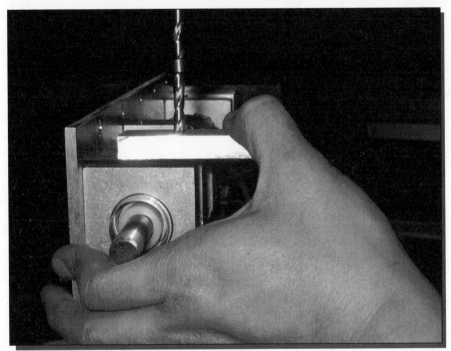

FIGURE 10.17: Using the block to drill straight holes.

8. Use the 1/4-inch hole on the block to guide the tap into the hole straight. Again, make sure to use lubricant.

9. Chamfer the tapped holes with a countersink in a cordless drill.

10. Reinstall the front and rear polycarbonate panels. Use 1/4-20 × 1-inch long button-head cap screws on the new holes. Optionally, you can install 1/4-20 × 1-inch long socket-head cap screws on the upper holes only, leaving the button-head cap screws in the lower holes, which will prevent the robot from balancing on the front or rear panels, and make it fall back down onto its wheels, instead of becoming incapacitated.

Soldering Leads onto the Motors

The motors are supplied without leads, so you will have to either solder a lead directly to the motor terminal or use a crimp connector. Soldering is described in detail to give you some experience with that procedure. However, you may optionally use flag crimp terminals to make your motor connections, which is described in greater detail on the Kickin' Bot Web site at www.kickinbot.com.

Caution

Adequate ventilation is essential. Make sure not to inhale the solder fumes. Never blow on any solder joint to speed cooling. Allow the joint to cool on its own or you'll get a cold solder joint (bad bond), which may lead to intermittent failure. Good solder joints are bright, smooth, and shiny.

1. Cut off a 4-inch long pair of red and black leads of 12-gauge Deans Wet Noodle or Ultra Wire for each motor.

2. Apply solder to the tip of a hot soldering iron (called "wetting the tip").

3. Touch the tip to one of the motor terminals. After a short while, the solder on the tip should flow to the motor terminal; as this is happening, add more solder so that the terminal has a bright and shiny coating of solder.

4. Apply solder onto the lead, just as you did with the motor terminals.

5. Hold the lead in contact with the terminal using a pair of helping hands as shown in Figure 10.18. Then hold the soldering iron so that it touches both the lead and the terminal at the same time.

6. Add more solder as solder begins to flow between the terminal and the lead. Remember not to blow on the solder. Allow it to cool on its own.

FIGURE **10.18: Holding the lead in contrast to the motor terminal.**

Cutting Down the Axles

This series of steps focuses on cutting down the axle material into usable lengths for the project robot. Although the hacksaw will make it through the material just fine, a reciprocating saw can save you a lot of time (and effort).

1. Use a hacksaw or reciprocating saw to cut down the axles using the procedures described in Chapter 5. The plain keyed shafts should be cut to 5½ inches long, while the splined motor shafts should be cut to 6¼ inches long, including the spline.

2. Lightly clean up the edge on the disc sander.

3. Use the deburring wheel to put a slight chamfer on the end by turning the axle slowly while holding it into the wheel.

4. The splined motor shafts had some residue on them from the heat-treating process that interfered with the fit. I had to file them a bit to remove the coating before attempting to insert them in the bearing blocks.

Assembling the Drive Components

In order to prevent the shaft from simply pushing out of the bearing, you've got to lock it in place. Thanks to the excellent engineering by Dan Danknick of Team Delta, the bearings in these blocks are captured on opposite sides. This means that you can prevent the shaft from sliding by putting a shaft collar on either side of the block. Note, however, that you want to only touch the center of the bearing, which spins along with the shaft. Otherwise, you'll cause drag. To deal with this, you'll use some smaller-diameter nylon spacers to make sure that you only touch the center of the bearing. You'll also use them to correctly space the sprocket on the shaft, see Figure 10.19.

1. Use the 1/2" bore × 5/8" OD × 3/8" long nylon bearings as spacers on the inside and the 1/2" bore × 3/4" OD × 1/2" long nylon bearings as spacers on the outside towards the wheel.

2. In order to install the sprockets, you need to clamp the keys in the bench vise and cut them down to 3/4-inch long with a hacksaw.

3. Use the disc sander and deburring wheel to clean up the edge of the key.

4. Assemble the spacers, shaft collars, and sprockets (with keys) on each bearing block. The splined axle will be assembled exactly the same way, with the spline sticking out of the sprocket side. Do not install the wheel and wheel key yet. Those items will only get in the way, and will be installed just before the drive test.

5. Feed the axles through the axle holes in the side panels. Fit the motors onto the splined shafts. The motors should be mounted with 10-24 × 1/2" long button-head cap screws.

FIGURE 10.19: Bearing block drive component assembly.

6. You may have to tweak the position of the splined shaft if it jams up against the motor. Under normal conditions, the motor should have a little play, allowing you to rock the splined shaft back and forth. If it feels too tight, then adjust the shaft collars so that the tip of the splined shaft moves more towards the outside of the robot, away from the motor.

7. Apply Loctite to all screws once you're satisfied with the position of everything. Avoid getting any Loctite on the polycarbonate.

Wrapping Up

These components function as the "glue" that helps major mechanical parts fit together and function properly. They join drive components, and help secure them in place. They reduce friction and help transmit power. You won't use every one of these building blocks in your robot, but you'll definitely use at least a few.

Working with Roller Chain and Sprockets

Y ou've got the makings of a frame, a pair of drive motors, and some output axles. The next step is to get power from the motors to the axles, and that's where roller chain comes in. It's the choice of most robot builders because of its ease of use, relatively light weight, versatility, and tolerance to misalignment. The latter is important when compared to gears. If your sprockets aren't perfectly aligned, the roller chain won't mind at all. But if your gears are misaligned by a small margin, they may bind and cause friction, which results in power loss, and ultimately, a less powerful drive system or weapon. That's why I don't recommend making your own gearboxes unless you have access to more advanced machine tools. Let's get rolling.

What Is Roller Chain?

Roller chain is very much like bicycle chain. It wraps around a toothed wheel called a *sprocket*. But in this case, instead of your legs pushing pedals to move the sprocket, your drive motors will be doing the turning. The turning force on the sprocket results in a pulling force on the chain. To complete the system, there is another sprocket for the output, and the chain pulls on this output sprocket, which results in a turning force on the output axle. In the case of both the bicycle and your robot, the output happens to be a wheel, but you could also use roller chain to drive a rotating weapon, for example. As shown in Figure 11.1, what's important is that there is an input and an output joined by a chain. The chain wraps all the way around both and transmits power from one to the other by a pulling force.

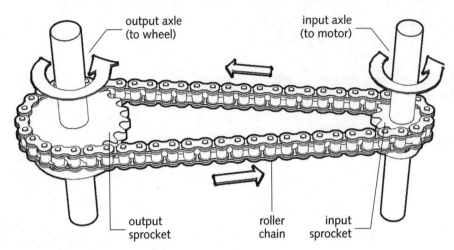

FIGURE 11.1: Typical roller chain setup.

What Can You Do with Roller Chain?

As mentioned earlier, you can use roller chain to transmit power between drive motors and wheels, or motors and weapons. One of the chief advantages of this method of transferring power is that you can change gear ratios by choosing different-sized sprockets for the input and output.

Why Can't You Just Use Bicycle Chain?

Although roller chain and bicycle chain are very similar, they're not the same, and you won't be able to mix chain and sprockets. The pitch, which is a measure of the pin-to-pin distance of the chain, is different for bicycle chain, and therefore incompatible. Besides, roller chain is available in many different mounting configurations and has a lot of accessories.

Designing with Roller Chain

In this section, you'll find out exactly what you should be looking for in roller chain and sprockets. When it comes to these items, there are lots of different sizes and specifications, and it can be confusing. In this section, I'll help narrow down the choices with my recommendations, and give you the information you need to select the roller chain and sprockets that are right for your job.

ANSI Numbers

The easiest way to talk about sprockets and chain is to refer to the ANSI number. ANSI stands for American National Standards Institute, an organization devoted to setting manufacturing standards for various mechanical components. By having a standard, sprockets from different manufacturers can be used interchangeably.

Each size of roller chain has a specific ANSI number. If you order sprockets with the same number, you can be sure that they'll fit together with your chain. Standard numbers are 25, 35, 40, 50, 60, and 80. We'll be using ANSI #35 roller chain and sprockets for the project robot. This size is appropriate for our needs for two reasons: The chain is strong enough to handle a lot of abuse without being too heavy, and the sprockets are widely available in a variety of sizes. ANSI #25 chain would be strong enough for most lightweight robots, but the sprockets are not available through standard suppliers such as McMaster-Carr or Grainger. Usually, the chain is sold in lengths of 10 or 100 feet. It's best to get a few 10-foot lengths to begin with. Buy more than you need in case you make a mistake in breaking the chain.

Sprocket Ordering Options

You can't just walk into a hardware store and ask for a sprocket. The same goes for calling McMaster-Carr or Grainger. You might have to answer half a dozen questions or so before you can order the sprocket you want. Following is a brief introduction to each of these options and what's important about them.

Finished Bore versus Plain Bore

Finished bore sprockets have two setscrews on the sides and a keyway already cut. Plain sprockets just have a hole in the middle with no setscrews or keyway. You'll need finished bore sprockets in almost all cases except where you're planning to use keyless bushings (described in detail in the "Mounting Sprockets to Motors and Axles" section). Since the bushing locks itself to the shaft, no keys or setscrews are necessary to transmit power. Figure 11.2 compares both types side by side.

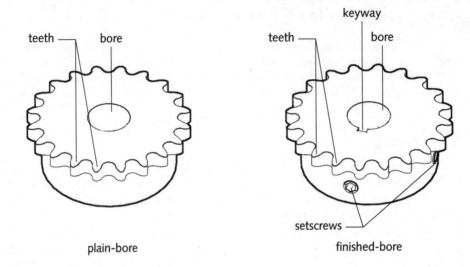

FIGURE **11.2: Plain-bore versus finished-bore sprockets.**

Bore Size

The bore size is the size of the hole in the center of the sprocket, as shown in Figure 11.3. It's also usually the size of the shaft that you'll be attaching the sprocket to. Read the specifications carefully to make sure that the sprocket you need is available in the bore size that you need. When you use keyless bushings (see next section), you will have to specify a larger bore size than the shaft you're mounting to.

Number of Teeth

As mentioned earlier, you can change gear ratios through your choice of number of teeth on the sprockets. Read the ordering page carefully to make sure that the sprocket with the number of teeth you need is available in your bore size. An important thing to remember is that the outside diameter of the sprocket increases as the number of teeth increases. This means that if you're trying to gear a motor down a lot with a single stage of sprockets (that is, a single input and output pair), then the output sprocket may become too large to fit in your frame. Fortunately, you can add a second stage of reduction, where the output sprocket of the first stage is connected to the input sprocket of the next stage. Bear in mind that you'll pay a price in weight, cost, and complexity. Also, it's one more thing that could break down. There is a table in Appendix F, "Tables and Charts," of standard sprocket sizes and outside diameters to help you in your designs.

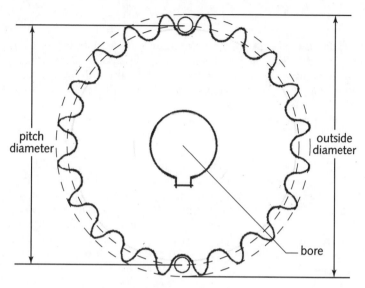

FIGURE 11.3: Sprocket parts and diameters.

 Note As shown in Figure 11.3, the outside diameter is measured from the tip of a tooth to the tip of a tooth. The pitch diameter is measured from the middle of a link (threaded into the chain) to a link directly opposite.

Choice of Material

Fiberglass-reinforced plastic sprockets are available from McMaster-Carr. These are suitable for lightweight robots with nonexposed internal wheels, but probably won't hold up to middleweight punishment. They give you a significant weight savings, but at the price of strength. We'll be using steel sprockets for our project.

Mounting Sprockets to Motors and Axles

In order to transmit power from your drive motors to the axles, you'll have to find a way of mounting the sprockets. Following are a few of the ways that this can be accomplished.

Setscrews

This method is cheap and easy, since the setscrews are already included with the sprocket, but it's not recommended. You are relying only on the force of the setscrew to transmit the large amount of power that your motors are generating. If you can't use any of the other mounting methods listed in this section, then you can help yourself by filing a flat on the shaft so that you don't gall it, or mar the shaft where the setscrew was attempting to dig in. There are better ways. Do you really want to lose for a reason as silly as this?

Pinning the Shaft

This is another relatively cheap and easy method, but still not the best. The idea is that you drill a hole through both the sprocket and the shaft. Then, you press a steel pin through the hole to join the sprocket and shaft together. Sounds pretty good, but consider the amount of shear force (slicing force) that this tiny pin will be subjected to under combat conditions. It's a huge amount of torque that could cause this pin to fail. It's actually used in some applications as a safety measure to prevent damaging other expensive equipment. There are still better ways of securing the sprocket, and your robot is worth just a little more effort.

Welding

Yes, given a steel sprocket and a steel shaft, you could weld the two together. The good news is that the sprocket will not loosen and move out of alignment—ever. But that's also the bad news, because removing the sprocket (from a bent or otherwise damaged axle) involves grinding the joint away. Very robust, but not too much fun when there are other ways around the problem.

Keys and Keyways

This is the most preferred method of mounting the sprocket, illustrated in Figure 11.4. As mentioned in Chapter 10, "Mechanical Building Blocks," the idea is that the finished-bore sprocket has a keyway cut into its center hole that matches a similar notch cut into the axle. A key slides into the square-shaped hole created by the sprocket and shaft. The key and keyway solution is easily installed, easily removed, and repositionable, yet incredibly strong.

The trick is that you buy pre-keyed shaft and appropriate-size keys from McMaster-Carr or Team Delta. (See the chart in Appendix F for standard key and keyway sizes.) Then, you push the key in the slot and slide your sprocket to where you want it, making sure the key is lined up. Tighten down the setscrews and you're ready to go.

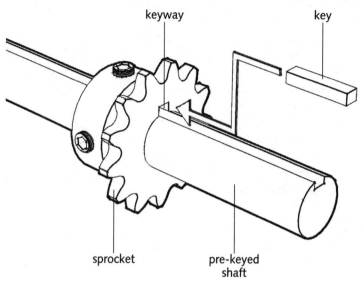

keyway   key

sprocket   pre-keyed
                shaft

FIGURE 11.4: Keyed sprocket and shaft.

Why use setscrews? In this case, the setscrews don't transmit the power, the key does. All the setscrews do is hold the side-to-side position of the sprocket. Still obsessed with *not* having setscrews? Okay. You can set the side-to-side position of the sprocket with a shaft collar on either side and get rid of the setscrews altogether. Or you can cut thin-wall tubing to the appropriate size and slide a piece on each side of the sprocket to relieve you of setscrews forever.

Keyless Bushings

Why do you need to know about keyless bushings if I've just told you the most preferred way of mounting the sprocket? Because this is a very interesting solution that has a few must-use applications, especially if you have a motor that doesn't have a keyway cut into its shaft. You could attempt to grind one with a rotary tool. That's pretty messy and since there are specifications on how deep and how wide keys have to be for maximum benefit, you could be setting yourself up for disaster. Here's where the keyless bushing comes in, as shown in Figure 11.5.

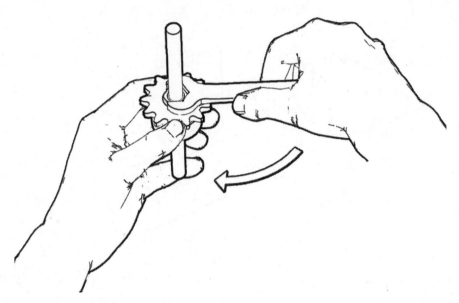

FIGURE 11.5: Keyless bushing.

The keyless bushing slides onto the shaft and inside the sprocket, so the shaft and sprocket don't touch. It has a threaded, tapered inside that simultaneously squeezes down on the shaft while pushing out on the sprocket as you tighten the bushing down, as shown in Figure 11.6. Pretty tricky, eh? Just as with the key solution, it's easily removed and installed, repositionable, and very strong, but you don't have to have a keyway cut into the shaft or the sprocket.

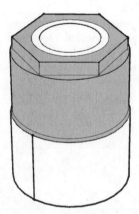

FIGURE 11.6: Keyless bushing installation.

How well does it work? It works great, but only if the shaft and bore size are within a few thousandths of an inch of specifications. You won't be able to get away with drilling out sprockets with smaller holes to use the keyless bushings. You should order plain-bore sprockets (no setscrews or keyway) with a much larger bore size than your axle diameter (see the keyless bushing bore size table in Appendix F). The catch is that keyless bushings cost about $30 to $50 each, depending on the bore size.

Note If you're running a fixed-axle system, then the sprocket won't be mounted to the shaft. It will be mounted to the wheel, as described in Chapter 10. You could drill your own mounting holes in a stock sprocket, but a better solution is to use aftermarket custom go-kart sprockets and hubs that are designed with fixed-axle applications in mind. These options are discussed in greater detail in Chapter 12, "Let's Get Rolling"

Sizing the Chain

Roller chain is sold in minimum lengths of 10 feet, and unless you've got a *really* big robot, you're going to have to make a few modifications in order to size it correctly. This involves determining the right chain length, breaking the chain using a chain breaker, joining it back together with a master link, and possibly tweaking the length using an offset link. Following is a detailed description of each of these procedures.

Where Should I Break the Chain?

As described at the beginning of this chapter, you'll be joining the two ends of your chain with a master link. It's important to realize that the master link needs two clean chain ends with no outer link plates attached to do its job. That means that when you size your chain, you must pick a pin that will produce a clean link. This limits your choices and may result in a chain that is longer than you really want, since it will be slack when installed. No problem. It's better to have a slightly longer chain than one that's too short. You'll take care of slack chains with a tensioner, described at the end of this chapter.

Using the Breaker

This is not as violent as it sounds. The tool that you use for this purpose just happens to be called a *chain breaker* (shown in Figure 11.7). The jaws of the breaker are spring loaded so that you can clamp them around the sides of the chain. Following are the steps you'll take in order to break (that is, shorten) the chain.

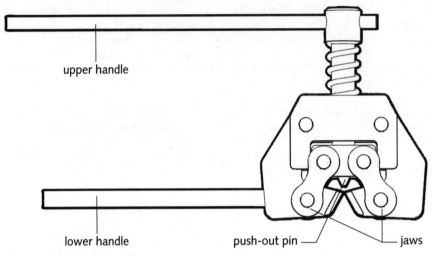

FIGURE 11.7: Chain breaker.

1. Spread the jaws by grasping both handles and squeezing, as shown in Figure 11.8.

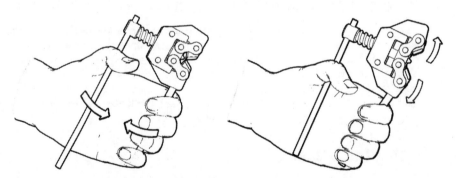

FIGURE 11.8: Opening the chain breaker's jaws.

2. Guide the spread jaws over the chain link that you want to break. Make sure that the V-shaped part of the jaw fits around the link's sides (see Figure 11.9), and the pointed push-out pin of the breaker is centered over the pin.

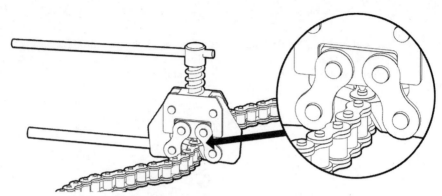

FIGURE 11.9: Positioning the push-out pin.

3. Release the jaws and let the breaker grasp the pin, as shown in Figure 11.10. If it doesn't line up, or catches the wrong link, squeeze the jaws open and try it again. If you can't get the jaws to close over the chain, then the push-out pin may be extended. Reset the tool by turning the upper handle counterclockwise, which pulls the push-out pin back in and out of the way.

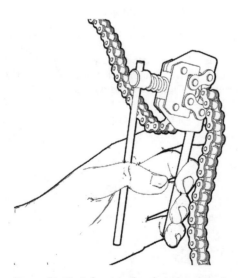

FIGURE 11.10: Releasing the chain breakers jaws on the roller chain.

4. One hand holds the lower handle while the other hand turns the upper handle clockwise (as seen in Figure 11.11). Keep turning until you push the pin about halfway through the link plate. Keep in mind that you're not going all the way through, just enough to loosen the pin. If the bottom part of the pin link starts to bend, you've gone too far, and you may have a difficult time getting the pin link out later.

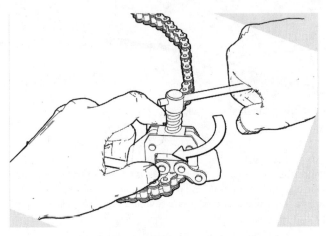

FIGURE 11.11: Turning the handle to extend the push-out pin.

5. Next, jump to the other pin on the same link and repeat Step 4 of this list, as shown in Figure 11.12. Remember to turn the top handle counterclockwise to reset the tool before each use. Otherwise, you won't be able to clamp down on the chain. Move back to the first pin and turn a little more so that the pin is pushed through the link plate. Then, jump to the second pin and push it through the link plate also. Both pins should have the pushing operation done twice to them and the link plate should pop off.

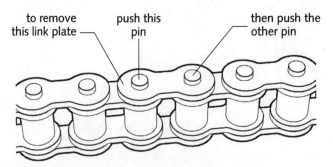

FIGURE 11.12: Link plate removal sequence.

6. The link plate should now be free from the chain. You should be able to easily pull the pin link from the chain. If it becomes stuck in the chain, use a pair of needlenose pliers to pull it out while gently swiveling the chain back and forth around the stuck link (see Figure 11.13).

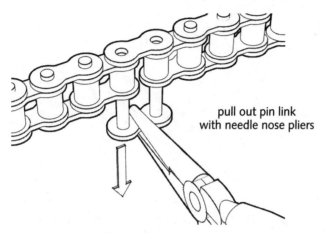

pull out pin link
with needle nose pliers

FIGURE **11.13: Removing the pin link.**

7. Make sure that both ends of the chain are "clean." In other words, they should have no plates hanging from them, as shown in Figure 11.14. If so, then it's on to the master link.

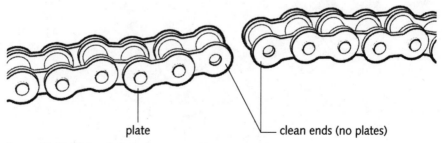

plate   clean ends (no plates)

FIGURE **11.14: Clean chain ends.**

Using the Master Link

Now that you've cut your roller chain to the correct length, how do you join it back together? Well, you need to add a *master link*, shown in Figure 11.15. This is a special link that's reusable. You can remove and install it quickly and easily without the use of the chain breaker. Here's how it works:

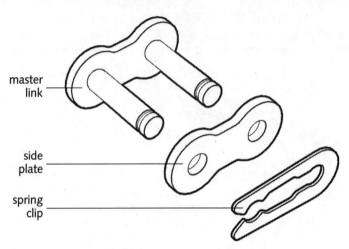

master link

side plate

spring clip

FIGURE 11.15: Master link parts.

1. Put each tip of the chain puller into the last link of each side of the chain you're joining, as shown in Figure 11.16, and twist the knob at the top until the two chain ends are almost touching.

FIGURE 11.16: Pulling the chain together.

2. Insert the master link by pushing it all the way through the holes (see Figure 11.17). If you can't get the link all the way through, then tighten the chain puller a little more and try again. Once you have the master link in place, you can loosen and remove the chain puller.

FIGURE 11.17: Inserting the master link.

3. Put the master link side plate over the tips of the pins that are sticking through the link, and push it down so that the pin ends come through the plate (see Figure 11.18). Note: The master link has a special plate that is different from the plates that you remove with the chain breaker and the two can't be interchanged. It will become obvious if you try it, but I thought I'd mention it here.

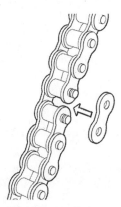

FIGURE 11.18: Installing the master link side plate.

4. Rotate the spring clip 90 degrees so that the side plate and clip form a cross (as seen in Figure 11.19). Snap the middle part of the clip onto the end of one of the pins.

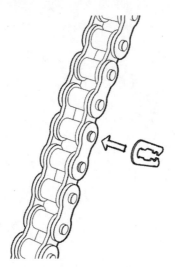

FIGURE **11.19: Positioning the spring clip.**

5. Rotate the spring clip so that it lines up with the side plate (see Figure 11.20). You may have to push the pin back a little bit as you swing it around.

FIGURE **11.20: Spring clip installation sequence.**

6. Insert the tips of a pair of needle nose pliers, as shown in Figure 11.21, and squeeze until you hear a click.

7. To remove the link later on, insert the tips of the needle nose pliers as shown in Figure 11.22, and squeeze to push the clip back. Continue with Step 5 and the rest of the sequence in reverse order.

clip moves this way

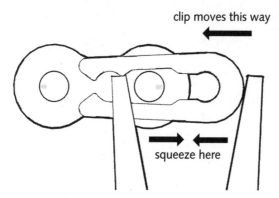

squeeze here

FIGURE 11.21: Locking the spring clip in place.

clip moves this way

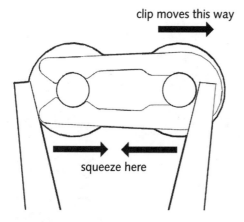

squeeze here

FIGURE 11.22: Spring clip removal.

Note If the chain hangs down a little bit after it's cut, no problem. We'll take care of that later in this section with a chain tensioner. If the chain is too tight to get the link in, then you've sized the chain a little too short, and it may have to be recut. First, however, try an offset link, which is described next.

Using an Offset Link

An *offset link*, shown in Figure 11.23, gives you half of the length of a standard link. This can give just enough slack to make a chain that's too tight just right. The offset link consists of three pieces: the link body, the offset pin, and a cotter pin, as shown in Figure 11.24. To assemble the pieces, follow these steps:

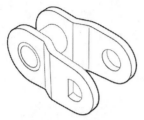

FIGURE 11.23: Offset link.

1. Remove the master link (if installed) from the chain that's too tight, as described earlier (see Figure 11.24).

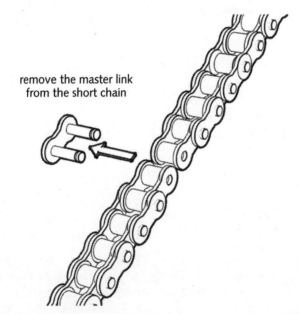

remove the master link
from the short chain

FIGURE 11.24: Removing the master link.

2. Carefully remove the parts from the package, keeping track of the microscopic cotter pin, as shown in Figure 11.25.

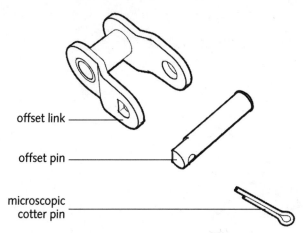

offset link

offset pin

microscopic
cotter pin

FIGURE **11.25: Offset link parts.**

3. Select one end of the chain and slide the offset link over the end (see Figure 11.26).

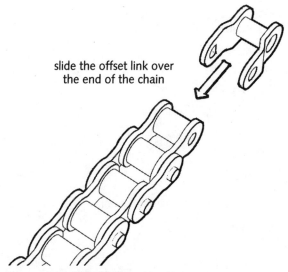

slide the offset link over
the end of the chain

FIGURE **11.26: Offset link positioning.**

4. One of the holes on the side of the offset link is round, and the other is squared off, as shown in Figure 11.27. You will slide the offset pin through the round hole towards the one that is squared off. (There should be only one way that the pin fits through both holes.)

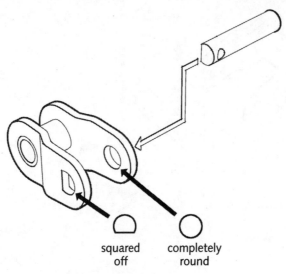

squared
off

completely
round

Figure **11.27: Offset pin installation.**

5. Push the cotter pin through the hole in the offset pin starting at the round side and pushing through to the flat side (see Figure 11.28).

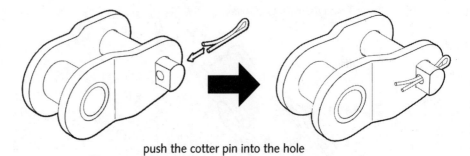

push the cotter pin into the hole

Figure **11.28: Cotter pin installation.**

6. Spread the legs of the cotter pin outwards (see Figure 11.29) to lock it in place.

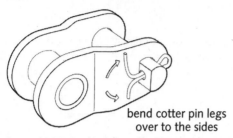

bend cotter pin legs
over to the sides

FIGURE **11.29: Locking the offset pin in place with the cotter pin.**

7. Insert the master link as described earlier. Figure 11.30 shows the final installation with both the offset link and master link.

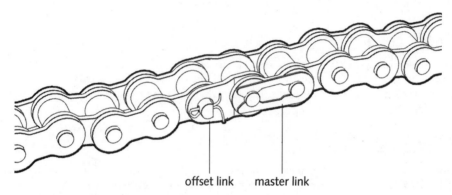

offset link   master link

FIGURE **11.30: Offset link and master link installed.**

Note If the chain is too loose, then it's time to add a chain tensioner, described next. If the chain is still too tight, then you should recut the chain to a slightly longer length and try again. (Good thing you bought extra chain, right?)

Why Chain Tension Is Important

There are a few reasons why chain tension is critical to the system. It can't be too loose *or* too tight. You're looking for *just right*, which I'll talk about at the end of this section. As I mentioned earlier, you can't always have the perfect length chain, and that's where the tensioner comes in to take up the slack.

Efficiency

Efficiency is a measure of how much of the source's power is actually transmitted to the load. In plain terms, how much of the motor's power will get to the wheels? Of course, it's not possible to get 100 percent efficiency, but you want as much as you can get. If the chain is too loose, your efficiency goes down because the chain slips a little as the teeth try to catch the rollers. If the chain is too tight, your efficiency will suffer because the sprocket will be fighting against the chain, which will be pulling in the direction of the other sprocket, resisting turning.

Skipping

If the chain is slack, it may jump out of the sprocket teeth when the system experiences a high torque. If it skips repeatedly, this can cause a lot of wear on the sprockets and chain. Also, a skipping chain means that your robot isn't moving. Finally, the chain can jump off the sprocket altogether, leaving you with a bunch of chain piled up inside the robot and an incapacitated drive system.

Shock Load

If the chain is too slack, when the wheel experiences a shock load (hitting another robot, the arena, or a weapon), it can actually tense the chain in a single spot and focus a huge amount of energy on one or two links, causing either a kink in the chain (a source of efficiency loss) or actual breakage.

Stretching

Your chains will stretch over time. As they wear in, you will see more and more slack. Having a chain tensioner will allow you to correct for that increasing slack over time.

What Exactly Is "Just Right"?

Just right, according to most manufacturers, is a slack amount that is 2 to 4 percent of the run length, as shown in Figure 11.31. It's about 4 percent under normal conditions and 2 percent for extreme loading (that is, combat) conditions. How do I measure this? Take a pad of paper and prop it up behind the tensioned chain. Pick a pin in one of the links near the middle as your reference. Pull up the chain and roughly mark the maximum height of the pin. Then, pull the chain down and mark the minimum height of the same pin. If the distance is somewhere between 2 and 4 percent of the tangent-to-tangent (run length) distance, then you're right on.

For example, let's say that the run length is 12 inches. Two percent of that distance is 0.24 inches and 4 percent is 0.48 inches, so I would put the "just right" tension somewhere around 0.375 inches.

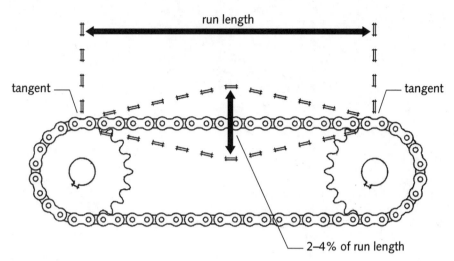

FIGURE **11.31: Measuring chain tension.**

Installing a Tensioner

In the previous section, you found out why correct chain tension is important. Okay, but what do you use to get correct chain tension? A *chain tensioner* of course. There are many different kinds of tensioners out there. In the following section, I'll describe some of the common off-the-shelf solutions, and some things you might want to think about if you make your own.

Getting a Good Wrap

One of the most important things to consider when adding a tensioner to your system is figuring out where to put it. You've got to make sure that all of the sprockets in your system have adequate wrap. In other words, does the chain go around at least 50 percent of the sprocket, as shown in Figure 11.32? If not, the chain may skip, even if you have perfect tension.

Figures 11.33 through 11.38 show some examples of various types of chain wraps. Note the position of each tensioner in these systems and how they help each sprocket get good engagement with the chain.

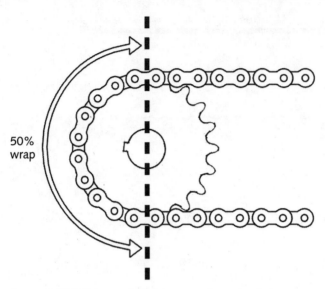

**50%
wrap**

FIGURE **11.32:** Minimum chain engagement on sprocket.

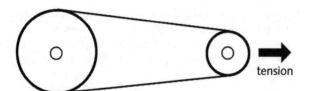

tension

FIGURE **11.33:** Motor bracket used as a tensioner.

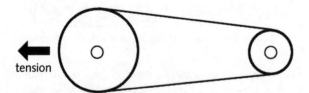

tension

FIGURE **11.34:** Wheel bearing bracket used as a tensioner.

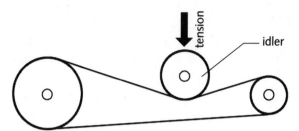

FIGURE 11.35: Standard idler.

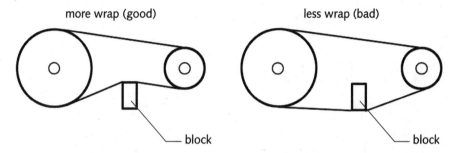

more wrap (good)   less wrap (bad)

block   block

in general, you want a tensioner that increases chain wrap around the sprockets

FIGURE 11.36: Delrin block.

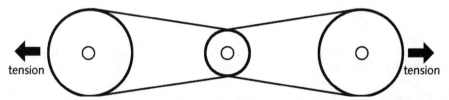

FIGURE 11.37: One motor driving two wheels with tensioner in each wheel bearing bracket.

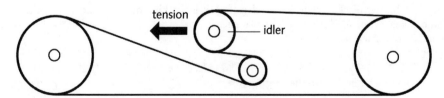

FIGURE 11.38: One motor driving two wheels with an S-wrap tensioner.

Design Considerations

You don't really need a fancy store-bought tensioner. In fact, many of those items have cast-iron bodies — much too heavy for our needs. You can design your own chain tensioner, keeping in mind some of the following points:

- Is it easily tensioned? Can you reach the tensioner to tweak it between matches? Are the retaining bolts easily accessed with an Allen wrench or ratchet?

- Does it have enough range? You've got to make sure that the slots are long enough and placed in a location that will give you enough slack to remove the master link and extra tension to take up any stretching in the chains later on.

- Is it serviceable? Can you remove the tensioner easily to service or replace it? Do you have to remove many other parts of the robot to access it?

Project 5: Installing the Chain

This chapter explains how chains and sprockets can be used to transfer power between rotating shafts. In this project, you'll do exactly that for the project robot. First, you'll size the chains and install them. Then, you'll add the tensioner blocks, and any spacers, if necessary. This will give you some experience in working with the chain tools and getting a feel for correct chain tension.

It's likely that the chains will be super oily, so you may want to get some latex gloves from the hardware store, or have lots of clean rags handy. Eye protection is required for all of these operations.

Review all of the general power-tool safety protocols described in Chapter 5, "Cutting Metal," as well as the sections that correspond to the specific tools used below.

Making the Chain Tensioner Blocks

The chain tensioner blocks will push the chain up and take up some of the slack, so that the chain is a bit tighter. The material selected for this job is Delrin, which is naturally slippery, and can be used in applications where you need low friction. The blocks will be cut into half-rounds out of a full rod.

1. Take the 1-inch diameter Delrin rod and place it on a 3/4-inch L-channel. Mark a 4-inch long line down the length of the rod.

2. Feed the rod straight into the bandsaw keeping the blade centered on the line. When you reach the end of the line, withdraw the rod, carefully pulling straight backwards, so that you leave a slot in the rod. Make sure to use water or WD-40 for lubrication.

3. Mark off two 1¾-inch lengths on the shaft.

4. With the slot parallel to the bandsaw table, feed the shaft into the bandsaw blade on the first line. This should produce two half-round shapes.

5. Feed the rod into the blade on the second line to produce two more half-round shapes for a total of four blocks.

6. Clean up the block on the disc sander, being careful to keep your fingertips away from the abrasive surface.

7. Flip all the blocks over so that the flat side is facing up and apply the hole pattern with spray adhesive.

8. Mark the holes with an automatic center punch.

9. The half-round may be difficult to clamp in the drill press vise, but if your vise is equipped with a horizontal notch, as shown in Figure 11.39, you can get one side of the half-round in there.

FIGURE 11.39: Setup for drilling holes in tensioner blocks.

10. Drill the two 1/4-inch holes in each block.

11. Install the blocks on the base using 1/4-20 × 3/4-inch long button-head cap screws.

Sizing the Chain

Below is a listing of steps to size the roller chain for installation. It's a bit unusual to install the tensioners before sizing the chain. Usually, it's the other way around. In this case, we need the height of the blocks to keep the chain up off of the base so that we can pass wires underneath it later.

1. Take the loose end of the chain and thread it around the sprockets. Note that the sprocket connected to the motor won't turn, so you'll have to pull the chain past and wrap it around when you have enough slack.

2. Keeping hold of the link that lines up for a decent fit, break the chain using the technique described earlier.

3. Use the chain puller and insert the master link as described earlier.

4. You may find that the chain is a little slack with the master link and tensioners in place, as shown in Figure 11.40. Not to worry. Next, we'll make some spacer plates to raise the blocks up high enough to correctly tension the chain.

FIGURE **11.40: Chain with master link and tensioners installed.**

Note   There's a calculation that you can use to find a length for an integral number of links. However, I usually dictate the position of the axles and use a tensioner to take up the slack.

What Do You Do If the Pin Link Gets Stuck in the Chain?

The best tool that I've found to deal with this is the SmithTool model #B-5035 Chain-A-Part (McMaster-Carr #6669K11) chain breaker. It consists of an outer barrel that clamps the chain, and an inner pin that actually forces the stuck pin out of the chain.

The first step to getting the stuck pin out is to pop off the other side of the link plate, so that you leave an exposed pin, as shown in Figure 11.41.

Figure 11.41 Exposed pin stuck in the chain.

Next, you tighten the outer barrel down and then you use a box wrench or adjustable wrench to turn the screw clockwise, which pushes the middle pin down. At some point, it will get much easier to turn the screw, which indicates that it's been pushed all the way out of the link. It may or may not fall out of the bottom, depending on how oily the chain is.

You may need a pair of slip-joint pliers to loosen the outer barrel. Since the middle pin is kind of floppy, you may have to push it back in by hand as you turn the screw counterclockwise.

Making the Spacer Plates

In this operation, you'll make four plates from scrap 3/16-inch aluminum, which will be inserted below the tensioner blocks to push them up to a height that gives correct chain tension.

1. Begin by marking a 1-inch-wide strip at least 7 inches long on a scrap of 3/16-inch aluminum from the base.

2. Use the band saw to cut the strip.

3. Clean up the strip on the disc sander, being careful to stay on the side of the disc that rotates down, as described in Chapter 6, "Shaping and Finishing Metal." You will have to flip the piece over to sand the entire length of the strip.

4. Finish the edge on the deburring wheel.

5. Mark four 1¾- inch plates on the strip. The kerf of the bandsaw blade is relatively small, and the length of these plates isn't that critical, so you can mark them all at once.

6. Cut the strips and clean them up on the disc sander and deburring wheel.

7. Next you'll drill the plates. You can save time by clamping all four plates together, as shown in Figure 11.42, and drilling all the holes at once. Use spray adhesive to apply the hole pattern to the top plate, and mark the hole positions with an automatic center punch before drilling.

FIGURE **11.42: Drilling multiple plates at once.**

8. Install the plates and check the chain tension. You should be able to move the chain at least 1/8 inch, but not so much that the upper part of the chain comes in contact with the lower part. Correct chain tension should appear as shown in Figure 11.43. You can also use washers to tweak the height of the block.

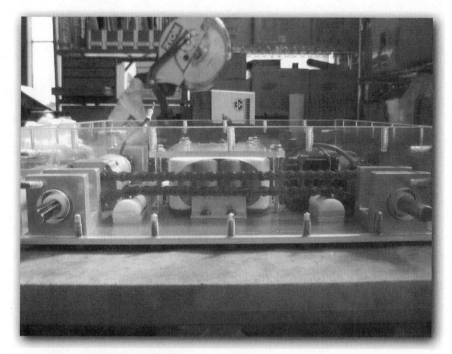

FIGURE **11.43: Correct chain tension.**

 Note Don't be surprised if over time, your chains begin to stretch and sag. It's natural and an example of why it's a good idea to have a tensioner in the system. You may have to make another set of spacer blocks in the future, using the previous technique to maintain tension.

Wrapping Up

This chapter has provided an introduction to the tools and techniques for working with roller chain. Although not every robot uses roller chain, every robot builder should be aware of how to use it, because it's so versatile. You can change sprockets to get a different gear ratio, and in most cases, all you have to do is cut a different-sized chain. Through this introduction and the project robot, you should develop a clear understanding of how roller chain and sprockets can distribute mechanical power over long distances, and how it can be applied in your own designs. Best of all, you don't need any fancy, expensive machine tools with absolute precision to use roller chain and sprockets. If the sprockets are a little bit out of true, no problem. The roller chain won't mind at all.

Let's Get Rolling

Wheels are the final step of the drive system. Without the appropriate wheels, your performance could be just as bad as if you'd incorrectly geared your motors. There are more choices than you might imagine, and this chapter will discuss the importance of traction and durability, as well as different factors that will affect your selection, such as drive setups, wheel construction, and materials. You'll find out about the top wheels used by the pros and why they're so popular. Finally, I'll show you how to find your own wheels, if you choose, and how to enhance your traction by using special chemicals.

What Makes a Good Wheel?

There are many good wheels to choose from out there. Perhaps a more important question than "What makes a good wheel?" is "What is the most *appropriate* wheel for *your* robot?" The traction and durability of the wheel, as well as the weight class of the robot must be taken into account, because each of these factors affects how your wheels will perform in actual combat conditions.

Wheel Anatomy

In the discussion of wheels, it is helpful to define a few terms first. Manufacturers have specific names for different parts of a wheel. A *standard wheel* has a *tire*, which is the outer surface that contacts the ground. The *rim* (or also, confusingly, *wheel*) usually comes in two halves and bolts together, squeezing around the tire, and providing a solid mounting surface. The *hub* connects the whole wheel to the axle by bolting to the rim. Figure 12.1 shows a typical wheel system.

Bonded wheels are a bit different. They have a *core* that performs the same function as a rim, but the tire is molded (bonded) directly to the core, and not sandwiched between two rim halves.

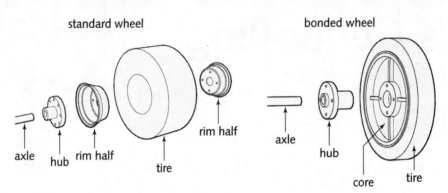

FIGURE 12.1: Anatomy of a typical wheel system.

Why Is Traction Important?

Traction is the ability of the tires to grab the arena floor. Without traction, a combat robot might simply spin its wheels when trying to push an opponent. You could have a huge amount of torque, but without traction, it's wasted. Worse yet, you could be pushed all around the arena by other robots.

The tires are the key to traction. They should be slightly compressible (squishy), but not enough to deform permanently under the weight of your robot. The tire material and pattern will determine the traction for a given wheel. How these factors interact with the playing field will determine your ultimate success.

Tire Materials

By definition, the compressible materials in tires are flexible. At the same time, you want the tire to be resistant to tearing or becoming delaminated or disconnected from the hub or core. It's a difficult balance for most materials to achieve, but rubber, foamed polyurethane, and thermoplastic elastomers have all been proven in battle (see the "Top Wheel Choices" section).

A *durometer* reads the hardness (and thus the compressibility) of a material. As mentioned in Chapter 4, "Selecting Materials," the *Shore-A* scale was created to measure the hardness of soft (pliable, compressible) things. In combat robots, you should be looking for wheels and tires with a durometer rating of no more than 75 Shore-A. Experience has shown that a durometer rating of about 65 Shore-A is appropriate for robot combat because it has the right amount of "squish" to give you good traction in the arena, while being able to stand up to punishment. The *Shore-D* scale is also used by manufacturers to evaluate some harder materials used in solid wheels. Which scale is used is up to the manufacturer, although the Shore-D scale gives better resolution with harder materials. To compare a wheel rated in the Shore-D scale, add 50 and you will get an equivalent hardness in Shore-A. However, if the wheel or tire is rated in Shore-D, it's a good bet that it's way too rigid (not enough "squish") to provide you with good traction on the arena floor, and will probably slide rather than grab.

Tire Patterns

Another important choice is what tire pattern to use. Arena floors will vary, and you should try and find out as much as possible about the surface you'll be playing on before the competition. If you can't get this info, then see if you can get tapes of previous competitions, and see how the robots handle in the arena. Usually, you can find a competitor's Web site and discover the wheel system that they used. You may be fighting on steel (coated with a textured traction paint), concrete, or even asphalt. Different patterns have different reactions to the various surfaces. In general, the more textured the surface, the more textured the tire should be. As shown in Figure 12.2, the patterns you may have to choose from are smooth, rib, diamond, lug, bell, jag (sawtooth), and knobby.

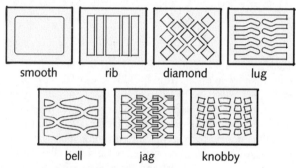

FIGURE 12.2: Different tire patterns.

Tip

Removing the very outside skin of a tire is a good idea. Since most tires are molded, the outer layer is usually pretty smooth. By sanding off this smooth outer skin, you can give the tire a rougher surface (called "tooth") to grip the floor better. Some builders recommend removing the outside layer with a hand-held sander while running the drive wheels. This is done so that you don't sand down the tread too far in one spot, and you keep the tire thickness uniform. Remember, you're only taking off enough to rough up the surface, not weaken the tire.

The Importance of Durability

The wheel has to be able to carry the weight of your robot without collapsing. It should also be capable of taking high shock loads from impacts with the arena walls, hazards, or other robots. You should count on your wheels getting attacked and abused in the arena. As the mechanical connecting point between the wheel and the rest of the drive system, the hub (and/or core) will be absorbing most of the shock, so their materials will have a direct effect on the overall durability of the wheel.

Core and Rim Materials

The first thing to realize is that you won't be supporting thousands of pounds on your tires, so cast iron cores can be ruled out immediately. A wheel with a higher load capability isn't necessarily better, and may be much, much worse because of weight issues. Remember that robot parts used in multiples add up quickly, and if each one of them is heavy, you'll be out of weight in a hurry. What is important is that the core or rim won't rupture when hit.

Usually, aftermarket rims are made out of steel, aluminum, or nylon. Steel and aluminum are certainly up to the abuse, but nylon rims (while a good lightweight solution) should be backed by an aluminum hub, and an aluminum plate on the outside of the wheel if it's fully exposed, to armor them and increase their durability. The core material used on the bonded caster wheels is usually tough enough to take through-hole bolting. It's resilient enough to stand up to abuse, but not brittle like plastic lawnmower wheel cores, which are prone to shattering with shock loads.

Hub Materials

Aluminum is the preferred material for hubs because it's light yet strong enough to take a beating. There are many different aftermarket solutions from NPC Robotics, Team Delta, Robot Marketplace, and other online vendors. Plastic hubs (such as Delrin) should be avoided in all but lightweight robots with fully protected wheels. However, bonded caster wheels with aluminum hubs have been successfully used in many weight classes, even 340-pound superheavyweight robots. Steel hubs are usually reserved for aftermarket pneumatic tires. There are several steel go-kart hubs that have heavy-duty bearings to carry the load and that allow you to bolt the wheel onto one side and a sprocket onto the other, for a convenient drive wheel solution for fixed-axle systems.

Dimensions

As mentioned in Chapter 9, "Selecting Drive Motors," the diameter of the wheel will help determine the speed of the robot and the pushing force. Larger-diameter wheels give the robot a higher speed, but less pushing force. The thickness of the tire will affect the surface contact area, and thus the traction.

Weight Class

As always, every part on the robot has to be evaluated for weight. A single wheel and hub can weigh up to 15 pounds. With four wheels, that would already be 60 pounds — half the weight of a middleweight class robot, only leaving another 60 pounds for armor, weapons, drive, and batteries. By scaling the wheel to the robot, you can leave some weight for the *rest* of the basic systems.

Drive Setups That Affect Your Wheel Choices

The type of drive system you are using should be considered in your wheel decisions. Described next are some of the secondary factors related to your drive system that will influence your choice of wheels, including number of wheels, their exposure, and axle configuration.

Two-Wheel versus Four-Wheel Drive

Is there such a thing as "too much" traction? Yes, if you're dealing with a four-wheel drive slip-steer robot. As its name implies, turning a four-wheel slip-steer robot means that all the wheels have to slip in order for the robot to turn. Unfortunately, the higher the traction, the more torque it will require to turn, which means more battery power as well. Two-wheel drive robots don't suffer the same penalty since their wheels don't have to slip.

Why have the four-wheel drive robots become so popular, then? Because they will inherently go much straighter at higher speeds than a two-wheel drive system. Also, if you're lifted up, the wheels are closer to the edge of the robot, and you usually have at least one drive wheel on the ground to get away. With a typical two-wheel drive system (with casters, for example, as shown in Figure 12.3), you're usually at the mercy of the lifter. More wheels give you more traction, so you're less likely to get pushed around with four, as opposed to two, of the same wheel. In addition, the friction during turning in a four-wheel drive system (from slipping) slows down your turns, which can help you compensate for overturn, giving you added control.

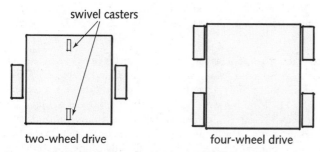

swivel casters

two-wheel drive   four-wheel drive

FIGURE 12.3: Two-wheel versus four-wheel drive configurations.

Note Adding more wheels driven by a chain will *not* give you more pushing power. Although you will have more traction, the torque is divided by the number of wheels that you're driving. You can only get more power with higher power motors, or by driving the additional wheels with separate motors.

Wheel Exposure

Another variable to consider in wheel selection is whether your wheels are fully exposed, fully protected, or partially exposed (see Figure 12.4). Fully exposed wheels are big targets for spinner robots and arena hazards, and need to be more durable than protected wheels. In particular, the hubs and rim (or core) should be as tough as possible. For example, the wheels on my robot Deadblow had nylon rims, so I machined some aluminum plates for the outside that bolted into an aluminum hub, sandwiching and armoring the rim. Note that partially exposed wheels are subject to less direct damage, but need to be larger in diameter to exit the top and bottom of the frame.

| fully exposed | fully protected | partially exposed |

FIGURE **12.4: Exposed, fully protected, and partially exposed wheel examples.**

Live or Fixed Axle

This refers to whether the axle turns or not. In either case, you've got to spin the tire, but this factor will help you determine what kind of rims and hub you'll need, and if your bearings will be in the wheel, or in bearing blocks in the robot, as described in Chapter 10, "Mechanical Building Blocks."

In the live-axle case, which is the most popular solution, the wheel is firmly attached to the axle, and as the axle turns, the wheel rotates along with it, as shown in Figure 12.5. Usually a sprocket is connected to the axle to transmit power from another sprocket on the motor (via a chain, as described in Chapter 11, "Working with Roller Chain and Sprockets"), and the axle itself transmits power to the wheel. (Gears can also be used in place of the chain and sprockets.) The axle should be prekeyed, or have a keyway cut into it (no set screws), so that you have a robust mechanical connection. Technically, any wheel that is directly connected to the motor (such as an NPC T64 and 10-inch NPC flat-proof tire) is a live-axle system.

Cross-Reference

For sources for prekeyed shafts (a real convenient thing), see Chapter 10.

In the fixed-axle system, shown in Figure 12.6, the axle is firmly attached to the robot's frame, and does not rotate with the wheel. The sprocket is bolted directly to the wheel, and the hub (or rim) has bearings that allow the wheel to turn around the axle. The axle is just supporting weight and does not transmit any power.

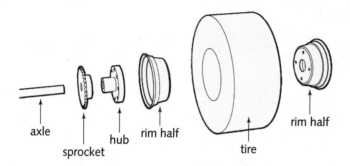

The rim halves sandwich the tire and bolt to the hub. The hub and sprocket firmly attach to the shaft and all components rotate along with the shaft.

FIGURE **12.5: Live-axle example, exploded view.**

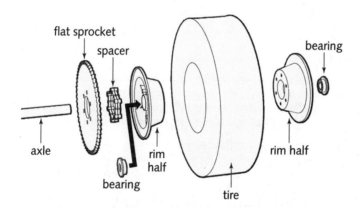

The rim halves bolt together and sandwich the tire. The sprocket bolts directly to the wheel on top of the spacer. The wheel spins on the axle using bearings. The axle does not rotate.

FIGURE **12.6: Fixed-axle example, exploded view.**

Tip

Azusa Engineering, an online go-kart distributor, carries rims (wheels), custom hubs, and aluminum sprockets for fixed-axle systems. They have rims manufactured out of aluminum, steel, and nylon, with their own bearings installed. They have dozens of bolthole patterns for custom aluminum sprockets for many different chain sizes. You can visit the Web site at www.azusaeng.com or call, fax, or e-mail a design to them.

Top Wheel Choices

The top wheel choices used by builders today fall into a few different categories: solid, foam-filled, or bonded. This section will talk about the strengths and weaknesses of each of these types of wheels.

 Note Air-filled tires are not recommended for the simple reason that if the tire and tube are punctured or otherwise ruptured, the tire will deflate, instantly losing its effectiveness. Some competitions (such as BattleBots) have arena hazards such as circular saws that pop up out of the ground. No matter how well you protect the tire, you still have to leave a part of it exposed to contact the ground. Thus, foam-filling tires are highly recommended, and are described later in the "Finding Your Own Wheels" section. Also see the "Foam-Filled Wheels and Tires" section.

Solid Tires

Solid tires are available in many different materials. Most of them, however, are too heavy and/or rigid for use in combat robots. Solid rubber, hard polyurethane, phenolic, and polypropylene wheels fall into this category and should be avoided. The good news is that there is one solid tire that is outstanding for use with combat robots, the Carefree BattleTread.

Carefree BattleTread Tires

When they say carefree, they mean carefree. No tubes, no rubber, no flats. Unlike other molded or solid polyurethane tires, this line is lightweight and compressible, due to its unique microcellular polyurethane foam. The foam is dense enough to support a load, but has millions of tiny closed cells that squish a little, giving you better contact with the arena floor.

The foam allows the tire to have larger dimensions without the added weight that a comparably sized nonfoamed polyurethane tire would have. Because they're solid, you can drill holes through them to lower the weight even more, as some competitors have done. You will have to source your own rims and hubs from online suppliers for this tire (see the "Finding Your Own Wheels" section at the end of this chapter for vendors).

BattleTreads (see Figure 12.7) are available in several different tire patterns from Robot Marketplace. Make sure to remove the slick outer layer of a stock tire to expose the foam underneath.

Foam-Filled Wheels and Tires

These tires are generally standard rubber, intended for use with inflatable tubes, that have been filled instead with a lightweight foam for rigidity and resistance to failure due to puncture.

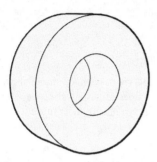

FIGURE **12.7: Carefree BattleTread-type solid tire.**

NPC Flat-Proof Wheels

NPC Robotics has a line of tires prefilled with foam and mounted to steel rims with mounting holes for a hub. There are currently three fairly large tire sizes, which produce a good combination of speed and torque when coupled to lower-RPM motors, such as the gearmotors they sell (see Chapter 9).

NPC also carries hubs that are custom machined to fit both the wheel rims and the output shafts of the motors they sell. This is a great solution, because it's an all-in-one package that's designed to work together, as shown in Figure 12.8. The wheel sizes are good combinations because you usually have more torque than you need, giving you accurate positioning (the robot stops when you tell it to), but also a wide enough diameter to give you above average speed in the arena.

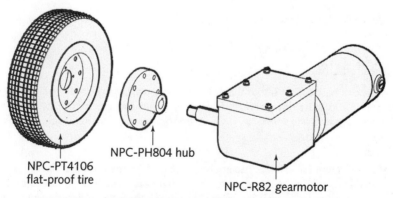

NPC-PH804 hub

NPC-PT4106
flat-proof tire

NPC-R82 gearmotor

FIGURE **12.8: NPC-type wheel and motor system.**

Note You can use the NPC hubs to connect NPC gearmotors to any aftermarket wheel that has four boltholes in a 2¾-inch or 2¹³⁄₁₆-inch-circle.

American Airless Airfree Innerthanes

American Airless is a company that takes rubber tires and fills them with a lightweight foam. In fact, they're the company that NPC uses to fill their tires. Specializing in the wheelchair and mobility markets, they offer rubber tubeless caster wheels that were originally intended for wheelchairs, but work great for combat robots.

Their Airfree Innerthanes are rubber tires with the foam filling, as shown in Figure 12.9. They have several choices in the 6- to 10-inch range with different tread styles, and they fit standard size rims. The Superlite series wheels come with a nylon rim and bearings. You'll need to add an aluminum hub to connect them to the axle.

FIGURE **12.9**: American Airless-type tire.

Bonded Wheels

Bonded wheels are a unique solution to the problem of preventing the tire from spinning on the rim. Instead of a separate rim, the material is molded directly on a *core* of plastic or aluminum. The core connects the wheel to a hub, and in some cases, functions as the hub as well.

Caster Wheels

Caster wheels (see Figure 12.10) have become staple items in combat robots. Their light weight combined with good traction properties and toughness make them an excellent choice for lightweights as well as larger weight classes. The Colson Performa series of caster wheels has a solid thermoplastic elastomer tire (a flexible rubber replacement) bonded to a tough plastic-like polyolefin core. When used with an aluminum hub and through-hole bolted (as is the case with Team Delta or Robot Marketplace hubs), the core stands up well in the arena. The 65 Shore-A durometer hardness rating on the tire is just about perfect for this sport.

Deadblow and American Airless Tires

I heard about American Airless in the Robot Wars days through Donald Hutson of Mutant Robots (Diesector and Tazbot) fame. I was lucky to start out with these tires. My robot Deadblow has used American Airless tires for the past three years. I've never had to worry about killsaws popping out of the ground. I've had tires cut completely through. Even if you lose a tread, the foam itself has enough grip to get you through a match. All that you have to worry about is getting the tire material caught between a wheel well and the wheel, which can jam the wheel and prevent it from turning. (Since I had fully exposed wheels, this wasn't a problem for me.)

In a 3-minute match, you won't be going through many tires. I usually carry two spares and change them out if they get too chewed up. The only time I've ever had them burn through was when I was running four-wheel drive with four independent drive motors (no chains) and one of the motor/speed control packages blew up, leaving that wheel dead. The rest of the robot was dragging it along as if nothing had happened, but it wore a ring around the tire, and it eventually separated.

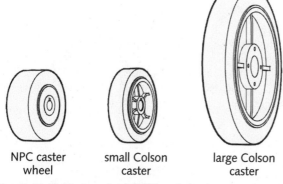

| NPC caster | small Colson | large Colson |
| wheel | caster | caster |

FIGURE 12.10: Caster wheels (different sizes).

NPC Robotics takes similar caster wheels from another manufacturer and adds their own custom Delrin inserts to them. The inserts are keyed for use with similarly keyed shafts. To lock the insert in place, they've also added a taper pin drilled at a 45-degree angle. While this isn't as solid a scheme as through-hole bolting, it represents a significant price advantage for smaller robots that don't need to deliver huge amounts of torque. I recommend fully protecting these wheels and double-supporting them. NPC also plans to release a line of aluminum hubs soon.

Andrus Engineering Wheels

The Andrus wheels shown in Figure 12.11 are a newer product. They offer really wide wheels for a large contact area. This wheel has an aluminum core with a bonded polyurethane tire.

The polyurethane is more rigid than I would like (75 Shore-A), but otherwise this appears to be a nice wheel. You will need to add your own hub to connect the wheel to an axle.

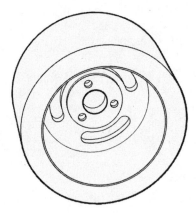

FIGURE 12.11: Andrus-type wheel.

Team Whyachi Wheels

The Team Whyachi wheels (see Figure 12.12) are aluminum-core wheels with a polyurethane tire bonded directly to the core. They wisely have a similar durometer rating to the highly successful Colson casters. Unfortunately, they have a hexagonal center bore, which is a strong, but not a very convenient choice for the average builder. Fortunately, their TWA 40 and T-Box gearboxes are available with a matching hexagonal drive shaft. This is a very sturdy drive solution, and should be an effective combination with their gearboxes.

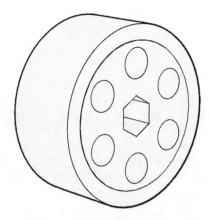

FIGURE 12.12: Team Whyachi-type wheel.

Other Configurations

It's true that your basic two- or four-wheeled robot isn't the only choice you have out there. Following are some of the other configurations, and why they may not be the best (or most efficient) choice for your robot.

Track Drive

Very few combat robots use tracks. Tracks have a high coolness factor, but you've got to consider the issues of expense and durability. You don't have many off-the-shelf solutions to choose from. It's not to say that some famous robots don't use them. On the contrary, Peter Abrahamson's Ronin and Scott LaValley's DooAll are both successful tracked robots. They're also both in the superheavyweight class. My recommendation is to steer clear (pun intended) from these as a locomotion solution, unless you're prepared to custom make CNC track parts, and you have the weight to spare.

Car-Type (Ackerman) Steering

While it's intuitive for anyone who drives a car, Ackerman (car-type) steering has a few problems when applied to combat robots. First and most importantly, when compared to four-wheel slip-steer or two-wheel drive, it gives your robot an inherently wide turning radius. In this game, being able to perform a tight turn could mean the difference between firing your weapon or being fired upon. (Think how long it takes to perform a three-point turn in a car.) Car-type steering is harder to build. With a two- or four-wheeled system, all you have to do is mount a motor and wheel on either side of the robot. With car-type steering, you actually have to use high-power servos to swivel the front wheels. This is mechanically more complicated. It's best to avoid this type of steering.

Walking or Shuffling Robots

While also very high on the coolness factor, walking robots are very difficult to implement. Because walkers usually garner a weight bonus, restrictions on exactly what qualifies as a walking robot are tight. In addition, there are issues with durability and power consumption, because it requires a lot of energy to move without the efficiency of rolling wheels. Shuffling robots have been successfully implemented, but currently, a shuffler will probably not qualify for a weight advantage. From a competitive standpoint, it's difficult to justify a walker. However, if you're more interested in building something cool rather than winning matches (as some builders are, such as Mechadon's creator Mark Setrakian), then a walker is a sure-fire way to create some excitement.

Finding Your Own Wheels

Maybe you don't like the look of the previous wheel solutions. Maybe you want to use something else. This section will introduce some aftermarket distributors and talk about what you need to know to order from them.

What Do the Tire Sizes Mean?

The outside diameter (OD) is the final outer dimension of the wheel, as shown in Figure 12.13. The model number of the tire doesn't always give you a clue as to the tire's OD. In fact, the rim size has more influence on the OD. For example, a standard model 4.10/3.50 tire can have an OD of 10.5 inches, 11.5 inches, or 12.5 inches, depending upon the rim size that's specified. The bottom line is that you should always check to see if the manufacturer has specified the OD of the tire. That way, you'll know the final diameter that you should see in the wheel.

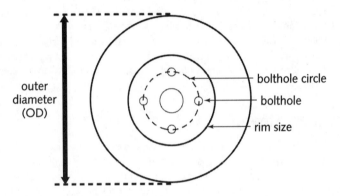

FIGURE 12.13: Tire sizing and bolthole diagram.

You will also need to know the bolthole pattern on the rims that you order. The bolthole pattern consists of the actual hole diameters, number of holes, and the size of the circle that they're aligned on, as shown in Figure 12.13.

Azusa Engineering

This online distributor caters to the go-kart market and sells rubber tube tires, rims (wheels), custom hubs, and aluminum sprockets. They have rims manufactured out of aluminum, steel, and nylon, along with matching keyed steel and aluminum hubs to mount them on. They have dozens of bolthole patterns for custom aluminum sprockets for many different chain sizes. You can visit the Web site at www.azusaeng.com or call, fax, or email a design to them. You can purchase a tube tire and rim and have it foam filled by American Airless, as described in "The (American) Airless Solution" section below.

Gokartsuppply.com

This vendor has a good selection of tube tires, rims, and hubs. You may see *tubeless tires* mentioned on this site and others. In the go-kart world, a tubeless tire does not mean that it's foam filled. It means that it's intended to be used with a single-piece rim. The rim and the tire form the air-filled cavity. The problem is that the rims are generally welded together, and in many cases have no mounting holes for sprockets. For our purposes, I would suggest buying a regular tube tire with standard split hubs and having it foam filled.

Can I Foam-Fill My Own Tires?

Sure you could save a few bucks by performing dentistry on yourself, but why? There are commercial products made for foam-filling tires (polyfill), but they're usually heavier and more rigid (not as squishy). They're meant for filling hand-truck tires and are optimized for durability, not weight and performance. Stick with the American Airless guys; you'll be much happier.

The (American) Airless Solution

How do those guys on TV get those cool wheels? Aren't they worried about the circular saws popping out of the ground? No. They're regular tube tires, but they're foam filled by American Airless. You can use the same vendor as NPC Robotics to get your custom setup filled. They have excellent service, quick turnaround, and good prices. All they ask is that you advertise with a sticker. Apparently, they can fill almost anything, which frees you to use whatever rim, hub, and tire combination you want.

Better Traction through Chemistry

It's possible to enhance your traction with aftermarket chemicals developed by the radio-controlled (R/C) car racing industry. Some manufacturers have separate compounds for rubber and foam tires. The following compounds can be applied to your tires to increase traction by attacking the tire material:

- Trinity Zip Free (foam)
- Trinity Death Grip (rubber)
- Paragon Ground Effects
- Racer's Choice FXII (foam)
- Racer's Choice TQ Sedan Grip 'Blu' (rubber)

Hobby distributors who specialize in R/C car racing should carry these compounds. If you order by mail, place your order early, since the flammable compounds have to travel by ground and cannot be shipped by air. Tower Hobbies carries these products in its online store.

Caution Most of these compounds contain octyl and/or methyl salicylates, which are pretty nasty and strong-smelling chemicals. Make sure that you have adequate ventilation so you're not found collapsed over your robot with your face stuck to a pair of incredibly grippy tires.

Wrapping Up

There are a lot of wheel choices. The trick is to match the wheel to the arena, drive system, and robot size. A mismatched wheel can make an excellent drive system perform horribly, and result in bad maneuverability and no pushing power, while a well-matched wheel will ensure that all your horsepower is transformed into movement, bringing out the full performance of your drive system.

Choosing Your Control System

I f the motors are the muscles of the robot, then the control system is the *brain*. This section will tell you what the brain has to do, and what your choices are. This chapter begins with standard radio controls, including the controls on a standard R/C transmitter, and some basic R/C concepts, such as servos, PWM signals, and mixing. Different radio features are discussed, as well as price and receiver size. I'll also talk about the IFI control system technology and features, and give you some important tips on correctly using the system.

Control System Basics

In order to compete, every robot must have a *wireless control system*. It's just too fast and violent a sport to have a wire running from the robot to your control station. A wireless system consists of a *transmitter* and a *receiver*. The transmitter is what you use to send radio signals to control the robot. It's either hand-held or has joysticks or other controls that plug into it. The *receiver* is the companion piece to the transmitter. It rides inside of the robot, picks up the radio signals and interprets them, and distributes your commands to speed controls, servos, and relays.

As far as wireless systems go, you can choose one of two major routes: off-the-shelf standard radio control and IFIrobotics controls. The reason that there are only two choices is that competitions may have dozens of robots, and it's easier for event organizers to deal with radios that fall into one of these two well-defined categories. Event safety is a major concern, and sticking to these technologies ensures that your signal won't end up inadvertently controlling someone else's robot.

Standard Radio Control

Standard radio control (or R/C) is the easiest and cheapest way to control a robot. There's a huge R/C hobby industry out there that predates robot combat, and you can buy R/C equipment at most hobby stores or online hobby retailers. Depending on the complexity of the radio, prices can range from moderate to expensive.

The Radio Signal

An R/C transmitter sends out radio signals on a certain frequency, which is designated by the *band* and *channel*. The band is used to distinguish the base frequency. These bands have been set aside in the United States by the FCC (Federal Communications Commission) specifically for model applications. Flying models use 72 MHz, and surface (ground) models use 75 MHz. (See the "Switching to 75 MHz" sidebar in this chapter.) The band is further divided into channels, so that each plane flying at a competition can have its own distinct channel and won't interfere with the other planes. Because the individual models (planes, robots, cars, and so on) in an R/C system are *noncoded*, they may pick up signals from any transmitter, and *interference* happens when the signal from one transmitter interrupts the signal from another, causing unpredictable results. Since unpredictable results could spell disaster at a competition (personal injury), event organizers have gone to great lengths to prevent interference. (See the description of transmitter impound in Chapter 20, "Going to a Competition.")

Cross-Reference For a complete listing of frequencies sorted by band and channel, see Appendix F, "Tables and Charts."

The system is completed as shown in Figure 13.1 by having a radio receiver in the robot on the same channel (frequency) as the transmitter. Both the transmitter (in the hands of the operator) and the receiver (in the robot) need to be tuned to the *same* frequency. This is done with crystals. The transmitter and receiver each have a crystal that they use to lock in the correct frequency.

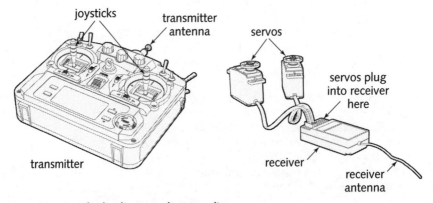

FIGURE **13.1: Standard radio-control system diagram.**

When the system is working properly, a change in one of the controls on the transmitter should produce an equal change in one of the outputs (also called channels) on the receiver. Note that these channels are different from the radio-frequency channels mentioned earlier. The *output channels* of the receiver are numbered and an aircraft receiver can have as many as nine total. Normally, these channels correspond to specific controls on the transmitter, as shown in Figure 13.2.

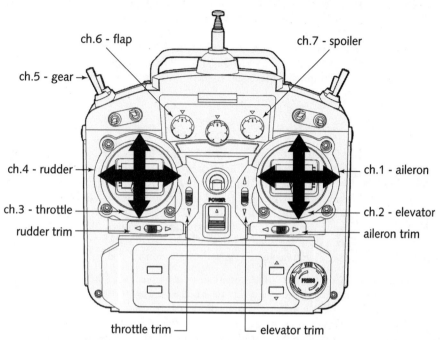

ch.6 - flap

ch.7 - spoiler

ch.5 - gear

ch.4 - rudder

ch.1 - aileron

ch.3 - throttle

ch.2 - elevator

rudder trim

aileron trim

POWER

throttle trim

elevator trim

FIGURE 13.2: Standard R/C transmitter control and channel assignments.

The channels have standard names and a standard order that comes from the hobby aircraft world. For example, moving the right stick from side to side will move a *servo* connected to the channel 1 (aileron) output of the receiver. (See the sidebar entitled "What's a Servo?" for a description of a servo.) Since each *axis* of a stick is considered a *different* control, moving the right stick up and down will move a servo on channel 2 (elevator). Some high-end transmitters (such as the Futaba 9C and 9Z series) allow you to change (remap) the default channel assignment to a different control, which is handy for some situations.

What's a Servo?

In order to control model aircraft, the hobby people had to come up with a small but powerful motor that was capable of holding a commanded position. This became the servo that we know today. It's basically a small motor with a gearbox and electronics to help it compare the commanded position from the receiver and its own position, and make the two the same. All this comes in a small, lightweight package.

Most of the hobby grade servos are small. If you need big power, then check out Chapter 18, "Choose Your Weapon," for a discussion of making your own big servos.

The actual signals that come out of the channel outputs are pulse-width modulated, as described in the sidebar titled, "The PWM Signal." All speed controls, servos, and R/C switches interpret their commands based on standard radio-control pulse-width modulation (PWM) signals.

Mixing Your Signals

Sometimes it's advantageous to combine multiple controls to produce an output. This process is called *mixing*. A prime example of mixing for the robot builder is putting the drive controls on a single stick, also called "tank turn" mixing. Push the stick forward and the robot moves forward, pull it back and the robot backs up, and move it to either side and the robot turns in that direction. Sounds pretty simple, right? Here's the kicker: Moving the stick into the upper-right corner should cause the robot to move forward *while* veering to the right. But, as mentioned previously, each axis of the stick is a separate control. Therefore, you've got to read *both* of these controls at the same time and figure in *both* positions to generate a correct output.

Since most robots have a left-side drive and a right-side drive, both sides need to be controlled at the same time. Now it's suddenly starting to sound a bit more difficult. Don't worry. Most midlevel radios have mixing built in. In fact, you can easily use the *elevon* preprogrammed mode to perform the single stick mixing described earlier, as shown in Figure 13.3. Just activate elevon mode, and plug your speed controls into channels 1 and 2. The right stick will control the drive.

You can also use programmable mixing to create special modes for inverted driving if your robot gets flipped over, for example. (More advanced mixing is discussed in Appendix A, "Advanced R/C Programming.") If your radio doesn't have mixing, you can also purchase an external mixer made by a third-party company such as Ohmark, Watt-Age, VeeTail, or RobotLogic that connects between the receiver and the speed controls and performs the same function. Finally, some two-channel speed controls (see Chapter 14, "Choosing Speed Controls") have mixing built into them, so all you have to do is connect them to the receiver.

The PWM Signal

All the radio-control systems use the same standard for how they convey signals to their servos and speed controls. This is called PWM, or pulse-width modulation. In this type of signal, a pulse of 1 to 2 milliseconds long for each output channel is repeated 30 to 60 times a second out of the receiver. A pulse that's 1 msec long represents one extreme direction, while a 2-msec pulse is the other extreme. The neutral (middle) position is usually 1.5 msec. In a servo, this is interpreted as a position to go to and maintain. In a speed control, it corresponds to a speed and direction. One side of neutral is forward, and the other side is backward. The actual speed in that direction is proportional to the distance from neutral.

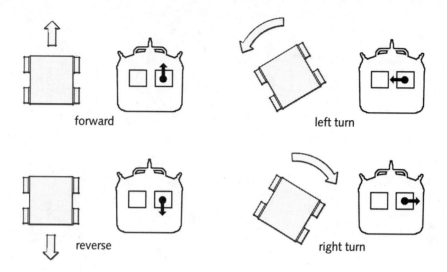

FIGURE **13.3: Tank turn steering with elevon mixing.**

Note It doesn't matter if you have two- or four-wheel drive (as described in Chapter 12, "Let's Get Rolling!"); the control signals are the same from a mixing standpoint. Also, if you have four speed controls (two per side), the signal for each side can be split into two using a Y-connector. That way, the same signal can go to both speed controls on the same side.

Maybe you don't want to have both controls on the same stick. That's okay too. In that case, you don't need any special mixing or preprogrammed modes. You can put the right speed control on channel 2 (elevator) and the left speed control on channel 3 (throttle) and simply use the forward-backward axis on both sticks to drive. To go forward, you move both sticks forward. To go back, pull both sticks back. To turn right, move the left stick forward and the right stick back, and to turn left, do the opposite.

Although I feel it's more complicated, some builders prefer driving this way, much the way some car drivers prefer a manual transmission. Besides, with practice, anything can become second nature. The really important thing to keep in mind is that if you intend to drive with this method, make sure that you can get and install the forward-backward centering spring on the left stick. In nearly all aircraft radios sold in the United States, this control is the throttle, and it doesn't normally come with a spring return. The throttle control usually comes with a ratchet installed so that it will stay wherever it's set. You can drive using the ratchet, but it will make coming to a stop more difficult, because you'll have to concentrate on centering the stick manually, instead of letting the spring do the work. Tank steering without mixing is shown in Figure 13.4.

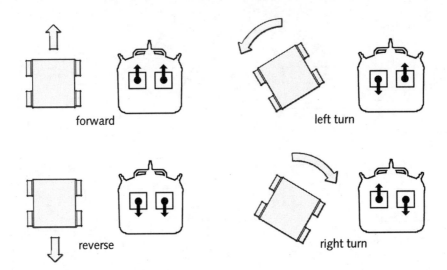

forward

left turn

reverse

right turn

FIGURE **13.4: Tank turn steering without mixing.**

Tip

Mixing can sometimes get pretty confusing. No problem. Here's a way to figure out exactly what signals are coming out of the mixing program. First, plug a servo into each output channel that the mixing affects. Put a small piece of tape on the servo heads to mark the center, and then slowly move the control stick back and forth. The movement of each servo is exactly proportional to the commands that the receiver is getting from the mixing program. This is the same signal that will be passed to whatever is plugged into those channels, whether they are servos or speed controls. This is a good visual check if you're trying to determine whether a problem is in the radio or the speed control, or if you just need to get a handle on what's going on.

What Kind of Radio Should I Get?

You don't need the top-of-the-line radio to do well in battle. However, there are a few features that you should take into account when you go shopping, such as transmission type, features, receiver size, and cost.

Transmission Type

When it comes to transmission type, there are three choices: AM (amplitude modulation), PPM (pulse position modulation, also commonly called FM), and PCM (pulse code modulation). This particular aspect of radio technology is not so much a convenience as it is a safety issue. AM radio equipment should not be used in any combat robot. It's too susceptible to outside interference and should be considered a safety risk.

This leaves you with PPM/FM and PCM. Technically, both PPM/FM and PCM are FM transmission systems. The difference is in how the data is encoded. PCM data is encoded digitally, and gives a higher degree of immunity from noise than PPM/FM, because it has built-in error identification. Invalid data packets are caught and thrown out immediately, before they can reach the outputs. PPM/FM data is analog, and incorrect data could be passed along to the output channels.

Because of these varying degrees of resistance to interference, you will see that most low-end radios are AM, the midrange models are PPM/FM, and the high-end units are PCM.

Brand and Features

There are four major brands in the R/C market: Airtronics, Futaba, Hitec, and JR. The biggest and most popular brand of radio on the market is Futaba.

The radio-frequency channels described previously in the section entitled "The Radio Signal" are an industry standard, and are the same for all brands, as are the default control assignments and corresponding receiver channel numbers.

The following are common features that you'll find offered among the different brands. Just as all cars have turn signals and windshield wipers, the actual implementation of controls varies slightly among brands, but the basic function is the same.

- **Number of output channels.** Usually four to nine in aircraft radios, and three in pistol-grip R/C car radios. You'll need at least two for the drive, and any weapon on/off switches or other controls will require additional channels.

- **Servo reversing.** Ability to reverse which way the servo turns relative to the control.

- **ATV/EPA.** Ability to adjust how far a servo can turn in each direction.

- **Dual rate/exponential.** Adjustment for stick sensitivity and linearity (described in greater detail in Chapter 19, "Troubleshooting").

- **User-programmable mixing.** Ability to create your own mixing setups.

- **Preprogrammed mixing modes.** Factory-programmed mixing setups, such as elevon.

- **Multiple models.** Lets you set up several different models (robots), so that each robot can have its own trim and mixing settings.

- **Failsafes.** Returns servos to safe positions when radio contact is lost (described in greater detail in Chapter 17, "The First Test Drive"). Only PCM radios can take advantage of this feature.

- **Control reassignment.** Ability to reassign controls to different channels (high-end Futaba radios only).

Most low-end FM radios have at least reversing and ATV/EPA. Mixing is usually reserved for medium- to high-end radios, which have small computers. Having a computer radio isn't necessary, but I highly recommend it.

Receiver Size and Weight

As with every component that goes inside of the robot, size and weight are concerns. However, because R/C receivers are designed to be used in flying models, they're very compact and light. (Component weight in an aircraft is much more critical than in a combat robot.) In fact, almost all R/C receivers weigh less than 2 ounces. The case sizes vary with manufacturer and model, but they're usually pretty small. For example, the largest Futaba receiver measures 2.5" × 1.4" × 0.88".

You should also figure into the weight a separate battery pack. Even though a BEC (battery eliminator circuit) would save weight by replacing the receiver battery, my personal feeling is that a separate battery is essential in preventing problems due to battery voltage dropouts, which can occur during pushing matches in the arena.

Cost

Generally, R/C transmitters and receivers are sold as packages, grouped together with a couple of servos, batteries for the transmitter and receiver, and a charger. Of course, features and brand affect price, but be prepared to pay anywhere from $150 to $800 for a package.

If you're on a budget, as most of us are, try to leave yourself an upgrade path. Midrange radios are usually FM/PPM, but the Futaba 6XA series can also transmit PCM, allowing you to purchase an FM receiver and upgrade to PCM later by buying a PCM receiver. An upgrade to PCM will give you the ability to use failsafes. Note, however, that unless you purchase the 6XAS-75MHz, you'll have to go through the retuning process.

My pick for a great radio value is the Futaba 9CAP. It essentially has all the features of their flagship 9Z transmitter series, but at half the price. What's more, it's already tuned to the FCC-compliant 75 MHz surface model band. Make no mistake; it was made for robot combat.

This will probably be one of the biggest single expenses associated with your robot. Think of it as an investment in your hobby. You can easily use the same radio with multiple robots. The same can't be said of most mechanical parts.

Tip You should purchase multiple crystals so that you have some backup choices if another robot is using your frequency. They can be easily swapped out to change the channel. Make sure that you purchase crystals that are made for your band. If you're running 75 MHz, then you need to buy 75 MHz crystals. Also, make sure that the crystals that you buy are compatible with your radio. Crystals from different brands of radios usually aren't interchangeable. You'll need to change the crystals in both the transmitter and receiver.

Cross-Reference This chapter is focused on introducing you to control systems and helping you decide which system is right for you. Several more technical concepts have been grouped into other sections, where the discussion is more applicable:

- Radio tuning, including trim adjustment and failsafe setting is covered in Chapter 17.

- Antenna placement and radio interference problems are discussed in Chapter 19.

- An example of exponential programming to improve drive response is described in Chapter 19.

- Advanced programming, including programming for inverted driving, is covered in Appendix A.

- Using your radio to remotely activate something (such as a weapon) onboard the robot is discussed in Chapter 19.

Switching to 75 MHz

In combat robots, you are strongly encouraged (and in most competitions *required*) to comply with FCC guidelines and use *ground* frequencies only. Unfortunately, most of the radios on the market are in the 72-MHz (aircraft) band. What if you've already spent a lot of money on a 72-MHz radio system? Not to worry. You can have any 72-MHz transmitter retuned by a professional for around $20 to $50, not including the new 75-MHz crystal. Unfortunately, the retuning is necessary—you can't change *bands* simply by changing the crystals, only the channel *within* a band. Listed below are a few places you can contact for professional retuning:

- **Hobby Services (www.hobbyservices.com).** A service center that's factory-authorized by Futaba and many others

- **George Steiner, Sacramento, CA (916) 362-1962.** A longtime radio-control guru and author of *A to Z - Radio Control Electronic Journal*, a collection of articles on various R/C topics, which I proudly have on my bookshelf

- **D & M Electronics (www.dnmelectronics.com).** A radio-control specialty service shop recommended by many competitors, including Dan Danknick of Team Delta

You will also need to retune your receiver to work on ground frequencies. Usually, however, it's more cost-effective to buy a new receiver than it is to buy a new crystal and go through the retuning process.

If you don't want to deal with sending out your radio, then you may have another choice. The high-end Futaba radios (5U, 7U, 8U, and 9C series) have frequency *modules* (not just crystals) that allow them to change bands simply by swapping out the module. No retuning is necessary, because the crystal is *inside* the module, which has already been factory-tuned to the correct frequency. Ground modules tuned to 75 MHz (called TJ modules) are available from major hobby retailers and cost about $55 each. The catch is that you've got to buy a complete module for every channel that you want to use, not just a crystal. And of course, you'll still have to purchase a new receiver specifically tuned to ground frequencies.

Also, Futaba has recently started coming out with radios aimed specifically at the 75-MHz market. The 6XAS-75MHz and 9CAP transmitters are both factory-tuned to 75 MHz, and include a matching receiver.

The IFIrobotics System

IFIrobotics has been providing control systems to the FIRST competition (a national high-school robotics contest) for years. Their equipment is very stable and robust, and is generally considered safer than R/C for large competitions.

Parts of the System

In this system, the transmitter unit is called the operator interface (OI), and the receiver unit is called the robot controller (RC). You plug joysticks and other controls into the OI, which is a stand-alone (not hand-held) base unit. The speed controls and servos plug into the RC inside the robot. Unlike standard radio control, the IFI system transmits on 900 MHz using radio modems, as shown in Figure 13.5. This is a two-way signal, so the OI and RC transfer information back and forth to each other. For example, you can check the battery voltage on the robot while standing at your control station. This sort of information is impossible to get with a standard R/C system, which is a transmit-only, one-way system.

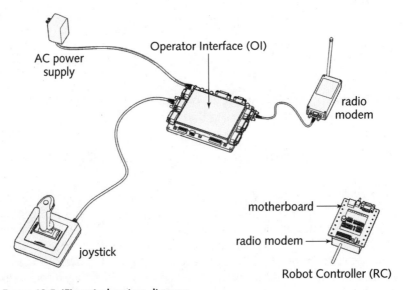

FIGURE 13.5: IFI control system diagram.

Each data packet passed back and forth by the OI and RC has a team number associated with it. The RC will only listen to valid data packets that have the right team number. This eliminates the possibility of other transmitters being able to control your robot (unless they're set to the same team number as you, they're on the same channel, and they're closer to the robot).

Note Usually, competitions require that you set your team number to your pit table number.

This inherent immunity to interference is a great safety factor. It also leads to another advantage for the robot builder: exemption from transmitter impound (described in Chapter 20). This doesn't mean that you can fire up your system at any time and run around the pits. On the contrary, you will be required to use a *tether*, which is a DB9 cable that connects the OI to the RC without transmitting any radio signals.

Flexibility

The IFI system is also different than a traditional radio-control system because it has an onboard processor that runs a program. What makes it flexible is that you can perform math calculations, read sensors, set up whatever mixing you want, and even allow the robot to make some autonomous decisions. You can also perform timed sequences. For example, if you have a series of pneumatic valves that have to open at specific times in a specific order, then you simply write it into your program. There's no easy way to put a pause of any length of time in an R/C system. You can change the way the control system responds with just a few keystrokes without rewiring anything. Loading your program into the RC can be done from any PC, and only takes a few seconds. It usually takes longer to connect the cables.

Tip

Make sure to leave yourself access to the program/tether port and the reset button when you mount the RC. If you don't, it may make debugging a lot more difficult. Also, make sure that you get the most recent version of the download software from the IFI Web site.

The onboard computer based on the Basic Stamp IIsx (also called the BS2SX) uses the PBasic language for programming, which is probably the most popular language among electronics hobbyists. Check Appendix D, "Online Resources for Builders," for a listing of Web sites that offer free downloadable support materials and tutorials for PBasic. Also, *Programming and Customizing the Basic Stamp Computer* by Scott Edwards (McGraw-Hill, 1998) and the *Basic Stamp Programming Manual*, published by Parallax (I have version 1.9) are the two best books that I can recommend. They are both easy to read and follow.

Tip

It's a good idea to bring a laptop PC to the competition with a copy of your code. This will allow you to make minor changes easily, or even reload your code into a loaner RC in the event of damage. Remember, a DB9 cable is required to connect from the PC's serial port to the programming port on the RC.

If you're not big on programming, don't worry. Since the RC comes preloaded with a routine for single stick tank turn mixing, most competitors don't even have to touch the software. You just plug speed controls into PWM1 and PWM2, and a joystick into Port 1, and you're ready to go for single stick control. If you prefer two-stick (nonmixed) control, plug joysticks into Ports 3 and 4, and speed controls into PWM4 and PWM5. If you have a weapon that requires a relay, the default code uses the Port 1 joystick button to control the relay 1 output and the Port 2 joystick buttons to control relay 2.

Cross-Reference

IFI has a series of relay switches (called Spike Relay Modules) that plug into the RC and can be programmed to turn on valves and large solenoids. These switches are discussed in Chapter 18, and the programming is addressed in Appendix B, "IFI System Programming and Troubleshooting."

Which System Should I Get?

IFIrobotics sells two control systems: the Isaac 16 and the Isaac 32. The Isaac 16 is more than enough for most combat robots. If you need more than 16 digital inputs, 16 analog inputs, 8 PWM outputs, and 8 relays, then you're building something other than a combat robot.

The Isaac 16 RC measures 4.5" × 3.5" × 2" and weighs about half a pound with the motherboard. You should mount it by the holes on the sides of the motherboard with shock isolation.

Neoprene sandwich mounts (as described in Chapter 4, "Selecting Materials,") are perfect for this job, or just buy the shock mount kit made specifically for the Isaac system and sold by IFI.

What to Expect at an Event

At larger competitions, you will actually have IFI personnel at the event to run the Competition System in the arena. During a match, the Competition System provides the competitors with frequencies that aren't normally available. It is the highest level of interference protection, and ensures that you're the only person transmitting on a secure frequency. Every time you go up to fight, you'll connect the competition port on your OI to the Competition System, which provides power for the OI and a radio modem that's locked to one of these frequencies.

At smaller regional competitions, IFI personnel may not be present. In this case, it's good to have a few things prepared to make competing easier. You may not have AC power at the arena, so you should rig a 12-volt DC battery for the OI. Also, you should prepare a *channel access adapter* for the competition port, as described in the *Operator Interface Reference Guide*, available from the IFI Web site. When installed, the adapter (which is basically a DB15 connector with two pins shorted) will allow you to access additional frequencies, so that competitors in the arena can choose from four additional channels. It doesn't offer quite the security of the Competition System, but at a smaller event, you'll have far fewer robots to deal with.

IFI Usage Tips

Here are some helpful tips to keep in mind when using the IFI system:

- The joystick inputs on the IFI systems are intended for *analog* joysticks only. Newer digital and USB joysticks will not work. You can use your own joystick, but it's best to order a joystick from IFI when you purchase your system. It's critical to have a really good spring return or else your robot will start to drift when it's supposed to be standing still. You could use the trims to correct this, but without good centering springs, it will just drift again in another direction the next time you move the stick and let it go back to center. The CH joysticks sold on the Web site have excellent springs and trim controls.

Note

Some people find it easier to maneuver their robots with a smaller control stick, like those found on R/C equipment. You can choose your own stick for use with the IFI system, with a few requirements: It has to be an analog joystick with a standard DB15 connector (no USB interface), there should be trims so that you can tune the idle position to prevent movement, and the centering springs should be good and tight. If you find that the robot is still drifting even after you've correctly set the trims once with the new joystick, then the centering springs aren't strong or precise enough. Unfortunately, IFI can only recommend the CH sticks they have on their Web site. Any other control stick will require some testing to make sure it's good enough.

- Once the joystick trims have been set correctly to prevent robot movement when idle, they should be taped over to protect them from being inadvertently bumped. If you're really worried about not getting bumped, then you can also use hot-melt glue.

- You should make a cover for the Isaac 16, or at least tape over the exposed PWM pins so that you prevent any random pieces of metal that enter the robot from shorting out these connections, causing damage to the RC.

- Usually, robots have metal armor on all sides, which essentially creates a sealed metal box, and makes it difficult to send and receive radio signals. You can help out your equipment by locating the RC near a slot or window in the top or side of the armor. You can cover this hole with Lexan for protection. The ideal situation would be a 4-inch diameter hole, but you can also get away with a 4" × 1/2" slot. Note that the antenna doesn't have to stick out of the robot. You just need a nonconductive path for the radio signals to enter and exit.

- The RC should be located as far away as possible from any (electrically) noisy motors which may interfere with the radio signal. Remember to leave access to the program/tether port and the reset button on the RC so that you can easily plug in and tweak your existing program, or run off of the tether in the pits.

- You should have a secondary power switch to be able to turn the RC on and off. The switch should be a shock-resistant one, as described in Chapter 16, "Wiring the Electrical System." If you really don't want a switch, then you *must* provide clear external access to the reset button on the RC, which is required so that you can reset the radio modem just before going to the arena, which will allow you to lock to one of the arena frequencies when you plug in your control system.

- Mount your IFI control equipment (OI, modem, joystick, and so on) to a piece of plywood or acrylic. That way it all stays together, and you don't have the modem hanging down and getting banged around. Make sure to leave access to the competition port. You may not have a shelf to rest the system on, so bring something to use as a stand. I use a portable musical keyboard stand.

Battle Hardened

Unlike hobby R/C equipment, the IFI stuff is battle-hardened. They have been constantly upgrading their equipment to meet the demanding needs of the combat robot.

There are only a few drawbacks to this system. IFI units are larger and heavier than standard R/C equipment. They are also as expensive as the high-end R/C radios, with a complete system running at about $900.

Cross-Reference For interfacing and troubleshooting information on the IFI system, refer to Appendix B.

Wrapping Up

This chapter introduced the two major choices you have for control systems: radio control and IFI. They represent two different philosophies. The R/C system is simple and straightforward with the ability to do some mixing if you want. The IFI system is more complex, but has a large amount of flexibility to program custom-timing sequences and automatic weapons triggering. The choice is up to you which way to go, based on the information in this chapter, including features, price, physical size and weight, and convenience.

Complete Control uses its powerful finger-like clamp and forks to capture and lift Subject To Change Without Reason.

T-Minus flips vertical spinner Heavy Metal Noise with its powerful pneumatic launching arm.

Author Grant Imahara tests
one of the R2-D2 units.

ILM R2-D2 fabrication/
restoration crew: (L-R) Crew
chief Don Bies, Scott
McNamara, Grant Imahara,
Nelson Hall, Preston
Donovan, and Erik Jensen.

A bruised and battered Deadblow (version 1.0) goes the distance and wins the Megabot (Middleweight) Rumble at the first BattleBots competition in Long Beach, CA in 1999.

T-Minus launches SOB into the air and upside down.

Grant Imahara's Deadblow pounding Derek Young's Pressure Drop on its way to becoming Middleweight Runner-up at BattleBots San Francisco 2000.

Deadblow wins the Middleweight Rumble at BattleBots Las Vegas 2000.

Choosing Speed Controls

Now you've got your control signals going out to the robot, but how exactly are you going to make the motors move? That's where the speed controls come in.

This chapter begins by giving you the basic run-down on these items, and where they fit into the whole system. I identify some important features to look for when selecting a speed control, and then show you the most popular choices on the market and how to use them effectively. Also included is a discussion of some other (newer) speed controls to consider and a few solutions that you should avoid using in a combat robot. I end by setting you loose on Project 6: Mounting the Electrical Components.

The Basic Idea

The job of the electronic speed control (ESC) is to transfer power from the batteries to the motors in an efficient and *controlled* way. It interprets the PWM signal coming from the receiver and, based on this information, decides the speed and direction to turn the motor. It accomplishes this control by regulating the *voltage* that's applied to the motor. The polarity determines the direction, while the average voltage level determines the speed. The higher the voltage goes, the faster the motor wants to turn.

Controlling voltage is only half of the battle, however, since you are transferring *power*, and power is proportional to *both* voltage and current. As the ESC raises voltage, the motor will ask for more current, and the ESC has to be able to supply enough to cover the demand. You run into problems when the motor wants more current than the ESC can handle. Your first instinct may be to use the speed control with the highest current-handling capability you can get. The tradeoff, however, is that the higher-performance controls are usually larger, heavier, and more expensive. So, what you really want is a speed control with enough rated capability to handle your motor, and a little bit extra, as a safety margin to help cover any unanticipated situations.

Where Speed Controls Fit in the System

As mentioned in the previous chapter, all of your commands are processed through the control system and come out of the receiver in PWM format. This goes directly to the speed controls. The ESC also has a direct link to the batteries on the input side and to the motors on the output side. Some speed controls are single-channel (one motor) and some are dual-channel (two motors). Both types are connected in the same place in the chain, as shown in Figures 14.1 and 14.2.

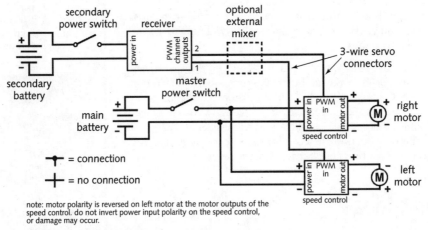

FIGURE 14.1: Typical single-channel speed control setup.

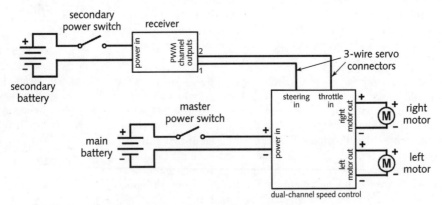

FIGURE 14.2: Typical dual-channel speed control setup.

Important Factors in Choosing a Speed Control

The following items — current handling, durability, size and weight, mixing capability, and cost — are characteristics that you should evaluate for the speed control that you intend to use. There are many tradeoffs, and you should be aware of the risks and benefits involved.

Current Handling

In general, more powerful motors will draw more current. However, most speed controls won't need to produce maximum current all the time. In fact, under normal conditions, the demand isn't that high. If the robot gets into a pushing match, or a wheel is jammed, this may create a stall condition, where the current demand goes through the roof, and then the speed control may suffer damage if it is not rated high enough. Given this, the ESC should ideally be chosen so that the maximum (intermittent) current-handling capability is greater than the motor's peak (stall) current.

High current-handling capability means larger size and higher cost. Therefore, your speed control should be scaled to both your budget and your motor. You can use a higher-performance speed control with a small motor, but the size and cost may be overkill.

Also, you should take into account the voltage that you're using with the motor. If you have a 36-volt motor, for example, but are only using a 24-volt speed control, then you will never be able to achieve maximum power. Almost all the popular speed controls can handle up to 24 volts, but as you get up to 36 and 48 volts, your choices become limited.

Durability

Although they're not directly in harm's way like armor or exposed wheels, speed controls are still subject to a large amount of damage if not properly shock mounted. Since they're full of tiny electronic components, they're one of the most fragile parts of your robot. Some controls have socketed chips that can come loose during a large impact, or output transistors that can touch, rendering your otherwise robust drive useless. The way to avoid this failure is to protect and isolate your speed controls from large and violent jolts using neoprene sheet (old mousepads can be used) to cushion around the case. Be careful not to block fans and airflow pathways for cooling, or you'll cause the ESC to overheat, and eventually fry. Neoprene sandwich mounts (described in Chapter 8, "Fasteners — Holding It All Together") are also handy for shock isolation.

You should also ask yourself if the speed control in question is battle-proven, and if you are willing to take the risk of proving it. Check out Appendix D, "Online Resources," for the Web site addresses of several builders and the BattleBots Builder's Forum, where you can see if someone has used a particular speed control in battle, and how it performed.

Size and Weight

In most speed controls, weight isn't a problem, because they're basically composed of electronic components on circuit boards. Some controls, however, have built in *heatsinks*, which add to the weight dramatically, but help improve performance. A heatsink is a piece of metal (usually aluminum) that is thermally coupled to the power transistors. As the transistors generate heat, the heatsink draws it out and radiates it to the environment, helping to prevent overheating and potential damage.

The size of an ESC can also be an issue, especially in the lower weight classes. Fortunately, smaller robots usually need less power than larger ones, and that means that the ESC can also be smaller. You should carefully check the size of the control that you want to purchase and make sure that you've got enough space for it in your frame.

Mixing Capability

As mentioned in Chapter 13, "Choosing Your Control System," mixing is the ability to take signals from two sticks or controls and combine them to produce an output. It's required if you want to use single-stick tank-style steering. While most mid- to high-end radios have the capability to mix internally, some people prefer the ease of mixing that's built into the speed control. There are a few controls that have multiple mixing programs built into them, and all you have to do is plug into the receiver. The programs are selected by buttons or jumper settings on the ESC itself.

If you've got a radio that's capable of internal mixing, then you can get away with a less expensive speed control without built-in mixing. I personally prefer using the mixing on the transmitter, so I can change the settings without having to open up the robot (and in some cases, the speed control).

For instructions on how to use your radio for internal mixing, see Appendix A, "Advanced R/C Programming."

Cost

Most of us are pursuing this hobby on a budget. Compared to the radio, the speed control is probably the second biggest investment you'll make in the robot. In general, you've got to pay for performance, but there are some great values available.

Also factor in the likelihood of damage. It's rare that someone will blow a radio during competition, but it's all too common that someone will blow a speed control. You can never predict what's going to happen in the arena. Having spares on hand during a competition is an excellent idea. Imagine barely winning a match and then having to go into the next round with only *half* your drive.

Popular Speed Control Choices

Most robot builders stick with the battle-proven top three brands: Vantec, IFIrobotics, and 4QD. This section will help explain their differences and discuss the strengths and weaknesses of the big boys in the speed-control market.

Vantec

Vantec has been around since the beginning. In those days, choices were limited, and if you didn't have a Vantec, then you were stuck using an inferior hobby speed control. Their large and loyal following continues to this day. The RDFR (Director) series is the flagship of the Vantec line. These are high-power dual-channel speed controls with many mixing options built into the controller. They provide a convenient solution for competitors who can't mix in their radios and don't want to deal with an inline mixing unit. A typical RDFR control is shown in Figure 14.3.

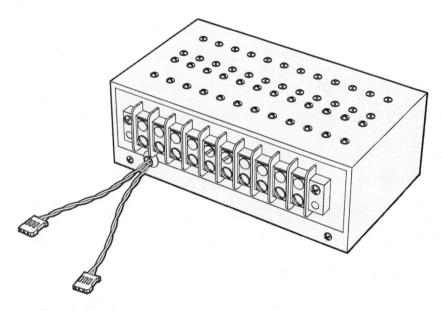

FIGURE 14.3: Sample Vantec RDFR controller.

The RDFR line has an impressive range of voltages and current ratings, as listed in Table 14.1. Bear in mind when comparing size, weight, and cost that these are *dual* (two-channel) speed controls, as opposed to the size, weight, and price listings for the single-channel speed controls in the rest of this section. (Current ratings are for continuous and intermittent conditions.)

Table 14.1   Vantec Dual-Channel Speed Controls

Model	Voltage (volts)	Current (amps)	Size (inches)	Weight (ounces)	Price
RDFR21	4.5-30	14/45	4.25 x 2.9 x 1.4	7	$250
RDFR22	4.5-30	20/60	4.25 x 2.9 x 1.4	9	$275
RDFR23	4.5-30	30/60	4.25 x 2.9 x 1.4	9	$350
RDFR33	9-43	35/95	6.25 x 2.2 x 4	27	$450
RDFR36E	9-43	60/160	6.25 x 2.3 x 4.5	39	$525
RDFR38E	9-32	80/220	6.25 x 2.3 x 4.5	43	$750
RDFR42	32-60	20/54	6.25 x 2.3 x 4	27	$500
RDFR43E	32-60	35/95	6.25 x 2.3 x 4.5	39	$550
RDFR47E	9-43	75/220	6.25 x 2.3 x 4.5	43	$785

Source: Vantec Web site: www.vantec.com.

Generally, lightweights use RDFR22 and 23, middleweights use RDFR23, 33, and 36E, and heavyweights use RDFR36E, 38E, and 47E. However, actual control sizing should be done according to the motors used as per the recommendations on their Web site at www.vantec.com. The Vantec controls have been battle-hardened over several generations. Their size and weight are moderate, and they tend to be expensive (but remember, these are dual-channel controls).

Vantec Usage Tips

- Order early, since experience shows that delivery may take a while, but everyone says, "it's worth the wait."

- Always use spade or ring crimp connectors on leads going to the terminal block. The terminal screws are too close to each other for bare wires, and a stray wire could case a short-circuit, damaging the unit.

- Always shock mount the speed control (which should be the case for all speed controls, but especially the Vantec because of the dual-board construction). They have been known to reset on large impacts.

- Most Vantec speed controls have open cases. You should fashion a cover out of Lexan or thin metal, or at least wrap the case with electrical tape to keep bits of metal from shorting out the exposed boards.

Note Something to consider is that while a dual controller is much simpler to wire, and the built-in mixing is convenient to use, a major drawback is that if you blow one side, you'll have to replace the whole unit. This can get expensive, considering that the dual units typically cost a lot more to begin with. Fortunately, Vantec has an excellent replacement policy. This won't do much good while you're at the competition, though, so think about bringing spares if you can afford them.

IFIrobotics

IFIrobotics is newer to the combat arena, but has gained a reputation for bulletproof controls that are well designed and handle heaps of abuse, all in an affordable, compact package. The Victor 883 controller is the flagship of the IFI line. They also have a higher-power version called Victor 885 (previously known as the Thor 883). It has twice the continuous current rating, but fits in the same package, which has the smallest size and weight of all the high-performance controls listed in this section. An IFI controller is shown in Figure 14.4, and their specifications are listed in Table 14.2.

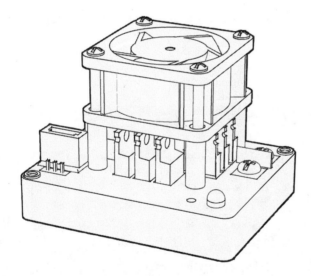

FIGURE **14.4: Typical IFI controller package.**

These are some of the least expensive and most available controls out there, with immediate delivery at most times, even when the demand is high, right before major competitions. This price allows competitors to keep a spare or two on hand without breaking the bank.

Unlike most of the other brands, IFI speed controls feature active cooling. This allows them to get the most out of their electronics. (Current ratings are for continuous and intermittent conditions.)

Table 14.2 IFIrobotics Speed Controls (Bidirectional)

Model	Voltage (volts)	Current (amps)	Size (inches)	Weight (ounces)	Price
Victor 883	12 or 24 (30V max)	60/200	2.7 x 2.2 x 2	4	$150
Victor 885	12 or 24 (30V max)	120/300	2.7 x 2.2 x 2	4	$299

Source: IFIrobotics Web site: www.ifirobotics.com.

At press time, the high-voltage version of the Victor (up to 60 volts) has not been released yet, but should be available soon.

Single-direction speed controls such as the Victor SC and Thor SC that are intended for rotary weapons will be discussed in Chapter 18, "Choose Your Weapon."

Signal boosters are also available from other vendors, such as Team Delta. Check Appendix D for a listing of specialty electronics vendors.

IFIrobotics Usage Tips

- Be careful inserting the PWM cable into the shroud on the speed control. The pins are easily bent. The connector should be pushed all the way down. Strain relief is recommended to help keep the PWM cable in place. Hot-melt glue can also be used to make sure that the connector doesn't move.

- Make sure to leave access to the calibration hole on top of the controller. Although unlikely, you may have to recalibrate the controls to compensate for an offset in your control system.

- Make sure to wire the fan into the *battery* side of the speed control.

- The fans in past versions have been reportedly susceptible to high shock loads. Current models ship with a Rotron fan, which is much less susceptible to damage. As with other controls, shock isolation should prevent this problem.

- When using an IFI speed control with a standard radio control, you will probably need a special interface cable to go between the speed control and the receiver, which boosts the receiver's output pulse. It runs about $10 and is available in all places that sell the IFI controllers. Get a few extras if you can.

4QD

The 4QD controls are the most popular on the British robot fighting circuit. Their controls are geared towards golf carts and other small electric-powered vehicles that carry full-size adults. They have plenty of power to control robots that are small by comparison. Figure 14.5 shows a 4QD controller.

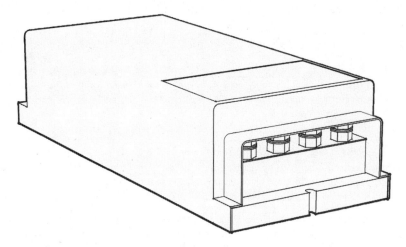

FIGURE **14.5: A 4QD 200 series controller.**

While 4QD has many speed controls in their line, the most popular among robot builders are the NCC, Pro, and 4QD series. The NCC is the economy line, with 35-amp and 70-amp versions; the Pro is the midgrade line, with 110-amp performance; and the 4QD series are the big boys, with 150-, 200-, and 300-amp versions. The models and their various characteristics are shown in Table 14.3. (Current ratings are for continuous and intermittent conditions.)

Table 14.3   4QD Speed Controls

Model	Voltage (volts)	Current (amps)	Size (inches)	Weight (ounces)	Price
NCC 35	12-36	35/55	6.8 x 3.2 x 1.6	9.2	$100
NCC 70	12-36	70/110	6.8 x 3.2 x 1.6	9.2	$150
Pro 120	12 or 24	110/145	6.3 x 4 x 1.6	11.5	$200
Pro 120	36	110/145	6.3 x 4 x 1.6	11.5	$215

Continued

Table 14.3 (continued)

Model	Voltage (volts)	Current (amps)	Size (inches)	Weight (ounces)	Price
Pro 120	48	110/145	6.3 x 4 x 1.6	11.5	$265
4QD 150	24 or 36	120/160	9.8 x 4 x 2.4	45.9	$305
4QD 150	48	120/160	9.8 x 4 x 2.4	45.9	$320
4QD 200	24 or 36	210	9.8 x 4 x 2.4	45.9	$320
4QD 200	48	210	9.8 x 4 x 2.4	45.9	$335
4QD 300	24 or 36	320	11 x 4 x 2.4	59.4	$400
4QD 300	48	320	11 x 4 x 2.4	59.4	$415

Source: 4QD Web site: www.4qd.co.uk.

The sizes and weights are provided for the board only on the NCC and Pro series. (You've got to make your own case.) The 4QD series comes complete with a case and (massive) heatsink. These controls are more on the heavy side compared to most controllers (thanks to the built-in heatsinks), but they are well designed and have tons of internal protection devices.

Unlike most speed controls, the NCC and Pro series controls use relays to switch between forward and reverse. Some builders believe that having this mechanical element lowers the reliability of the control. As the flagship of the line, the 4QD series is fully solid state, and the philosophy of this speed control is to make something so robust that the idea of failure is ludicrous. (Really, read the 4QD manual.) When you see a 4QD controller, you'll know why. They're beefy.

Interfacing to the 4QD Controllers

Since the 4QD products are geared towards the vehicle market, they do not come with an R/C interface. You will need a special interface board to translate the receiver's PWM signals to something that the 4QD controller can recognize.

For some time, the standard interface has been an analog board called the DCI (Dual Channel Interface), which costs about $30. The idea is that you connect two servos to two pots that plug into the board. The DCI has built-in two-channel mixing for tank steering, and provides the necessary interface signals to command two NCC, Pro, or 4QD controllers. This setup is electrically isolated from the controller, so it prevents any noise from being coupled back into the receiver, but as with any physical servo interface, if the receiver loses power, the servo will stay exactly where it was last, even if this is full forward. (Note that this is different from a *failsafe* condition, where the receiver loses valid *signal* from the transmitter.)

4QD has just introduced a line of digital interfaces called the SMR (single-channel, $30) and DMR (dual-channel, $60). Unlike the DCI, which has a servo mechanically coupled to a pot, these boards are fully electronic. Unfortunately, the SMR and DMR boards only offer "64 states of resolution." That's pretty choppy, and you will probably see a stepped response, rather than a typical smooth response as you move the joystick up and down. It is, however, proportional. The good news is that 4QD is designed in a failsafe mode that shuts down the controller if the receiver loses power, or otherwise fails to generate the PWM signal.

4QD Checklist for Robot Builders

The following features of the 4QD speed controls are intended to enhance safety for vehicles, but in practice, tend to get in the way of robot builders.

- The default state of each control is to go half speed in reverse. Make sure to disable this feature, giving you full reverse speed (also called symmetrical reverse).

- Another protection feature is a low-battery shutdown monitor. When the supply battery gets too low, the controller shuts down. It should be defeated since momentary dips in the battery supply could cause the speed control to cut out — usually at the worst time.

- Lastly, a safety feature of the 4QD called HPLO (high pedal lock out) ensures that the control won't fire up if the signal to a channel is at maximum. Unfortunately for the robot builder, this may also cause an inopportune shutdown during a power fluctuation and should be disabled.

Other ESC Choices

You don't necessarily need to go with the big three mentioned previously. There are quite a few other choices out there, including the OSMC controllers, the Robotic Sporting Goods RSGSS, and the Roboteq AX2500. This section will describe some of their benefits and drawbacks.

Open Source Motor Controllers (OSMC)

The Open Source Motor Controllers produced by Robot Power (OSMC) and Robot Solutions (MC1-HV) are based on the concept of making very high power yet affordable kit-based speed controls. They have separate power and logic boards. The power boards are single channel and run on 13 to 50 volts DC with a continuous current rating of 160 amps and a peak of 400 amps. The logic boards (called Modular OSMC Brains, or MOBs) are required to interpret the signals from the receiver and generate the control signals for the power board. The separate logic board allows for more system flexibility, and this system is packed with features. Since this is a newer solution, few competitors are using the OSMC and MOB at this time. A fully assembled and tested combo of an MOB and two OSMC boards (dual-channel controller) goes for about $400.

Robotic Sporting Goods

Robotic Sporting Goods has introduced a controller called the RSGSS. This dual-channel speed control runs on 15 to 36 volts DC, and can source 60 amps continuous and 300 amps peak (pulsed). There are two built-in mixing modes available and the controller interfaces directly with standard R/C equipment. Fully assembled, the controller weights 20 ounces and measures 6" × 4.2" × 1.5". It's available from Robotic Sporting Goods and Robot Marketplace for around $250. Kit versions are available for significantly lower prices.

Roboteq AX2500

Created by a company called Roboteq and available through NPC, the AX2500 is a dual-channel speed control that runs on 12 to 40 volts DC and can source 120 amps continuously, with a peak of 250 amps. These specs put it in the heavyweight (and up) category of controls. This is a new speed control to the combat robot market, but a promising contender. Since it was only introduced at press time, I haven't seen any competitors use it yet.

It conveniently features a direct interface to a radio-control system, as well as an RS232 communication port. The real strength of the AX2500 is the powerful built-in processor that offers mixing for tank-style steering, joystick calibration, deadband adjustment, and exponential curves. In addition to the standard open-loop speed mode, the controller has a closed-loop position mode, which means it can be used to create your own high-power servos (with an external position-sensing potentiometer).

It's on the large side, with basic dimensions of 6" × 5.5" × 1.6", and a weight of 32 ounces. The $485 price tag puts it in direct competition with the upper-end Vantec controls. I am interested to see if it catches on with competitors.

Hobby-Grade Speed Controls

Using a speed control made for the hobby radio-control market used to be the cheap route. Since the introduction of the IFI controller, there is much less incentive to use hobby controllers, since the prices are about the same. In fact, most hobby controllers have several disadvantages. First, going in reverse is standard for a combat robot, but in the R/C world, where they race around a track, it's pretty rare. Even if you do find a controller that goes in reverse, chances are good that it has a momentary delay. This delay is meant to protect the small gearboxes in R/C cars, but can make it very difficult to maneuver a robot, especially when you want to spin, where one motor goes in forward, and the other goes in reverse. That delay means that after a second, the other motor will kick in and drastically change the rate at which you're turning. Save yourself the difficulty and get an IFI controller. It's about the same price, and you'll be much happier.

Nonproportional Speed Controls

You may be tempted to try using binary (on/off) type speed controls. Resist that urge. What this means is that you will be jamming maximum voltage into your motors, slamming them back and forth. It's very hard on the motors and gearboxes. All commercial speed controls have some sort of ramp to protect your motors.

The only situation where this might work is if you have low-RPM, high-torque gearboxes and small wheels. This will make for a slow but positionable robot that won't spin out when commanded to go forward. The problem is that while your driving may be manageable, the other robots will probably run circles around you.

Binary speed controls (relays) are good for one thing: turning big motors on and off. Don't try to drive your robot with a relay. Save it for the weapon.

Home-Built Speed Controls

Some of you may be disappointed to learn that I highly discourage making your own home-built speed controls, or any other custom electronic circuits, for that matter. There are three reasons for this. First, even for an experienced engineer, it takes time to design, assemble, test, and debug a circuit. And in this sport, time is a real limiting factor. Second, each electrical component that you make is a potential failure point in the arena. Proven reliability is a huge advantage. Third, there are a number of good speed controls already on the market, for very reasonable prices. There are a million things that can go wrong in the arena. Unless you believe that you can make a bulletproof speed control on your own, don't make it a million and one.

Cross-Reference For a discussion of relays and high-power one-direction speed controls, see Chapter 18.

Project 6: Mounting the Electrical Components

By now, the foundation of your project robot is in place. In this project, you'll apply the motor, electrical, and control concepts learned here and in the last few chapters to install brackets to hold the receiver and its battery. You will also mount the speed controls and the main battery. You'll get some experience with extruded aluminum material, which is available in many useful profiles, such as U and L. Velcro (along with a secondary restraint) will be used as shock-absorbing and mounting material for these sensitive electronic components.

Caution Eye and ear protection are mandatory for cuts using the miter saw. It is incredibly loud, and pieces of metal will be flying everywhere. When using the rest of the tools in this project, you will need eye protection.

Caution Review all of the general power-tool safety protocols described in Chapter 5, "Cutting Metal," as well as the sections that correspond to the specific tools used below.

Making the Receiver and Battery Brackets

The receiver and its battery will both be mounted to tall L-shapes that you'll make by modifying some U-shaped extruded aluminum. Both components will be mounted with both Velcro and cable ties, since either method alone won't be enough to hold them in place during a large impact.

1. Cut two 2-inch long pieces of the 1" × 2" extruded aluminum U-channel with the miter saw. Since the stock is provided in 8-foot lengths, you will need extra support as shown in Figure 14.6. Make sure that the stock is flat on the table in the area around the blade. Make sure to use WD-40 or another lubricant when cutting.

FIGURE **14.6: Support for cutting long stock on the miter saw.**

2. Attach the hole pattern using spray adhesive and drill the 1/4-inch holes in each of the brackets. When holding the U-channel in the drill press vise, it's better to use a scrap piece of metal to clamp the middle of the U, as shown in Figure 14.7.

3. Use the bandsaw to cut off the top of both U-channels so that the battery and receiver can both ride up a little higher when installed on the bracket.

4. Use the disc sander to true up the cuts and then take the pieces to the deburring wheel to finish off the edges.

FIGURE **14.7: Clamping the U-channel in the drill press vise.**

5. Mark a 1/8-inch deep by 1/4-inch long notch on each side of both brackets about 1 inch up from the base to keep the cable tie from slipping off the bracket.

6. Use the bandsaw to cut out the notches, and clean them up on the deburring wheel.

7. Apply Velcro to both brackets on the inside of the remaining L-shape and to the back of the receiver and its battery.

8. Install the brackets with 1/4"-20 × 3/8-inch long button-head cap screws using Loctite, and attach the battery and receiver. Secure each with an 8-inch long cable tie, as shown in Figure 14.8.

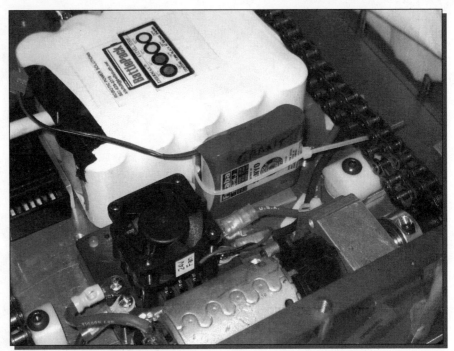

FIGURE **14.8:** Receiver battery and bracket installed.

Installing the Speed Controls

In this series of steps, you'll connect electrical leads to the speed controls using the stripper and crimper and mount them to the baseplate using Velcro. (Other shock-mounting options are discussed in Chapter 8.) Since the Velcro won't be enough to hold the speed control in place, we'll use screws as a secondary restraint.

1. First, cut down the fan leads to 4 inches and crimp on a 22-18 GA #8 stud ring terminal to each lead. Because the wire is smaller than 22 GA, you should strip the end twice as long as you normally would, and then fold it back to double the diameter before crimping.

2. Cut a 4-inch and 16-inch length of both red and black 12 GA Deans Ultra or Wet Noodle wire. Strip the ends and crimp on a 12-10 GA #8 stud ring terminal to one end of each lead.

3. Screw the 4-inch pair onto the speed control that will be mounted closer to the master power switch, along with the fan leads. Twist the red and black leads together on the 16-inch pair and connect it to the other speed control along with its corresponding fan leads. Both the 12 GA wires and the fan leads should be installed on the "power in" terminals of the speed controls. Make sure to connect red to +V and black to GND.

4. Apply Velcro to the bottom of both speed controls and to the baseplate.

5. Cut down the 6"-32 × 1¼-inch long alloy steel socket-head cap screws with a crimper that has a built-in bolt cutter, as described in Chapter 8.

6. Attach the speed controls to the Velcro on the baseplate and secure them with the screws you cut down in Step 5. Don't tighten the screws all the way down or you'll overcome the shock-mounting properties of the Velcro; they're just a secondary safety measure. You should, however, put Loctite on the screws to make sure they don't work themselves loose due to vibration. Route the longer pair of leads around where the battery would be.

7. Once the speed controls are mounted, you can pull the motor leads over to them and trim them off, leaving a little slack. Strip the ends and crimp on a 12-10 GA #8 stud ring terminal to each lead.

8. Screw the terminals on to the motor output of the speed controls, making sure to flip the polarity of one set of leads. Remember that one of the motors has to turn backwards for the robot to move forward.

Moving a Component after Installation

What if you find the need to move a component after everything is already mounted? Do you have to remove everything from the base to do this? No, that would be very time-consuming. Instead, you can make a template that you apply to the outside bottom of the robot, and drill it with a cordless drill. The trick is that the template should include through-holes for the existing tapped holes, and guide holes for the new positions. By bolting the template to the robot with the through-holes and existing tapped holes, you can guarantee correct alignment.

1. Templates should be made out of spare 3/16-inch plate. Since you're *only* interested in the holes, you can rough-cut the plates on the bandsaw. It doesn't have to be really precise.

2. Apply a drill pattern to the rough-cut plates that includes through-holes for the old tapped hole positions as well as tap drill holes for new hole positions. Remember that the drill sizes for through-holes and tapped holes are different.

3. Remove all electronics (speed controls, receiver) and put them in a plastic bag before doing any drilling. The chains and all mechanical components can remain as long as you wrap rags around the chains to keep metal chips from sticking to them and vacuum out the robot after you're done.

4. Flip the robot over and bolt the templates on, as shown in Figure 14.9. Make sure that the plate is flat and tightened securely to the robot's base. The holes provide a guide for the drill bit.

Continued

Continued

FIGURE **14.9**: Drilling through the bottom using a template.

5. After drilling, the template can also double as a tapping block. Loosen it from the base and hold it in place so that the through-holes in the template are directly over the new holes that you're about to tap. Make sure that it doesn't ride up on any of the screws poking out of the bottom of the robot, and that it sits flat on the surface.

Making the Battery Rails

The battery rails will help prevent the heavy main battery from sliding around on the inside of the robot. They will be simple lengths of L-channel bolted to the baseplate. The receiver battery bracket and the switch bracket (see Project 7) will also be mounted around the battery so that it's captured on all four sides.

1. Using the miter saw or the bandsaw, cut two 2-inch long pieces of 3/4-inch extruded aluminum L-channel. Note that when using the miter saw, you'll need a scrap block of metal to apply the clamping pressure to the L-channel, which sits too far back for the clamp to effectively hold it in place.

2. Clean up the cuts on the disc sander and deburring wheel. You may need gloves if the part gets too hot. You can cool the aluminum in a bowl of water when you're done.

3. Apply the hole pattern with spray adhesive and mark the holes with an automatic center punch.

4. Use a scrap block of aluminum to get a better hold on the L-channel, similar to the way you clamped the U-channel before.

5. Make sure to deburr all holes with a countersink. Install the rails on the base using 1/4"-20 × 3/8-inch long button-head cap screws using Loctite.

Making the Battery Plate Using the Bandsaw

Next, you'll be making a Lexan plate that holds down the main battery so that it can't jump over the rails or otherwise come loose. This is an example of an operation where the bandsaw really shines for cutting out parts. It's much easier to manipulate the plate by hand for intricate cuts than it is to wield the bulky jigsaw. You'll be using a paper template from the Kickin' Bot Web site to create the right shape and hole pattern.

1. Rough cut a 6" × 6" piece of 1/4-inch polycarbonate from the scrap left over from cutting the top armor.

2. Apply the pattern for the battery plate using spray adhesive.

3. Drill the 1/4-inch holes as indicated on the drill press.

4. Use the bandsaw to cut out the true outline of the plate.

5. Remove the pattern and use Goo Gone to get rid of any leftover sticky residue.

6. Clean up the edges on the disc sander, being careful to stay on the side of the disc that's rotating down into the table, and not crossing the middle of the disc to the side rotating upwards, as described in Chapter 6, "Shaping and Finishing Metal."

7. Use a flat file to remove any sharp edges around the plate.

8. Use the countersink to chamfer the holes, both on top and bottom.

9. Place the main battery on the base and install the plate with four 1/4"-20 × 2½"-long bolts using Loctite, so they don't loosen during combat. Optionally, if you would like to place a bit of foam underneath the battery (which is a pretty good idea), then you will need slightly longer screws. Unfortunately, the next size up is 2¾-inch long, so you may have to cut your screws down using a hacksaw or Dremel tool with an abrasive disc. When using the abrasive disc, a full-face shield is mandatory. Make sure to put a slight chamfer on the edges of the screws you cut down with the disc sander and clean up the edges with a deburring wheel.

Wrapping Up

Like most components for the robot, there are a lot of choices out there, but at least you can learn the important things to look for when selecting a speed control. There a lot of issues to consider, and hands-on work with the control systems and electrical components is an excellent way to learn them, as demonstrated in Project 6. The main thing to remember is that your speed control should be able to handle the requirements of your motors, be reliable under extreme combat situations, and be proportional, so that you don't limit your mobility.

Choosing Batteries

Batteries are the heart of your electrical system. They've got to provide all the power that you need for the entire length of a match. Batteries come in all shapes and sizes, and you've got to choose carefully, since they are potentially some of the heaviest and most expensive items on the robot.

The goal of this chapter is to help you match your battery to your robot by estimating how much power you'll need for a typical fight. I also discuss basic battery types used in robot combat, and their strengths and weaknesses, weights and prices, and how to estimate exactly how much power you'll get out of a particular type of battery. Finally, there's a discussion of safety when handling batteries and how to secure your battery in the robot, along with Project 7: Wiring the Electrical Components.

Battery Capacity

The first step in selecting batteries is to determine how much power you need. If you underestimate, you could be stuck out in the middle of the arena, crawling along and getting pummeled by your opponent. Not a very appetizing scenario. If you overestimate, you could end up spending a lot more money than you might want, and more importantly, wasting precious weight that could go towards armor or weapons.

How Much Power Do You Need?

By now, you should have chosen your motors for drive and weapons (if you've got electric weapons, that is). You'll be referring to the spec sheets for those motors to find important information about their rated voltage and current consumption. These are the pieces of information that you'll need to roughly determine how much power will be required out in the arena. I say, "roughly determine" because anything can happen out there, and you can only make an educated guess.

When you estimate how much power you'll need, what you're actually going to find is the amount of current that's required. Since electrical power is the product of voltage and current, and the voltage will remain fairly constant, current is the thing that will change to meet the increase in power requested by the motors.

Two popular techniques have evolved to help the robot builder calculate how much current is required to complete a match: the load profile method and the torque constant method.

Cross-Reference The stall current and torque constant (K_t) for many popular robot combat motors can be found in Appendix F, "Tables and Charts," to help you estimate battery requirements using these methods.

Load Profile Method

The load profile method is based on the idea that you're constantly moving during combat, and that this movement can be divided into five basic types, each with a different demand for current. These types are: *stall, pushing, accelerating, cruising,* and *stopped*. *Stall* is attempting to push with the motors, with the robot held at a dead stop. This is the most demanding condition. *Pushing* is the next most demanding condition and happens when you manage to move against another robot. *Accelerating* is when you move the robot from a stop, or change direction during a turn, which is the third-highest demand condition. *Cruising* is just driving around, which requires a relatively small current. *Stopped* is a pause in the action, where the robot is standing still, with virtually no current draw.

You assign an amount of time in seconds to each category, based on your guess of how long you'll spend doing each one. You can't assume that you'll be just cruising all the time, because every time you maneuver, that counts as acceleration. Also, you should figure in more or less pushing based on your robot's configuration and strategy. For example, a wedge may spend a lot more time pushing than a flipper robot, which may spend most of its time maneuvering.

Most motor manufacturers will provide you with the stall current. You have to scale everything from the stall current, figuring that the pushing is about 80 percent of stall, acceleration is about 30 percent of stall for a two-wheel drive robot and 40 percent for a four-wheel drive slip-steer robot (because of higher friction in a turn), and cruising is about 15 percent. Assign a zero current to the stopped condition. Multiply each segment's time by the scaled current, and add them all up to get the total number of amp-seconds required. Since large batteries are rated by the manufacturers in amp-hours (Ahr), you'll have to convert to amp-hours by dividing the total amp-seconds by 3,600. This will be the theoretical amount of battery capacity in amp-hours (more on amp-hours later) needed per motor to complete a match. Finally, multiply by the number of motors to get the total required capacity.

Overvolting Your Motors

It's true that you can drive motors at a higher voltage than the manufacturer's rating. This is an old robot-builder trick, and it's safe, within reason. Each time you double the voltage, you quadruple the power. However, you risk overheating and burning out the motors, especially if you get into a pushing match. Also keep in mind that the current consumption goes up as well. Some motors won't tolerate overvolting, but the manufacturer usually has an explicit warning.

The problem with this technique is that since every match is different, it's really hard to come up with an accurate estimate for how long a robot will be in any one of these states. The accuracy depends heavily upon your statistical guess for each of the time periods. To cover this, you should multiply the result by 1.5 to give yourself a safety factor of 50 percent.

For example, a four-motor, four-wheel drive slip-steer robot with a hammer weapon:

Given: Stall current = 100 amps, total time = 180 seconds (3 minutes)

Stall: 10 sec × 100 amps = 1,000 amp-seconds

Pushing: 35 sec × 80 amps = 2,800 amp-seconds

Acceleration: 30 sec × 40 amps = 1,200 amp-seconds

Cruising: 90 sec × 15 amps = 1,350 amp-seconds

Stopped: 15 sec × 0 amps = 0 amps = 0 amp-seconds

Total = 6,350 amp-seconds / 3,600 seconds/hour = 1.76 amp-hours per motor

Total required capacity = 1.76 amp-hours × 4 motors = 7.04 amp-hours

Safety factor adjustment = 7.04 amp-hours × 1.5 = 10.56 amp-hours

Torque Constant Method

While more of a brute-force technique, the torque constant method provides a suitable safety factor to cover those situations that simply can't be predicted ahead of time. The technique goes like this:

1. Divide the robot's weight by the number of motors you're using (not the number of wheels, since multiple wheels can be connected to the same motor by roller chain and sprockets, for example). This will give you a number in pounds that each motor will be pushing.

   Example: 120 lbs robot with four drive motors, 120 lbs / 4 motors = 30 lbs

2. Multiply the adjusted weight by the wheel radius in inches to get the required torque in pound-inches. (Since torque is the product of force and distance.)

   Example: 7.5-inch wheel dia. = 3.75-inch radius; 3.75 inches × 30 lbs = 112.5 pound-inches

3. Divide the torque by the gear ratio (if you have an external gearbox or chain and sprocket system) to get the torque at the motor in pound-inches. You can skip this step for gearmotors that are specified with the gearbox attached.

   Example: 8.33:1 gear ratio, 112.5 in-lbs / 8.33 = 13.5 in-lbs per motor

4. If the motor's torque constant K_t is specified in ounce-inches per amp, then convert the torque in pound-inches to ounce-inches by multiplying by 16. If it's specified in pound-inches per amp, then skip this step.

   Example: K_t = 6.6 oz-in per amp, 13.5 in-lbs × 16 = 216 oz-in per motor

5. Divide the torque at the motor by the motor's torque constant K_t (in ounce-inches per amp or pound-inches per amp) to get the number of amps for each motor.

Example: 216 oz-in / 6.6 oz-in per amp = 32.7 amps per motor

6. Multiply the number of amps per each motor times the number of motors to get the total number of amps required to push your robot.

Example: 32.7 amps per motor × 4 motors = 131 amps total

7. Multiply the total number of amps required by the match length in minutes, and divide by 60 to convert to hours. This will give you a figure for amp-hours, which is the unit of measure for battery capacity.

Example: 3 min. match, 131 amps × 3 min / 60 min = 6.55 amp-hours

Note Although the above two examples have been calculated with regulation 3-minute matches, I try to have some reserve capacity, since in BattleBots, the free-for-all rumbles, which come at the end of regulation competition, have a 5-minute duration, and tend to involve a lot more pushing and stall conditions than a regular match because there are so many robots in the arena at once.

Figuring in Weapons

Remember that you've also got to consider any electric-powered weapons or self-righting mechanisms in your calculations. You can use the load profile method to calculate the required current. In the case of a spinning weapon, you'll be concentrating on the acceleration and cruising segments.

Estimating a Battery's True Capacity

All commercial batteries are rated by the amp-hour. It's an industry standard that also happens to be a major source of confusion for the robot builder. Theoretically, this establishes a figure that you can scale by the rate of consumption. For example, a 10 amp-hour battery can deliver 10 amps for 1 hour, 20 amps for half an hour, and 5 amps for two hours, with the capacity theoretically remaining the same. In practice, however, this rating is usually only true for a 20-hour discharge rate in sealed lead acid batteries, and a 1-hour discharge rate in nickel-cadmium and nickel-metal hydride batteries. The capacity won't remain the same as you increase the consumption rate, and in fact, it will begin to decrease at an exponential rate. (You can tell because roughly linear-looking discharge data curves are published on log-log graphs — if you're truly interested in predicting the response, you can use a mathematical power series to get the equation.) Since the discharge rate we're interested in is going to be in the minute range (a really high rate), the true amp-hour capacity will be only a fraction of the rated capacity.

To get a rough idea of the actual capacity of a battery in time segments that are more appropriate for robot combat, you can use a process called *de-rating*. You can de-rate a battery by multiplying the manufacturer's amp-hour rating by a correction factor, which will give you a more realistic capacity for the 5-minute range. You'll see from the tables in the "Basic Battery Types" section that SLA batteries should be multiplied by a de-rating factor of 30 to 40 percent, while NiCad and NiMH cells should use a 85 to 90 percent factor.

Increasing Voltage and Current

Many off-the-shelf batteries come in a standard 12-volt size. What if you want to run your motors at 24 volts? What if the batteries don't have enough capacity? Not to worry. You can increase the voltage and current of your system by arranging the batteries in different configurations. A series connection will increase the voltage without affecting the current capacity, and a parallel connection will increase the current without changing the voltage. A combination of the two can yield an increase in both voltage and current.

Series Connection

A series connection, shown in Figure 15.1, means that the positive terminal of one battery is connected to the negative terminal of another battery. The positive output lead is connected to the remaining positive terminal, and the negative output lead is connected to the remaining negative terminal to create a stacked battery. This only increases voltage by the sum of the batteries in series, not the current capacity. For example, you can put two 12-volt, 7-amp-hour batteries in series to make a 24-volt, 7-amp-hour battery, and three batteries in series to make a 36-volt, 7-amp-hour battery.

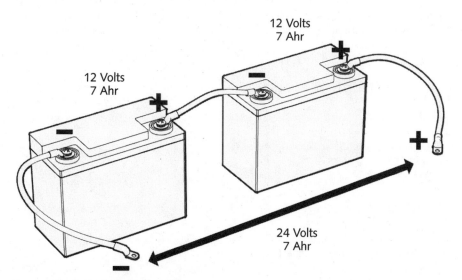

FIGURE **15.1: Series battery connection.**

It's possible to *tap off* an intermediate voltage in a series configuration to run a receiver or other lower voltage item. However, I strongly recommend that whatever you connect to the tap should be very low-draw. If you have a 24-volt drive system, but a 12-volt weapon motor, then I would not suggest tapping off one of the 12-volt cells in series. It will draw too much current down that leg, and may result in uneven charging and possible damage to the battery.

As mentioned in Chapter 13, "Choosing Your Control System," it's not recommended that you power the receiver from the drive batteries. You should use a separate, small-capacity battery instead. This will help prevent any glitches due to dips in the main battery's voltage. The power consumption is pretty low for a receiver, so the battery doesn't have to be huge. However, your power consumption may be higher if you're running several servos.

Parallel Connection

With a parallel connection, as shown in Figure 15.2, the positive terminals of both batteries are connected together and form the positive polarity output. The negative terminals are connected together to provide the ground reference. This will increase the overall battery capacity as the sum of the individual capacities of all the packs in parallel. For example, you can put two 12-volt, 7-amp-hour batteries in parallel to make a 12-volt, 14-amp-hour battery. Three of these batteries would produce a 12-volt, 21-amp-hour battery, and so on. It is essential that the batteries you connect in parallel be the same voltage and current capacity, as well as state of charge. They should be the same brand, and ideally, a similar age. When you connect two cells in parallel, they will attempt to equalize voltage. If the cells are significantly different in any way, it will cause current to flow between the cells, as the one with the higher voltage will attempt to charge the one with the lower voltage. Since the resistance in that line can be very small, this can cause a dangerously high current to flow, which may lead to a fire.

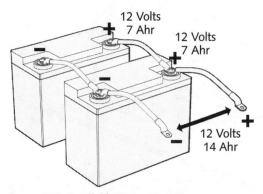

FIGURE **15.2:** Parallel battery connection.

Batteries can be discharged in parallel, but should always be charged separately to make sure that you get a full charge, because different batteries charge at slightly different rates.

Combination

You can put batteries in both series and parallel combinations to get higher voltages and currents, as shown in Figure 15.3. As mentioned above, the batteries in this arrangement should all be the same capacity and voltage.

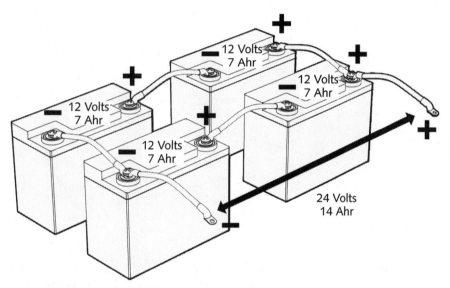

12 Volts
7 Ahr

12 Volts
7 Ahr

12 Volts
7 Ahr

12 Volts
7 Ahr

24 Volts
14 Ahr

FIGURE **15.3: Series and parallel connection.**

Basic Battery Types

While it's true that all the batteries in this section provide electrical energy through a chemical reaction, that's where the similarities end. There are a few popular choices, such as sealed lead acid, nickel cadmium, and nickel metal hydride. Each type of battery has its strengths and weaknesses. Just like everything else, there are tradeoffs, and there's no perfect battery for combat.

Sealed Lead Acid (SLA)

In general, sealed lead acid (SLA) batteries (see Figure 15.4) tend to be large and heavy, but relatively inexpensive. They have larger capacities and are capable of handling very high surges (peaks) of current, sometimes over 1,000 amps, thanks to their low internal resistance. The voltage stays relatively constant over the course of discharge, tapering down slowly. Many of the SLA batteries that are appropriate for combat are intended for the mobility (wheelchair and scooter) market. Though SLA batteries are generally sold in 12-volt packages, they are actually made up of six cells, each with a nominal voltage of 2 volts, connected in series inside the battery's case.

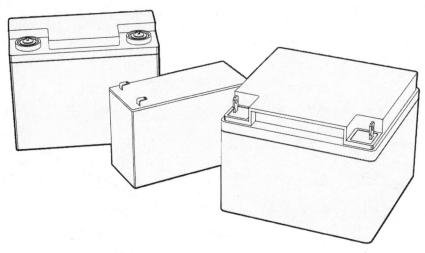

Figure 15.4 Typical SLA batteries.

You'll be looking for AGM (absorbed glass mat) SLA batteries. Fortunately, most of the SLA cells sold today are of this type. The reason for their popularity is that the electrolyte (battery acid) is absorbed into fiber separators within the cell. If the case is cut or otherwise breached, sulfuric acid won't come spilling out onto the floor. As a side benefit, the AGM cells are also invertible and can be used sideways. They are completely sealed, which makes them nonspillable and leakproof. Finally, AGM technology provides higher sustained current output than other types of lead acid batteries, including car batteries, which are meant to put out a huge surge for a few seconds to start a car, and then must be recharged immediately. In combat, you'll need sustained power if you're going to make it through a full match.

The Difference between Lead Acid Batteries

Traditional Lead Acid Batteries

These are regular car batteries, also called *wet*, or *flooded electrolyte* batteries. The *electrolyte* is a sulfuric acid solution that fills the battery and interacts with the plates in a chemical reaction that produces electrical energy. As a part of the recharging process, oxygen must recombine with the plates. But in flooded cells, there's no easy path for the oxygen to recombine because the plates are surrounded by the liquid electrolyte. As a result, the oxygen for recombination is drawn from the water in the electrolyte (with hydrogen as a byproduct) and must be replaced by adding water, which is why these batteries have a filler cap. The oxygen that could not recombine is released along with the hydrogen gas through a vent tube.

Unfortunately, these batteries are unsuitable for robot combat because the electrolyte is stored in liquid form. The vent tube and filler caps are a leak point, and if the battery case is punctured or ruptured, the electrolyte (acid) is likely to spill in a horrible mess.

Sealed Lead Acid (SLA) and Valve-Regulated Sealed Lead-Acid (VRLA or VRSL)

These are sealed cells, with an *immobilized* electrolyte. This means that the electrolyte is prevented from moving around inside the battery. This has two benefits: It keeps the electrolyte in contact with the plates while still allowing an oxygen recombination path, and it obviates the need to add any additional water to the battery to keep the process going. Since the oxygen is reused and stays within the cell, no water is needed to replace it. This also prevents the hydrogen byproduct from forming, so very little (or no) gas is released. A one-way relief valve vents pressure during charging, which should only be necessary if the battery is being overcharged.

Two special cases of this technology are *gel cells*, which use the addition of silica to form a semisolid gel, and *absorbed glass matte* (AGM), which uses a porous, fiber-like material to soak up and suspend the electrolyte. Gel cells allow recombination through cracks (called voids) in the semisolid electrolyte. The paths in an AGM battery are provided by the porous structure of the glass mat.

These batteries are great for robot combat because the semisolid or absorbed electrolyte won't spill like a flooded cell if the case is ruptured. Also they have higher output currents and lower internal resistances because of the more efficient oxygen recombination path.

Note While all SLA cells are commonly referred to as gel cells, in fact most of them today are AGM cells. Unfortunately, Hawker's Odyssey brand batteries have coined the confusing Drycell™ name. These are still AGM batteries, but they call them *dry* because almost all of the electrolyte is contained in the matte. Technically, all SLA batteries are considered "wet" because of the electrolyte, as opposed to "dry" primary cells.

Transportation

AGM batteries are easily transported. Under DOT standards (DOT-CFR Title 49), AGM batteries are classified as a "wet nonspillable electric storage battery" and are exempt from regulation in surface transport. They are also safe for marine transport under IMDG Amendment 27 and air travel under IATA/ICAO Special Provision A67.

Comparing the Favorites

Popular AGM batteries under the Interstate, PowerSonic, SVR, Genesis, and Odyssey brand names are compared in Table 15.1. Explanations of each of the columns follow the table.

Table 15.1   Popular SLA Batteries for Robot Combat

Manufacturer/ Model	Rated Ahr	5-Min Amps	Peak Amps	Actual Ahr	Diff/ Rated & Actual	Weight (lbs)	Price
Interstate SLA1075	7.2	25.94	80	2.2	30%	6.1	$25
Interstate SLA1105	12	44.09	80	3.7	31%	9.4	$62
Interstate SLA1116	18	61.37	250	5.1	28%	13.7	$90
Interstate SLA1146	26	89.3	288	7.4	29%	20.1	$75
Interstate SLA1156	34	115.5	280	9.6	28%	26.5	$84
Interstate DCS-33	33	121	2150	10.1	31%	27	$76
Interstate DCS-50	50	134	2500	11.2	22%	40	$96
Interstate DCS-75	75	212	3100	17.7	24%	54	$120
Interstate DCS-88	88	252	3300	21.0	24%	64	$155
PowerSonic PS-1270	7	21	70	2.5	-	5.7	$26
PowerSonic PS-12120	12	36	120	4.2	-	10.3	$48
PowerSonic PS-12180	18	54	180	6.3	-	13.1	$62
PowerSonic PS-12260	26	78	260	9.1	-	18.7	$82
PowerSonic PS-12280	28	84	280	9.8	-	19.4	$92
PowerSonic PS-12330	33	100	300	11.7	-	26.5	$79
PowerSonic PS-12400	40	120	400	14	-	30.9	$112
PowerSonic PS-12550	55	165	550	19.3	-	40	$122
PowerSonic PS-12750	75	225	750	26.3	-	58	$150
NPC-B1412 (SVR 14)	14	90	300	7.5	-	11.9	$80
NPC-B1812 (SVR 18)	22	100	350	8.3	-	17	$99
NPC-B2812 (28-12)	28	160	600	13.3	-	23	$139
Genesis G13EP	13	65.5	1400	5.5	42%	10.8	$77
Genesis G16EP	16	83	1600	6.9	43%	12	$65
Genesis G26EP	26	129.9	2400	10.8	42%	13.5	$102

Manufacturer/ Model	Rated Ahr	5-Min Amps	Peak Amps	Actual Ahr	Diff/ Rated & Actual	Weight (lbs)	Price
Genesis G42EP	42	193	2600	16.1	38%	14.7	$143
Genesis G70EP	70	247.4	3500	19.8	28%	22.3	$194
Odyssey PC 545	12	70.8	1200	5.9	49%	12.6	$118
Odyssey PC 535	13	71.9	1000	5.8	44%	12	$117
Odyssey PC 680	16	90	1800	7.5	47%	15.4	$124
Odyssey PC 625	16	91.6	1800	7.6	48%	13.2	$139
Odyssey PC 925	27	143.4	2400	12.0	44%	26	$160
Odyssey PC 1200	40	212	2600	17.7	44%	38.2	$192
Odyssey PC 1700	65	342.4	3500	27.4	42%	60.9	$263

Sources: Interstate Battery individual data sheets, PowerSonic *SLA Battery Technical Handbook*, PowerSonic individual data sheets, *Hawker Genesis Application Manual*, 5th edition, *Hawker Odyssey Starting, Lighting and Ignition (SLI) Drycell Battery Guide*, 5th edition, publication #ODY-BR-101, and NPC Robotics.

Note: All batteries are 12-volt AGM sealed lead acid. Quoted discharge rates are for batteries drained to 1.67 volts per cell (10 V terminal voltage), except for Odyssey series, which were drained to 1.75 Vpc (10.5 V terminal voltage). The listed PowerSonic 5-min rate is actually the 7-min rate. The listed SVR 5-min rate is actually the 3-min rate.

The *Rated Ahr* column is the manufacturer's nameplate capacity rating for that battery in amp-hours. The *5-Min Amps* column is the amount of current the battery can source for 5 minutes continuously. The *Peak Amps* column shows the surge or peak amount of current that the battery can produce in a short burst. The *Actual Ahr* column shows the true capacity of the battery in a 5-minute runtime. *Diff* is the percent difference between the rated and actual amp-hours. (The Diff value is what is averaged to estimate the de-rating factor previously mentioned in the "Estimating a Battery's True Capacity" section.) The *Weight* is specified in pounds by the manufacturer. The *Price* is an approximate street price for the battery in U.S. dollars.

The Interstate DCS series are deep-cycle batteries aimed at mobility applications, while the rest of their line (the SLA series) is very similar to the PowerSonic standard SLA batteries. Both of these lines should be easy to find. SVR is a relatively new battery brand available through NPC Robotics.

Perhaps the most popular SLA batteries among competitors are made by Hawker Energy Products. Odyssey is the consumer brand, while Genesis is the OEM (original equipment manufacturer) brand. They are sold through different vendors, and you should consult their Web site at www.hepi.com for a list of authorized dealers. The Odyssey consumer brand is commonly available through many battery retailers, and is used in automobiles, motorcycles, and watercraft. The Genesis brand is a little bit harder to acquire, since they are not intended for sale directly to the end consumer.

Types of Chargers

You're making a pretty big investment here, and you should do what you can to protect it. Although chargers can be pretty expensive, you don't want to trust your SLA batteries to an average car battery charger. To do so would be risking serious damage.

What you want is a fully automatic charger, with a multistage charge profile. There are specific charge profiles for different types of lead-acid batteries. A charge profile specifies the required voltages and currents at specific times (stages) in the charging cycle. If you deviate from those requirements, cells could be ruined, or at least the lifetime and capacity will be shortened.

Though expensive, both the Deltran BatteryTender series of chargers and the ChargeTek chargers come highly recommended among competitors.

Nickel Cadmium (NiCad)

Nickel cadmium cells, or NiCads, are used extensively in a variety of different applications. You can find them in most cordless power tools, for instance. This type of cell averages around a 90 percent efficiency versus the rated amp-hour specification, and the voltage remains remarkably constant throughout the discharge cycle. The individual cells have a nominal voltage of 1.2 volts, with a range of different sizes and capacities. They can be purchased individually, or in premade packs (see Table 15.2), as described later in this chapter. Unfortunately, they have relatively high internal resistance, which limits their ability to source large surge (peak) currents. The individual cells have a small resistance, but to achieve voltages in the ranges that we're looking for, multiple cells must be placed in series to form a battery pack, and all of those internal resistances add up. Of course, you can get around this by putting multiple packs in parallel, which is usually required to meet battery-capacity needs anyway.

Table 15.2   Custom NiCad Battery Packs

Manufacturer/ Model	Volts	Rated Ahr	5-Min. Amps	Peak Amps	Act Ahr	Diff	Wt (lbs)	Price
BattlePack CP2400-12	12	2.4	60	100	2.2	91%	1.3	$58
BattlePack CP2400-18	18	2.4	60	100	2.2	91%	2	$87
BattlePack CP2400-24	24	2.4	60	100	2.2	91%	2.6	$115
BattlePack CP2400-36	36	2.4	60	100	2.2	91%	4	$173
BattlePack 3000-12	12	3.0	80	100	2.7	90%	2	$74
BattlePack 3000-18	18	3.0	80	100	2.7	90%	3	$111
BattlePack 3000-24	24	3.0	80	100	2.7	90%	4	$148
BattlePack 3000-36	36	3.0	80	100	2.7	90%	6	$222
BattlePack CP3600R-12	12	3.6	80	100	3.3	92%	2	$79

Manufacturer/ Model	Volts	Rated Ahr	5-Min. Amps	Peak Amps	Act Ahr	Diff	Wt (lbs)	Price
BattlePack CP3600R-18	18	3.6	80	100	3.3	92%	3	$119
BattlePack CP3600R-24	24	3.6	80	100	3.3	92%	4	$158
BattlePack CP3600R-36	36	3.6	80	100	3.3	92%	6	$237

Source: Robotic Power Solutions Web site: www.battlepack.com.

R/C Racing Packs

The hobby radio-control market has yielded some handy packs for use in combat (see Figure 15.5). Packs assembled in six- and seven-cell arrangements (7.2 and 8.4 volts, respectively) are common, so you'll have to do some creative series arrangements to get the voltages you want. Also, you may end up having to put packs in parallel to get the current capacity you need. These packs can be darn expensive, too, ranging from $100 to $200 each.

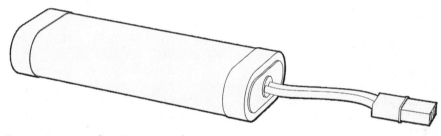

FIGURE 15.5: Premade R/C racing pack.

Commercial Cordless-Tool Batteries

NiCads are light and they have excellent capacity. It's no wonder that they're used in commercial cordless power tools (see Figure 15.6). This is a good source of batteries for the robot builder, and in the past, many builders have chopped up cordless power drills to get the motors, batteries, and battery connectors (or even left the body intact and mounted it to the frame!). The difficulty is that buying more batteries for spares can get expensive because power-tool replacement packs tend to be expensive. One advantage is that if you buy a power tool, you usually get two battery packs and a charger that's made for them.

Note

You may have to make custom mounts or chop up their cases to get rid of extra weight. Also, you don't get to choose what physical arrangement you want the cells to come in—you're stuck with what you get.

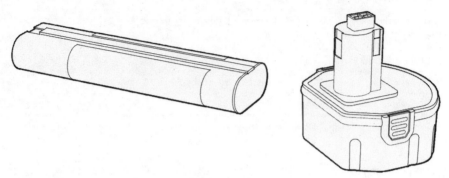

FIGURE 15.6: Various cordless power-tool NiCad packs.

DeWalt standard series batteries are rated for 1.3 Ahr, XR series are 1.7 Ahr, and XR+ series are 2.4 Ahr. Makita standard batteries are 1.3 Ahr, and the high-capacity batteries come in three sizes: 1.7 Ahr, 2.0 Ahr, and 2.6 Ahr. Not much is known about the true capacity of these batteries, only the rated amp-hours.

Custom Solutions

BattlePacks (made by Robotic Power Solutions) are by far the most popular custom NiCad packs in the sport. It's no wonder, with their commitment to constantly improving their products and excellent customer service. They currently stock premade packs (see Figure 15.7) complete with connectors in 12, 18, 24, and 36 volts, and a range of capacities. They are a complete power supplier, with packs, connectors, wire, and chargers, all in one place. The premade packs are well thought out and insulated. There are also higher-performance packs available with cooling fans built in.

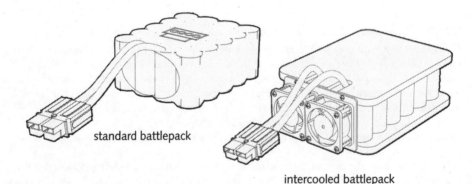

standard battlepack

intercooled battlepack

FIGURE 15.7: A custom BattlePack.

Rolling Your Own Cells

Yes, it's possible to make your own battery packs. It involves trying to match cells, which means testing their real capacity, and finding several in a batch with similar characteristics. Then you've got to solder tabs to individual cells (which are heat sensitive to begin with). You'll have to make sure that the cells are shock-isolated from each other with some foam. Then solder on some leads, find a really big piece of heat shrink, and you're ready to go. Wow, sounds easy, right? Wrong. It's very frustrating and time-intensive, and you could easily end up damaging the cells you're trying to solder. I really can't recommend this, unless you come from an R/C racing background and have done it before. I classify it in the "performing surgery on yourself" category. It's worth the extra money to have someone else do it. Believe me.

Ballistic Batteries is another custom pack manufacturer that teams are beginning to use. They don't carry stock packs like Robotic Power Solutions, but can make packs to your specifications.

Taking Care of Your Packs

The number one rule with NiCad battery packs is: *heat kills*. That's why BattlePacks come equipped with a temperature-sensitive sticker that helps the builder track the maximum temperature that a pack reached in service. If your battery packs are too hot to handle when you remove them from the robot, then you're attempting to draw too much current too fast, and you should add another pack in parallel. If you don't, then you're risking failure in the arena, and possibly fire. At the very least, you'll be shortening the lifetime of your batteries dramatically. Never, ever charge a hot pack. Always let the battery pack cool down before attempting to charge it. Remember, heat can kill a battery pack.

In some applications, *cell memory* is an issue with NiCad packs. Cell memory is a loss of capacity that occurs when the cells are consistently drained to the same intermediate level (not full discharge) and then recharged. However, robot combat is a high-drain application, and your packs are usually spent by the time you get out of the arena, so cell memory isn't an issue. Besides, it takes many, many cycles to develop a memory problem, and you're more likely to experience degradation in performance due to heat than anything else.

Having the incorrect charger can kill your batteries. Your charger must be able to sense when the cells have reached full charge and shut down. If you fail to do this, you will cook your cells, damaging them, and possibly causing them to rupture.

The AstroFlight 110D and 112D chargers (see Figure 15.8) are the only units that I know of that can handle 24-volt (20-cell) packs. Most hobby chargers designed for the R/C racing industry usually charge 7.2 to 9.6 volt (6 to 8 cell) packs. The AstroFlight chargers also keep track of how many amps they put back into the cell, and have both a fast and slow charge setting with adjustable current limits. The only drawback is the expense of both the charger and the power supply that's required to run it.

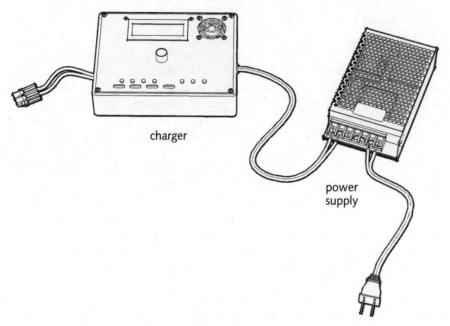

FIGURE **15.8**: AstroFlight battery charger and power supply.

Nickel Metal Hydride (NiMH)

The efficiency of these cells is about 90 percent. They have a lower peak current sourcing capability, but a better power-to-weight ratio, since they are generally lighter than NiCad cells of equivalent capacity. There's always a price. With NiMH cells, this price is a higher internal resistance than NiCad cells, which means a lower peak current (as you'll see in Table 15.3) and a lower continuous current rating. As a side benefit, NiMH cells are environment-friendly, since they contain no cadmium.

There aren't many power-tool or R/C racing packs featuring NiMH cells yet, so custom packs are the best solution at this time. Of course, Robotic Power Solutions (the BattlePacks people) have a line of premade NiMH packs for the robot builder, as shown in Table 15.3. They can be made in 12- to 36-volt packages like the NiCad packs mentioned earlier. They have a similar packaging to the NiCad BattlePacks. Ballistic Batteries also advertises the capability to create custom NiMH packs. They don't carry a stock, but can create packs to order.

Table 15.3 Custom NiMH Battery Packs

Manufacturer/ Model	Volts	Rated Ahr	5-Min Amps	Peak Amps	Act Ahr	Diff	Wt (lbs)	Price
BattlePack 2600N-12	12	2.6	45	60	2.3	88%	1.5	$63
BattlePack 2600N -18	18	2.6	45	60	2.3	88%	2.3	$67
BattlePack 2600N -24	24	2.6	45	60	2.3	88%	3	$126
BattlePack 2600N -36	36	2.6	45	60	2.3	88%	4.5	$189
BattlePack 3300-12	12	3.3	45	60	3	91%	1.5	$72
BattlePack 3300-18	18	3.3	45	60	3	91%	2.3	$108
BattlePack 3300-24	24	3.3	45	60	3	91%	3	$144
BattlePack 3300-36	36	3.3	45	60	3	91%	4.5	$216

Source: Robotic Power Solutions Web site: www.battlepack.com.

Charging NiMH Packs

Don't assume that a charger made for NiCads will work for NiMH packs as well. While many of the high-end chargers (like the AstroFlight 110D and 112D) will handle both, you've got to check the manufacturer's specifications and make sure that the charger specifically says that it can also handle NiMH packs. If it can't, then putting a NiMH pack on that charger will most certainly damage the pack. And remember, the same precautions about heat for NiCads also apply to NiMH packs. *Heat kills.*

Other Battery Technologies

Lithium Ion batteries may have a weight advantage even over NiMH cells, but they are currently regarded in the R/C community as too difficult to implement because of their sensitivity to excessive charge and discharge currents (requiring active onboard protection circuitry) and fondness for thermal runaway. High internal resistance may limit their current source capability, which isn't a problem for cell phones, but they don't routinely draw 100 amp peaks like combat robots. Also, the cells are not yet commonly available to the average user. Who knows? Perhaps this will become the next big breakthrough.

Technology Comparison

Sealed lead acid batteries are capable of very high short bursts of current that are much higher than NiCad batteries can produce, which can give you the edge in a pushing match.

Over the course of discharge, the voltage of an SLA battery has a gentle downward slope. You may see the robot get gradually slower and more sluggish towards the end of a match. With NiCad cells, the voltage stays pretty much constant all through the length of discharge, and you will probably not notice much of a performance degradation if they're sized properly.

SLA batteries are more suited for larger robots (120 lbs and up) in higher weight classes, due to their bulky size and heavy weight. Also, in heavier robots with bigger motors, you will end up using so many NiCads in parallel that the cost both in batteries and chargers will probably catch up to you.

The following examples will compare NiCads to SLA batteries in two different (extreme) situations:

Lightweight Example

Let's assume that a lightweight robot has a weight limit of 60 lbs. The robot has two EV Warrior drive motors being run at 24 volts, and an estimated battery requirement of 2.5 amp-hours. In this example, my NiCad selection is a 24-volt BattlePack (PN-3000-24) with a 2.7 Ahr actual capacity and a weight of 2.64 pounds (4.4 percent of total weight). The cost of the BattlePack is $115. The SLA choice is an Interstate SL1105 battery, which has an actual capacity of 3.7 Ahr and a weight of 9.4 pounds (15.7 percent of total weight). The cost of the SL1105 is $62. To get 24 volts, however, you will need two SL1105 batteries (12 volts each) in series, for a total of 18.8 lbs, and a cost of $124. Clearly, the NiCad selection wins out in terms of weight, and it has a comparable cost and peak current at this capacity.

SuperHeavyweight Example

Assume that the superheavyweight robot has a weight limit of 340 pounds. It is using four Bosch 750 GPA motors being run at 24 volts, and has an estimated current requirement of 10 amp-hours. The NiCad choice for this example would be four 24-volt BattlePacks (PN-3000-24) in parallel, for a total capacity of 10.8 Ahr and a weight of 16 lbs (4.7 percent of total weight). The peak current capability of this NiCad arrangement would be 400 amps. The SLA choice would be two 12-volt Hawker Genesis G26EP batteries in series, for a total voltage of 24 volts, a capacity of 10.8 Ahr, and a weight of 27 pounds (7.9 percent of total weight). The peak current capability of the Genesis batteries would be 2,400 amps. The cost of the four NiCad packs would be $592, and the cost of the two SLA batteries would be $204. Multiply those dollar amounts by three to include two spare sets of batteries (see the tip later on multiple sets of batteries), and the totals become $1,776 versus $612, not including battery chargers, which may require multiples to charge all those batteries in a timely fashion. If you have the resources to purchase NiCad packs, then you can save the weight. However, it's not as big a weight savings for a superheavyweight as it is for a lightweight, and the SLA has the clear advantage in cost.

Battery Safety

Given all this talk about battery characteristics and performance enhancement, it's easy to forget the basic safety issues with batteries. You've got to keep in mind that these are electrochemical sources capable of easily generating currents that could cause burns, fire, or explosion.

Batteries are dangerous, and their potential for destruction should be respected. Following are a few guidelines for battery safety:

- Never leave exposed battery terminals, whether the battery is in the robot, or out on the worktable.

- Always use polarized connectors on batteries, and especially in parallel configurations, where a wrong-polarity connection could spell disaster.

- Replace clip leads on battery chargers with a polarized connector that matches the batteries. Double-check it with a digital multimeter before connecting a battery.

- Use standard color codes to distinguish positive and negative polarity. Usually, red or white are positive, and black is negative.

- Use the correct size wire. Chapter 16, "Wiring the Electrical System," has important information about correctly sizing your battery cables. These should be the fattest wires in the robot, because they will be carrying all the current that needs to go to hungry motors.

- Always check your robot's power wiring before first connecting a battery. It only takes a few seconds to double check all the hookups, which will prevent damage to expensive speed controls and other electronics.

- In both series and parallel systems, the batteries should be removed and charged individually. Every battery is just a little bit different because of manufacturing differences or the like. It's best to separate the batteries so that you don't confuse the charger and potentially damage one of your packs.

Securing the Battery

The battery is one of the heaviest single components in the robot. When you start subjecting your robot to high shock loads in the arena, this large weight is going to want to move. If it breaks loose, you could be causing yourself all kinds of internal damage. Here are a few examples for securing the battery in your robot:

- Don't use cable ties (zip ties) to secure your batteries. They're too flimsy for SLAs, and NiCads can generate enough heat to melt them, or at least make them pliable enough to allow the battery to move around. You can and should use cable ties (or tape) to secure the battery's connector to the robot's main power connector, which will prevent it from popping loose during a big impact.

- Hose clamps (as discussed in Chapter 10, "Mechanical Building Blocks") can be used to hold batteries down, but you've got to make sure that the metal band is properly insulated. Hose clamps can exert a lot of pressure as you tighten them down, so be very careful not to crush your battery packs.

- Fabric and nylon ratchet straps have been used in the past, but you've got to make sure that the batteries can't move from side to side.

- Lexan plates have been used to hold down cells with long screws. You must make sure that the batteries cannot move from side to side, and that you have enough screws to keep the batteries in place in the event of a large shock load.

- Perhaps the best variation is to build a separate box of Lexan or (well-insulated) aluminum for your batteries. Make sure that any box has adequate ventilation all around (remember, heat kills) and active cooling would be an excellent addition.

- Also, leave a little slack on the connections so that if the battery does move during combat, it won't pull the connector out or damage anything else.

Tip

You need to have a battery that won't move around inside the robot, but you also need to design your hold-down so that the battery is easy to change. In combat, you can't always count on having a lot of time in the pits to charge batteries. That's why it's best to have multiple sets of batteries (three complete sets minimum) that you can swap into the robot. As you advance into later tournament rounds, the cycle time between matches gets a lot shorter, and you should be prepared to deal with this situation. One of the worst ways to lose a match is to simply run out of power from batteries that weren't completely charged.

Project 7: Wiring the Electrical Components

All the basic components of your project robot should now be in place. You now have to use the battery information found in this chapter to give your robot some juice.

You'll install a secondary power switch used to turn the receiver on and off, saving the battery between matches. For maximum safety, you need power indicator lights for both the receiver battery and the main battery to tell you when either system is powered. In this project, you will prepare and install the secondary power switch and power indicator LEDs (light emitting diodes) on a mounting bracket and plug the PWM leads into the receiver. This will be an opportunity to get some more practice with the soldering iron as well as the stripper and crimper. (This project ties in closely with the next chapter, Wiring the Electrical System, and its related project, Installing the Master Power Switch.)

Caution

Eye and ear protection are mandatory for cuts using the miter saw. It is incredibly loud, and pieces of metal will be flying everywhere. When using the rest of the tools in this project, you will need eye protection.

Caution

Review all of the general power-tool safety protocols described in Chapter 5, "Cutting Metal," as well as the sections that correspond to the specific tools used below.

Preparing the Power LEDs

For this project, I chose a red LED for primary power and a green LED for secondary power. Both had threaded housings to make mounting easier. The longer lead on an LED is the positive, and the shorter lead is the negative. It doesn't matter which of these legs you put the resistor on, as long as it's connected in series somewhere in the loop, although I usually put it on the positive leg out of habit.

Note

Some electronics vendors such as Radio Shack sell LEDs that are prewired for 12-volt systems, but they're not very bright, and can't be used with 24 volts or 6 volts. Actually, you can use any LED you want. The important thing is choosing the correct resistor. For 24-volt systems, use a 1k ohm, 1/2-watt, 5% tolerance resistor. (The resistors in the higher-voltage systems will dissipate a lot of power, so the 1/2-watt rating is important.) For 12-volt systems, use a 470-ohm, 1/4-watt, 5% tolerance resistor. For 4.8- to 6-volt receiver batteries, use a 220-ohm, 1/4-watt, 5% tolerance resistor.

1. Use a 12-inch long Futaba J-type servo extension for the secondary LED. It already has the right kind of connector to plug into the receiver.

2. Cut off the female end close to the connector, leaving the male end as long as possible. If you're unsure which is which, the male end is the one that can plug into the receiver.

3. Split the white lead off and use a hobby knife to lift the part of the plastic servo connector that holds the pin in place and slide the pin out, which will allow you to remove the white lead completely.

4. Separate the remaining red and black leads about 4½ inches down, and strip 3/8 inch of insulation from the black lead.

5. Cut a 3/4-inch long piece of 3/32-inch diameter black heat-shrink tubing and slide it over the black lead.

6. Wrap the black lead around the shorter (negative) leg of the green LED and apply the soldering iron and solder to the joint. Make sure that the solder left on the joint is bright and shiny. Do not blow on the solder to cool it. Let it cool down on its own. Cooling the joint too quickly will give you what's called a "cold solder joint," which can produce intermittent results.

7. Push the heat-shrink tubing up over the joint and use a heat gun or lighter to shrink the tubing in place.

8. Solder the 220-ohm, 1/4-watt resistor on the positive lead, as shown in Figure 15.9.

9. Cut a 2½-inch long piece of 3/32-inch diameter red heat shrink and slide it over the red lead.

10. Cut 1½ inches off the red lead and strip 1/2 inch of insulation from the end.

11. Wrap the red lead around one of the resistor's legs and solder it in place. Cut off any excess wire that extends past the joint on either side with a pair of flush cutters.

12. Slide the red heat-shrink tubing over the resistor and use a heat gun or lighter to shrink it into place.

13. You can test the secondary power LED by plugging it into the receiver and plugging in the receiver battery.

FIGURE 15.9: Soldering the resistor to the secondary LED.

14. For the main power LED, you will repeat the above steps, but instead of using a servo lead, use 24- or 26-gauge stranded hookup wire from the hardware store or an electronics vendor. Use red or white for positive, and black for negative. Cut each lead about 8 inches long. You will also use the red LED, a 1k-ohm 1/2-watt resistor, and 1/8-inch red heat shrink (for the larger diameter 1/2-watt resistor).

15. Let the ends of the LEDs hang loose for now. They will need to be loose for installation in the switch bracket, which is coming up in a few steps.

Making the Switch Bracket

The secondary power switch must be mounted to a bracket. The power LEDs can also be mounted to this bracket, which will be made of the same extruded aluminum U-channel you used in Project 6.

1. Cut a 2½ inch long piece of the 1" × 2" extruded aluminum U-channel using the miter saw.

2. Clean up the edge, if necessary, with light pressure on the disc sander.

3. Remove any sharp edges on the deburring wheel.

4. Attach the top and bottom patterns to the switch bracket using 3M Spray 77 spray adhesive.

5. Mark the holes on top and bottom with an automatic center punch.

6. Clamp the U-channel in the drill press vise using a piece of scrap stock as in Project 6.

7. Drill 1/4-inch holes for the two LEDs, and also as a pilot hole for drilling the switch hole. Make sure to use adequate lubrication.

8. Indicate the step on the large Uni-bit that corresponds to a 1/2-inch diameter hole with a marker.

9. Drill the hole for the switch hole with the Uni-bit.

10. Use a countersink in a cordless drill to clean up the small holes and the deburring tool on the large switch hole.

11. Clamp the U-channel in the vise again and drill the bottom two 1/4-inch mounting holes.

12. Use a countersink to clean up these holes. You can angle the deburring tool to reach the hole edges on the inside of the U-channel, where a countersink won't reach.

13. Remove the pattern and use Goo Gone or another solvent to remove any sticky residue.

14. When installed, the switch bracket will touch the bottom of the speed control, and may rub against it. To prevent this, you need to mark and cut 3/16 inch off the bottom of the bracket using the bandsaw as shown in Figure 15.10. Use plenty of WD-40, go slowly, and make sure to keep your fingers out of the way of the blade at all times.

15. Use the disc sander and deburring wheel to clean up the edge.

FIGURE 15.10: Cutting the switch bracket with the bandsaw.

Installing the LEDs and Switch in the Bracket

Now you can take the power LEDs that you wired and install them in the switch bracket, along with the secondary power switch.

1. Insert the secondary power switch into the switch bracket with the screw terminals facing outwards, as shown in Figure 15.11. Apply Loctite and tighten down the nut on the top with an adjustable wrench. Do not apply Loctite to the screw terminals. They will need an unobstructed path to conduct current.

2. Remove the servo connector from the red and black leads of the green secondary switch LED by using a hobby knife to lift the plastic that holds the pins in place, as you did with the white lead before. Pay attention where the leads go, since you will be putting them back in as soon as the LED is mounted.

3. Insert the wires in the LED hole closest to the secondary power switch, and pull the LED all the way down to the bracket.

4. Thread the lock washer and nut onto the wires as shown in Figure 15.11, and apply Loctite before tightening the nut on the mounting thread.

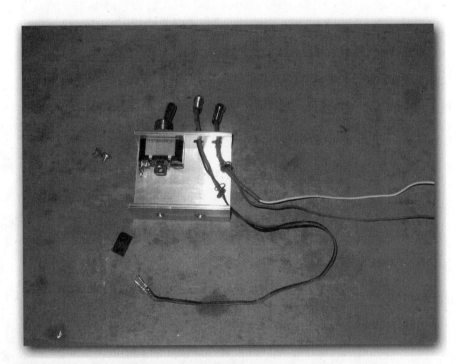

FIGURE 15.11: Power LED and switch installation.

5. Replace the servo connector on the end of the red and black servo leads.

6. Insert the leads of the red main power LED into the LED hole farthest from the secondary power switch.

7. You may have to do a little bit of shuffling the leads around so that you have enough clearance to get the large 1/2-watt resistor through the hole, lock washer, and nut.

8. Apply Loctite and tighten the nut on the mounting thread.

Wiring the Secondary Power Switch

For the secondary switch, you'll also use a servo extension. A 12-inch Futaba J-type servo extension (also from Tower Hobbies) will be just long enough to make it all the way to the switch and back.

1. Fold the 12-inch Futaba J-type servo extension in half and mark it in the middle.

2. Using a hobby knife, separate the red wire from the black and white wires in a 3-inch long area centered on the mark by carefully slicing the wire lengthwise along the joint where the colors meet.

3. Cut the red wire only at the mark.

4. As with the speed-control fans in Project 6, strip off twice the amount of insulation you normally would and fold it back to double the diameter.

5. Crimp a 22-18 GA #8 ring terminal onto each side of the red lead.

6. Screw the ring terminals onto the switch and tighten them down. You may want to bend the ring terminal farthest from the receiver over, as shown in Figure 15.12 so you get a little extra slack on the line towards the receiver.

Figure **15.12: Secondary power switch wiring.**

7. Paint all exposed electrical connections with liquid electrical tape.

8. When the liquid electrical tape is dry, you can install the switch bracket on the base using two 1/4-20 × 3/8" long button-head screws and Loctite. Since access to the screws is limited by the speed control, you should use ball end Allen wrenches to give you a greater working angle.

9. Route the secondary power LED servo connector and secondary power switch leads towards the receiver. The main power LED should be routed towards the closest speed control, near the position for the master power switch, where it will be connected in Project 8.

10. Apply cable ties as strain relief to make sure that the fragile LED wires don't get pulled when you change receiver batteries or work on the receiver.

Installing the Receiver

Next, you'll be plugging the PWM leads into the receiver and securing the cables with cable ties. Make sure to get the polarities right, or you could cause damage to the PWM signal driver cables.

1. Plug the secondary power LED and secondary power switch into the receiver. It doesn't matter what channel you plug these items into because power runs to every channel. You should not plug them into channels 1 or 2 because that's where your speed controls are going.

2. Plug the speed control cables into channels 1 and 2. When running IFI speed controls and R/C equipment, make sure to order the IFI PWM signal driver cable. It's not required for all R/C equipment, but it won't hurt.

3. Be careful when plugging the PWM signal driver leads into the receiver. Since they're not polarized, you can easily plug them in backwards.

4. Usually, it's a good idea to label the PWM leads at both ends with small number tape made by 3M. You may have to renumber them later on if the steering is reversed, so you can hold off labeling until the road test. In the meantime, you can put a temporary piece of masking tape on both ends of one of the PWM signal driver cables.

5. Make sure to provide adequate strain relief for the PWM cables and bundle the ridiculously long signal driver cables together. I usually use a ton of cable ties, as shown in Figure 15.13. Leave enough slack on the ends to swap receiver channels if you have to.

Wrapping Up

There is a lot of technical information in this chapter about the differences between battery types. Your choices should be based not only on performance, but also on weight, physical size, and cost.

Remember, this is all a guessing game. I've provided you with the means to estimate your requirements, but the only way to be sure is try your batteries as soon as possible with the drive system under real loads. That means driving hard and pushing around weights equivalent to the robots you'll be facing. Check your batteries after each test to see if they've gotten warm. Wait for them to cool, charge them, and test them again.

The project in this chapter provides a good preview for the more detailed information on wiring that you'll find in the next chapter. Keep both of these chapters — and the projects found within them — in mind when considering the electrical issues involved in creating a combat robot.

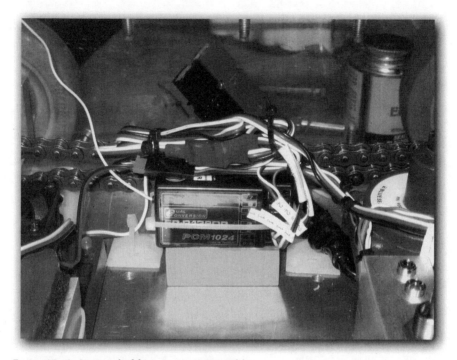

FIGURE 15.13: Strain relief for receiver PWM cables.

Wiring the Electrical System

In the last few chapters, you were introduced to the parts of your electrical system, which include the control system, speed controls, and the main battery. The next step is to connect all of those parts together with wire, switches, and connectors. Here, I show you what types of wire work best in robot combat, and identify the best switches, connectors, and crimp terminals to use that will carry the amount of power you need, while being resistant to the large shock loads that you might experience in the arena. Finally, there's info on how to avoid common combat failures by good wiring practice, which involves tips about selection of wire colors, insulating connections, and keeping wires tidy and well secured inside the robot. Project 8 will guide you through installing a master power switch.

Selecting Wire

Wire is where you need to start, because that is what you will use to connect the various electrical components of your robot, such as batteries, speed controls, and motors. Most types of wire will work, for a little while at least. This section will help you select wire that won't fail you in battle. You'll find out about what specifications are important, and where to get the best wire.

Wire Gauge

First and foremost, you should make sure that your wire is rated to handle the amount of current that you expect to put through it. There are a few factors to keep in mind when determining the current rating of a wire. In the real world, all wire has a resistance, although it's very small. Normally, you don't even notice it. However, when you run a very large amount of current through a wire, this small resistance results in a proportional voltage drop that you *will* begin to notice. This voltage drop means that the voltage at the end of the wire will be lower than the voltage at the source. With motors, less voltage means less RPM and less power.

With respect to the wire, however, this voltage drop takes the form of heat buildup inside the wire and can accumulate very rapidly. The heat buildup can melt the outer insulation (also called the jacket), the copper itself, and any solder that might be used to connect the wire to another conductor. Usually it's safe to say that the insulation will melt before the copper, and thus one of the guidelines for specifying for maximum current is the insulation material.

Since the resistance of the wire is what generates the heat, you should concentrate on ways to lower it, which will allow you to run higher currents with less voltage drop. Resistance can be lowered by making the wire fatter. The *wire gauge* number, also known as AWG (for American Wire Gauge), corresponds to the nominal diameter of the wire, and gets smaller as the wire's diameter gets larger.

Note   Resistance can also be lowered by making the wire shorter. As a rule, battery connections, which carry the most current, should be kept as short as possible.

In addition to the resistance, the ambient temperature should be taken into account in the evaluation of current capacity. If the wire is working in a high-temperature environment to begin with, less current will be required to bring the heat up to the limit. In the case of a combat robot, however, you can get away with specifying room temperature (30° C) as the ambient temperature.

Taking the above factors into account, Table 16.1 represents the current required to bring a multistrand copper conductor of the corresponding wire gauge at an ambient room temperature of 30° C to the temperature limit of the jacket. (In the case of PVC insulation, this temperature is 105° C, and in the case of a silicone jacket, it's 200° C.)

Table 16.1   Wire Gauge and Absolute Current Limits

Gauge	PVC Jacket Current (amps)	Silicone Jacket Current (amps)
2 AWG	170	240
4 AWG	125	180
6 AWG	95	135
8 AWG	70	100
10 AWG	50	75
12 AWG	35	55
14 AWG	30	45
16 AWG	15	32
18 AWG	10	24
20 AWG	7	17

Source: Alpha Wire technical data sheet, *Current Carrying Capacity of Copper Conductors.*

It's important to note that these figures should be taken as a guideline only. In general, if you can feel your wires getting warm, then they're probably on the edge of being too small and/or too long. It they're getting too hot to touch, then shut down immediately and do not continue testing until you've upgraded your wire. You'll probably notice a performance increase as well.

Wire Flexibility

Solid core wire (such as household electrical wiring) is not recommended, since you may end up repeatedly flexing your wire over the course of building, causing a fatigue point and eventual failure. Instead, multistrand wire is suggested, since it is extremely flexible and readily available in the larger gauge sizes that are appropriate for combat robots. The higher the strand count, the more flexible the wire. The more flexible, the better, since you may have to route your power leads through some tight spaces. Jacket material also affects the stiffness of the wire. Silicone is the best jacketing material, because it's very flexible and can handle momentary heating, if necessary.

Popular Choices

The following wire choices are popular among builders. You can also look for large gauge speaker wire from car audio manufacturers.

Silicone-Jacketed Wire

The most popular choice among builders by far is silicone-jacketed wire, available in 8, 10, and 12 AWG sizes. The high heat tolerance of the insulation combined with the excellent flexibility makes this wire ideal for our sport. Many silicone-jacketed wires are available for purchase by the foot from Team Delta, Robotic Power Solutions (makers of BattlePacks), and Robot Marketplace (these and other resources for this chapter can be found in Appendix D, "Online Resources," and Appendix E, "Catalogs").

Deans Wire

The Deans Wet Noodle wire is very popular among builders. It's available only in 12 AWG in red and black. The main advantage here is that the wire is extremely flexible, with a strand count well over 1,000. It's available from Team Delta, Robot Marketplace, and online radio-control hobby suppliers, such as Tower Hobbies.

Battery and Welding Cable (McMaster-Carr)

If you're looking for wire larger than 8 AWG, then McMaster-Carr has the solution. General-purpose battery cable is available in 6, 4, and 2 AWG sizes. It has a PVC jacket, and is available in red or black. High-flex battery cable (also known as welding cable) is available in 2 AWG and has a synthetic rubber jacket.

Machine Tool Wire (McMaster-Carr)

Machine tool wire (type MTW) from McMaster-Carr is overlooked by many builders, but is acceptable for most applications. It has a PVC jacket and is moderately flexible (low strand count compared to other selections). It's available in 8 and 10 AWG. The main advantage of this wire is that it's cheap.

Tip It's important to figure wire into your overall weight calculations. Depending upon the wire that you choose, this may end up adding a few unexpected pounds to your robot.

Selecting Main Power Switches and Disconnects

Every combat robot must have a means to disable all weapons and drive systems and render the robot safe for maintenance or handling. Some competitions require a physical disconnect that can be tugged to kill the system. In other competitions, you can use a main power switch. The catch is that you can't use just any disconnect or switch, since the current of all the drive and weapon motors will be carried through this link. If the equipment that you use is not rated for the load, it could become a bottleneck in performance for your whole electrical system.

Because most robots operate in the 12 to 48 V DC range, most of the popular switches come from either automotive or marine applications. In this section, you'll see the half-dozen most popular switches and disconnects used by competitors.

Caution The main power switch or physical disconnect should be a hard *cutoff*. There should be no relays that bypass the cutoff. When you turn off the switch or pull the plug, it should mean that the batteries are *completely disconnected* from the system and that there is no possibility of the robot becoming energized. A secondary path is a personal safety risk and has been outlawed in most competitions.

Hella Switch

The Hella switch (see Figure 16.1) has become a staple in robot combat because it's easy to mount and can handle 100 amps continuous and 500 amps for 10 seconds. It also weighs only about 5 ounces and can handle large impacts. The big red activation key is the only weakness, since you have to leave it in the slot during activation (it's held in place by a tab on the key), and there is a possibility that it may break lose in combat. Various builders have different solutions, such as altering or adapting the key. You'll have to decide for yourself what method you're comfortable with.

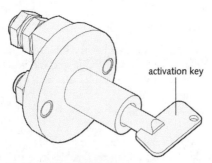

activation key

FIGURE 16.1: Hella swich.

The Hella switch is available from Team Delta, Robot Marketplace, and various automotive suppliers for about $17. Make sure not to fall for a Hella lookalike or imitation.

West Marine Mini Battery Switch

This marine switch (see Figure 16.2) can handle 250 amps continuous and 1,400 amps surge. It has a wide bottom with holes for base mounting. It can handle large impacts and is as light as the Hella switch. Unfortunately, just like the Hella switch, it has a big activation key. However, the Mini Battery Switch is available with both a removable knob (#3823325) and a nonremovable (fixed) knob (#3823333).

Both versions of the switch are available from West Marine for about $22.

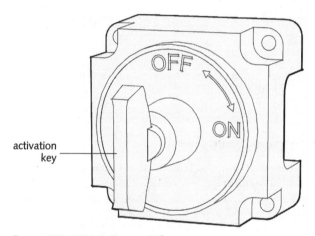

activation
key

FIGURE **16.2: Mini Battery switch.**

Team Coolrobots Switch

This is a custom aftermarket modification of a standard Hella switch, designed by veteran Christian Carlberg, builder of Toe Crusher, Overkill, and Minion. It has identical current ratings to the Hella switch. As shown in Figure 16.3, the outside housing is machined out of Delrin with 1/4"-20 tapped holes in a convenient 1.5-inch bolthole circle. The real trick is that the switch accepts a 1/4-inch hex key for activation. I have designed my robot Deadblow to use the same 1/4-inch hex key to activate the air system, so that I can use a single tool to make the robot ready for combat.

Since this is an item designed and custom manufactured for our sport, you can expect to pay a little more (in the $50 range). They are available exclusively from Team Delta.

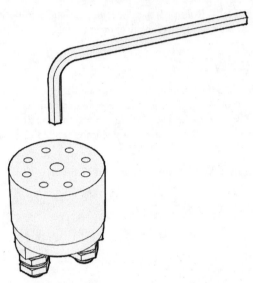

FIGURE **16.3: Coolrobots switch.**

Team Whyachi MS1 Power Switch

This is another custom aftermarket switch made by Team Whyachi (shown in Figure 16.4). It's more compact than the other switches and handles 80 amps continuous, 240 amps for 3 minutes, and 500 amps surge. It uses a 5/32-inch hex wrench for activation (isn't the custom market great?) and has a Delrin case with #10-32 tapped mounting holes.

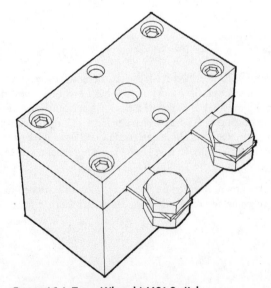

FIGURE **16.4: Team Whyachi MS1 Switch.**

You're looking at about $58 for this switch, but if you're short on space and weight, then it's worth the price. It's available from Team Whyachi and Robot Marketplace.

Cole M284 Switch

This popular switch is used in both automotive and marine markets as a battery disconnect (see Figure 16.5). It can handle 175 amps continuous and 800 amps surge. The difficulty here is that it's got a long mounting neck and a little knob on the end. You may be able to modify the knob, as other competitors have done, to use with your own activation tool.

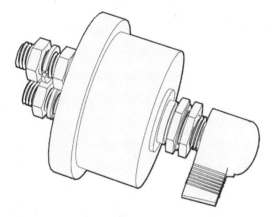

FIGURE 16.5: Cole M284 switch.

The Cole switch is available from West Marine (Model 542019) and Rebcoperformance.com (part #1601495) for about $25.

Anderson Multipole Disconnect

This is not a switch, but a physical link created by looping back a heavy-gauge wire such as 1/0 in an Anderson Multipole connector (as shown in Figure 16.6). When the connector is plugged in, it connects the batteries to the electrical system. When it's removed, the batteries are disconnected. It's pretty simple, and some people believe that this is the best way to implement a master switch, because it can be yanked out if something goes wrong. You don't need any special tools — just a strong finger to grab and pull the disconnect.

Make sure to check the competition rules to see if this type of master disconnect is required or optional. If you have the choice, I would suggest one of the other switches described previously, since they will allow for more protection from arena hazards and other robots.

Premade disconnects are available from Team Delta, or you can buy the parts from McMaster-Carr and make your own.

FIGURE 16.6: Anderson Mulipole disconnect.

Other Switches

Standard *heavy-duty* switches (see Figure 16.7) that you might find in the hardware store or through an industrial supplier are only rated for 10 to 15 amps. They may be able to handle the current demands of a lightweight robot, but robots in larger weight classes may need one of the switches listed previously to get maximum performance in demanding situations.

If you decide to use a switch like this, then make sure to purchase one that has screw terminals, not spade (slide-on) terminals. That way you can use the ring terminals described later in this chapter.

Note   This type of switch can be used as a secondary power switch to provide power to the receiver/radio system.

Selecting High-Power Connectors

Many popular connectors used in the wheelchair and forklift industries are also popular in combat robots. In addition, the radio-control hobby industry has yielded some connectors that may be valuable to the combat robot builder.

Anderson Multipole

These are single-piece plastic housings that have two conductors, as shown in Figure 16.8. These are the bulkiest connectors that you can use, but they are also rated for the highest amount of current, with models rated for 50, 175, and 350 amps. Note that the size increases as the current rating increases. It's likely that most combat robots will only require the 175 amp connectors.

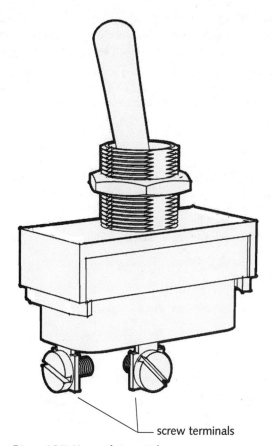

screw terminals

FIGURE **16.7:** Heavy-duty switch.

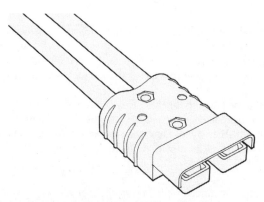

FIGURE **16.8:** Anderson Multipole connector.

The Multipole has the advantage of being genderless, which means that there is only one type of pin and one type of connector to keep in stock. Also, both ends of the connection are fully protected, unlike some other systems, where the male ends of the pin are fully exposed.

Anderson Multipole connectors are available from McMaster-Carr as *industrial battery connectors*, or from other online suppliers such as powerwerx.com or NPC. The pins may be crimped using any of the battery terminal crimpers from McMaster-Carr, including the economical hammer-style crimper.

Anderson Powerpole

This is another connector made by Anderson Power Products, but unlike the Multipole, the Powerpole is a modular connector. Each plastic housing holds only one conductor, but the housings can be locked together to form larger multiconductor connectors, as shown in Figure 16.9. This connector has the advantage of being fully configurable, since the individual connectors can be locked together to form multiple-conductor connectors, or split apart to be used conveniently in series applications.

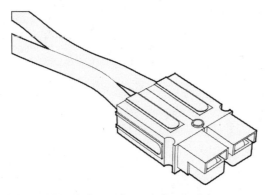

FIGURE 16.9: Anderson Powerpole connector.

The 75-amp version of this connector is the standard for high current NiCad BattlePacks, and is suitable for battery connections in all weight classes. The 45-amp version is useful for making moderate current connections. Lightweight robots that don't exactly need hundreds and hundreds of amps of capacity may be able to use these contacts as their main power connections.

 Tip
Although these connectors are designed to be plugged together, they can sometimes slide apart when you're plugging and unplugging them repeatedly. Fortunately, they also have a design feature that allows you to simply lock them together. By inserting the appropriate size roll pin into the hole that's formed by joining two connectors, as shown in Figure 16.10, you can prevent them from sliding apart.

Just like the larger single-housing Multipole connector, the Powerpole is genderless, giving you the benefit of single inventory stock. Both ends of the connection are fully protected.

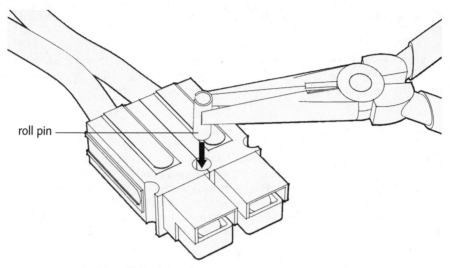

FIGURE **16.10:** Antislide roll pin insertion.

Anderson Powerpole connectors are available from McMaster-Carr as *modular connectors*, or from other online suppliers such as powerwerx.com or NPC. The pins may be crimped using any of the battery terminal crimpers from McMaster-Carr, including the economical hammer-style crimper.

AstroFlight Zero-Loss Connectors

These connectors feature gold-plated pins that fit into custom-molded plastic housings for a compact connector solution (see Figure 16.11). Their performance is good, and many robots have used these connectors in the past, although they are only rated for 60 amps continuous, which limits this connector to moderate current applications. Their main drawback is that they're difficult to use because the pins must be carefully soldered to install and unsoldered to be reused. I'll gladly trade this for the convenience of cutting the connector off and crimping on a new pin. Also, the solder cup is only large enough to accommodate 12-AWG wire at best.

FIGURE **16.11:** AstroFlight Zero-Loss connectors.

These connectors have a specific gender and are sold in pairs. The female side should always be installed on the battery or the side of the speed control that *supplies* the power, since the male side has exposed pins and is subject to being shorted.

AstroFlight Zero Loss connectors are available from AstroFlight and numerous online hobby distributors.

Tip If you happen to get solder on the outside of the barrel, don't try and push the connector into the housing. Instead, take a fine file and remove the solder bump so that the barrel is smooth again. Then you'll have a much easier time pushing the pin into its housing.

Deans Ultra Plugs

These are probably the most compact connection solution, as shown in Figure 16.12. According to the manufacturer, they are more efficient than an equivalent length of 12-AWG wire. Like the AstroFlight connectors, you have to solder leads to the pins. Since the housing does not cover the ends, heat shrink is a must for insulating the soldered connections. Because of the size of the connector and the closeness of the terminals that you will be soldering, I recommend using 12-AWG or smaller wire, which will limit this connector to moderate current applications.

These connectors have a specific gender and are sold in pairs. The female side should always be installed on the battery or the side of the speed control that *supplies* the power, since the male side has exposed pins and is subject to being shorted.

Deans Ultra Plugs are available from various online hobby distributors.

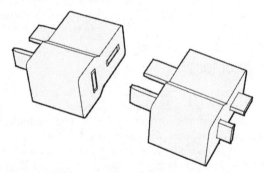

FIGURE **16.12: Deans Ultra Plugs.**

Selecting Crimp Terminals

As mentioned earlier, the wires in your electrical system may be subject to overheating if overloaded. Crimp terminals are a good way to ensure that your connections will stay together, even if the heat rises high enough to melt solder. You can also flow some solder into the crimp joint to get a better electrical connection. Unfortunately, the wire will try to draw solder into itself (through capillary action), which will make the wire hard at that location and may potentially lead to a stress failure if it gets bumped hard. The way to avoid this is to make sure not to get too much solder into the connection. If the wire near the crimp feels much less floppy than before, then you've got too much solder in the wire, and you should start over with a fresh piece of wire.

Battery Terminals

Some batteries, such as SLA cells, have screw terminals for their main battery connections. The appropriate connector for this application is a battery terminal, which is a large ring terminal with a crushable barrel for 8-AWG or larger wire, as shown in Figure 16.13. These terminals are also essential for connecting wires to most of the high-power switches listed previously, which have 3/8-inch and 1/2-inch studs with locknuts.

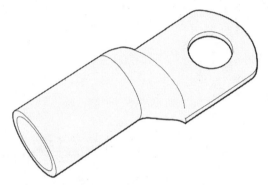

FIGURE **16.13: Battery terminal lug.**

McMaster-Carr carries a large selection of battery crimp terminals (battery lugs). They also have several crimpers, but for battery terminals, the hammer-style crimper is an economical and effective alternative to the $100+ crimpers. This can be used in a bench vise for ease of installation, as demonstrated in Project 8.

Low-Power Crimp Terminals

These connectors are used in low or moderate power applications, such as connecting to the secondary radio power switch and speed control terminals.

Crimping Connectors

A crimper is used to mechanically compress a connector onto a wire. If done correctly, this provides a strong mechanical connection that is not subject to failure from heat, like a soldered connection. One of my favorite sayings is, *you live and die by your crimps*. In the combat arena, the death is a bit more literal, but still valid. Invest in a good crimping tool, and check every crimp that you perform by yanking on it *every time*. It takes longer to test each crimp, but if the wire pops out, then you just discovered a combat failure waiting to happen.

There are three major types of crimp terminals, as shown in Figure 16.14: ring, spade, and quick disconnect. The ring terminal has a complete circle (ring) at the end. The stud size determines the size of screw that will fit through the ring. You also have a choice of wire gauge and ring width. The spade (or fork) terminal has a U-shape at the end, and is also specified by stud size and wire gauge. You have a choice of spade width. The quick disconnect terminal is meant to slip over a tab. Here you have a choice of tab width and wire gauge.

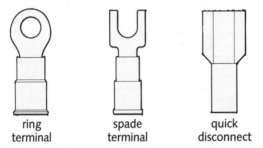

ring
terminal

spade
terminal

quick
disconnect

FIGURE **16.14: Various low-power crimp terminals.**

All terminals should have insulated barrels, and the quick disconnects should be fully insulated. These connectors should be crimped with a good ratcheting crimper made for fully insulated terminals, as shown in Figure 16.15.

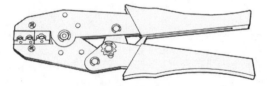

FIGURE **16.15: Ratcheting crimper.**

Good Wiring Practice

Now that you have an idea of all the parts of the system, you can connect everything together in a way that will keep you out of trouble at the competition, which includes appropriate component selection, uniform polarity colors, lowering noise by separating signal and power leads, adding power LEDs, properly insulating connections, and securing wires inside the robot.

Use Components That Are Rated for the Task

By now, you should have a good idea of what kind of current your motors will be drawing from the battery in a stalled condition, which might occur in a pushing match. Your current will also spike when you accelerate from a stop, or change from full forward to full reverse quickly. Normal running current usage won't be anywhere near the peak (stall) current, but you should design so that your connectors won't incinerate themselves if you happen to find yourself in a worst-case situation.

Keep Uniform Polarity Colors

Red or white should be used exclusively for positive battery polarity. Black should be used for battery and system ground only. Maintaining uniform polarity color throughout the robot will help prevent any unintended polarity reversals, which can instantly kill an expensive speed control or receiver. Keep in mind that this will help prevent disaster not only during the building process, but also during the *fixing* process in the pits, which is usually more frantic, and much less orderly.

For motors, which have a polarity relative to the motor's turning direction, other colors can be used as long as they're clearly marked.

Separate Power and Signal Leads

By running and bundling signal and power leads separately, you can reduce the chance of electrical noise interfering with your radio signals and causing strange or erratic behavior (in the robot, that is). Now is the time to pay attention to this. Going back after you've already wired the robot to separate these leads is a painful process that should be avoided.

Use Mechanical Connections Wherever Possible

As mentioned before in the "Selecting Crimp Terminals" section, wire can heat up in an overload and cause solder to melt. In high-power applications, crimping is the preferred method of installing a connector, and mechanically bolting that connector to a terminal is the safest technique.

In a power system, you will be faced with the situation of having to distribute power to multiple destinations. You can try soldering them, but a better solution is a star joint, which is composed of several ring terminals bolted together. Make sure that this joint is well insulated by several layers of electrical tape.

You can also use a terminal block or strip to distribute power, but these strips take up space and may become bulky as the current ratings increase. They will also need insulation to protect them from short circuits.

Only Use the Big Wire Where You Need It

You don't need to run 2-AWG wire all over the robot. The electrical system won't mind, but you will be adding unnecessary weight, which begins to add up when multiplied by several speed controls and motors. Batteries should get the biggest wire and connectors, because they source current to the rest of the system. From there, multiple speed controls and motors can use smaller gauge wire and connectors. The important thing is to carry the big wire up to the distribution point, which means also running it through the master power switch.

Keep wire leads as short as possible, especially for the higher-power lines, but plan ahead before you cut it to length. You should make sure that the components are going to work in those locations. I really hate adding extensions to high-power lines, because it adds resistance and creates a possible failure point.

Separate Your Receiver Power

Your receiver system should have its own battery and power switch. The reason is that during a surge load, a battery's voltage may dip slightly as it tries to keep up with the demand from the motors. If your receiver is connected to the same battery, the voltage may drop low enough to confuse it, reset it, or otherwise cause you problems. By having a separate battery, you eliminate that risk.

Adding a separate (secondary) power switch is also highly recommended. You may not know how long you have before a fight. Although the power drain is low, you may forget to unplug the receiver overnight, and end up with a dead battery in the morning.

IFI systems should definitely have a secondary switch for the receiver (RC) power. The reason is that the radio locks to the frequency that yields the first valid data stream. The frequency that you normally have available (channel 40) is different from the ones that they use in the arena. If you've been testing outside, and the RC is still powered, it will still be locked to channel 40, and your robot will ignore all of your signals when you're plugged into the official arena channels. Powering down the RC and powering back up will cure this. If you don't have a secondary switch, then this may make for some prefight scrambling.

For radio-control systems, it's acceptable to use a small slide switch. However, you should make sure that the slide axis is not aligned with the axis of your weapon, where the most shock will occur, For example, a horizontal spinner will generate a huge shock load in the horizontal plane when it hits something. In this case, the switch should be mounted vertically, so that the shock will not act in the direction that may turn off the switch. I personally like to use a larger switch with a heavier spring, such as the standard heavy-duty hardware store switch described earlier.

Add Power Indicator LEDs

The purpose of a main power LED (light emitting diode) is to show you from a distance whether the main power switch (or disconnect) is providing power to the rest of the system. It's a good idea and is now required at some competitions. The secondary power LED tells you if your receiver is turned on or not. Both LEDs should be located after their corresponding switches, so you can see if the switch is engaged, and whether the robot is actually receiving power or not.

LEDs have a polarity, which can be indicated in two ways. Usually, the bottom of the LED has a flange with a flat side on one edge. The flat side indicates which lead is negative. Also, LEDs usually have one lead that's shorter than the other. The shorter lead is the negative one.

You will also need to limit the current that flows through an LED with a small resistor. If you don't do this, the LED will go supernova and burn itself out in the blink of an eye. An appropriate resistor for 12 volts is 470 ohms. For a 24-volt system, use a 1k-ohm resistor. The resistor should be in series with one of the leads (it doesn't matter which one).

You can find LEDs that already have the resistor attached, which is very convenient, but they're usually only available in 12 volts. With any other voltage, you'll have to attach your own resistor.

Insulate Your Connections

Insulating all exposed electrical connectors in a robot is one of my *must-do* rules. Losing because a piece of metal shorted out a speed control or a battery is a really silly (and preventable) way to go out. This section will describe several options you have available to you to make sure that no part of the electrical system is left exposed to harm.

Electrical Tape

This is the basic vinyl insulating tape that you see in the hardware store. No surprises here. Just make sure that all the surfaces you apply the tape to are clean and dry.

Liquid Electrical Tape

While electrical tape is good for insulating *cylindrical* items, the majority of exposed connectors in robot combat are screw terminals, which aren't exactly cylindrical. Speed controls, main power switches, and solenoid terminals are all excellent candidates for liquid electrical tape.

Basically, liquid electrical tape is an air-dry flexible synthetic rubber coating that comes as a gooey liquid. You paint it onto the surfaces that you want to insulate. A few coats may be required for complete coverage. It sticks quite well and can be scraped off if necessary. It's flexible (won't crack if moved) and durable. I use several colors to help distinguish polarity. It's available in black, red, green, and white.

It's available at automotive supply stores and some hardware stores. If you can't find the Performix brand of liquid electrical tape, you can use PlastiDip (made by the same company).

Heat Shrink

Heat shrink is useful in finishing off your soldered electrical connections. (Deans Ultra Plugs require heat shrink to finish installation.) Heat shrink is basically a thin black vinyl tube that shrinks when you heat it.

You can use a lighter or portable butane torch, but be careful to keep moving the flame, or you'll end up burning the vinyl. It's best to use a low-power heat gun (not a high-power paint stripping heat gun) to shrink the vinyl.

Grommets

A grommet is a little rubber doughnut that is meant to insulate any metal holes that electrical wires pass through. Over time, vibration may cause a wire's insulation to scrape off if it's in contact with a metal edge. Grommets are a handy way to prevent any metal edges from harming your wires.

Clean up Your Wiring

If left unchecked, the wiring in your robot could easily become a nightmare of insulated spaghetti. With wires running everywhere, it's difficult to see where things are connected, which makes diagnosing a problem extremely difficult. You may also impede your ability to work on other (nonelectrical) parts of the robot. This section will give you a few tips on keeping the wiring tidy and organized.

Label Your Leads

Sure, it's obvious now. This is clearly the power for the left-side speed control. Now add three more speed controls and two relays, a power-on LED, and a master switch, and things can get confusing fast. Not to worry. The best way to handle this is to label everything. You should be able to quickly track connections and diagnose any problems that come up.

In situations where polarity isn't clear (such as motor leads), I put a tag on each side of the wire with a big plus symbol or a name like "Left +." Try to be descriptive with your tags. By putting a number on each side of the lead, you can keep track of where servo leads and extensions need to be plugged in. Sometimes you'll have to change out a damaged speed control or need to disconnect wires to access another part of the robot. Having these labels will make sure things go back together the right way.

Tip   You can buy a compact dispenser that holds ten rolls of numbered tape made by 3M. The tape is only 1/4-inch wide, which makes it large enough to see, but small enough not to get in the way, especially if you have a lot of leads terminating in one place, such as the receiver. The adhesive isn't very strong, so you should make a little tape flag, as shown in Figure 16.16.

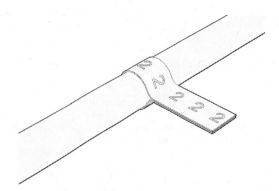

FIGURE **16.16:** Tape flag.

Tie Down Loose Ends

Tying down leads does two good things: First, it provides strain relief, which keeps wires from getting stressed where they connect, which could loosen the connector, or cause a failure due to fatigue. Second, it keeps wires from flying around inside and getting stuck in chains or gears, or other internal mechanisms.

You don't need a lot to add some strain relief to your electrical system. In fact, all you need is a hole in a support here or there that you can slip a cable tie (also called a zip tie) through. Cable ties are small nylon straps that can be tightened around a wire bundle as shown in Figure 16.17. They have a little ratchet-like tab in the head that keeps them tight. They're disposable and available in a variety of sizes. They can easily be trimmed with diagonal flush cutters for a neat appearance. Make sure to trim the cable ties flush to the head and not at an angle, unless you like stabbing your hands on small plastic daggers inside the robot.

Note It's good to keep things tidy and well secured inside the robot, but avoid tying everything down with zero slack. A little bit of slack prevents you from stressing connection points, which are more likely to break if they're constantly being pulled down. Sometimes, you may need to reposition an electrical component inside the robot, and having that slack means you won't have to start over and recut and wire a bunch of connections that are too short. Make sure no wires are flopping around inside the robot, but at the same time, give it a little slack.

If you don't have a handy bulkhead or support to tie into, you can use adhesive-backed hold-downs just about any place inside the robot, as shown in Figure 16.17. Make sure the surface is clean and dry.

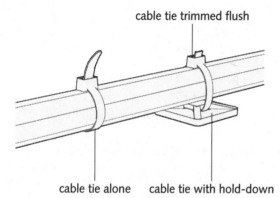

cable tie trimmed flush

cable tie alone   cable tie with hold-down

FIGURE **16.17: Cable tie alone and with hold-down.**

The previously-listed supplies are available from Grainger and McMaster-Carr (see Appendix E for more info).

Make a Map

A combat robot can be a pretty complicated place when it comes to wiring. Make a map of your electrical system (also called a schematic) so that you can quickly glance at an overview when you need to diagnose a problem. Along with a good set of identification tags on your wire leads, you should be able to navigate your power system easily.

Project 8: Installing the Master Power Switch

This project picks up right where Project 7 (Chapter 15) lets off. In this project, you'll apply your new knowledge of wiring to modify the West Marine switch that was selected to activate the project robot's master power, making it battle-ready by replacing the big red plastic key with a socket-head cap screw. This will allow you to use an Allen key to turn the master power on and off, which should make activation more convenient and reliable. You'll also be wiring the switch into the system by crimping battery terminals onto the battery leads and installing Powerpole connectors for quick battery changes.

Caution Eye protection is required for all cutting, drilling, and grinding operations in this project.

Caution Review all of the general power-tool safety protocols described in Chapter 5, "Cutting Metal," as well as the sections that correspond to the specific tools used below.

Modifying the Master Power Switch

The main drawback with the switch is the large red plastic key that's turned to activate it. This key doesn't make it easy to turn on the robot from outside of the frame. (You'd have to have a huge, gaping hole or some sort of hatch.) We'll fix this by replacing the head of the key with a socket-head cap screw, allowing you to use an Allen wrench to turn the robot on and off.

1. Remove the red key from the master power switch. It can only be removed by pulling it straight out when the switch is in the "off" position.

2. Cut the key on the bandsaw at the base of the head. If the cut edge is too ragged, you can clean it up on the disc sander.

3. Place the key stub in a drill press vise, making sure that the stub is perfectly vertical. Tighten a 1/4"-20 × 1" long socket-head cap screw in the drill chuck to give you a visual reference to line up the stub with the center of the chuck. Once you confirm that the spacing is equal all around the sides, clamp the drill press vise in place.

4. Remove the screw from the chuck and use a countersink to make a divot in the exact center of the key stub, which will act like a punch mark to make sure that the drill stays on center.

5. Drill a hole all the way through the key stub with a #7 drill, which is the correct size of a tap drill for a 1/4"-20 thread. Make sure to use adequate lubrication during the drilling process.

6. Tap the hole with a 1/4"-20 tap. Remember to use adequate lubrication. Make sure to tap all the way through the key stub.

7. Apply a drop of Loctite to the threads of a 1/4"-20 × 1" long socket-head cap screw and tighten it into the key stub. The Loctite should prevent the screw from loosening with repeated usage.

Installing the Powerpole Connector

As mentioned in Chapter 2, "Designing the Robot," changing batteries should be made as easy as possible, since you may not have enough time in the pits between matches to wait for your batteries to charge up in the robot. This means that you need a reliable connector to handle the load that's easy to connect and disconnect. For the project robot, we'll be using an Anderson

Powerpole connector as described earlier in the chapter. This will also give you a chance to use the heavy-duty hammer-style battery-terminal crimper.

1. Cut a piece of white 8-gauge wire 5½ inches long and a piece of black 8-gauge wire 4½ inches long. Strip both ends of each wire.

2. Use the hammer-style battery-terminal crimper in the bench vise, as shown in Figure 16.18 to crimp an 8-GA 3/8-inch stud ring terminal onto the white lead. Use the crimper to attach an 8-GA 1/4-inch stud ring terminal onto the black lead.

FIGURE 16.18: Crimping the battery lead.

3. Use the hammer-style battery-terminal crimper in the bench vise to crimp the Powerpole pins onto the white and black battery leads.

4. Insert the white lead into the red housing and the black lead into the black housing. Slide the two pieces together. Make sure that they are in the correct configuration to match the colors on the battery connector.

5. Using a pair of slip-joint pliers, insert a 3/16" × 1/2" long roll pin into the hole created by the interlocking Powerpole connectors to lock them together.

6. Place the connector as shown in Figure 16.19, and route the white and black 8-gauge battery leads under the chain towards the master power switch.

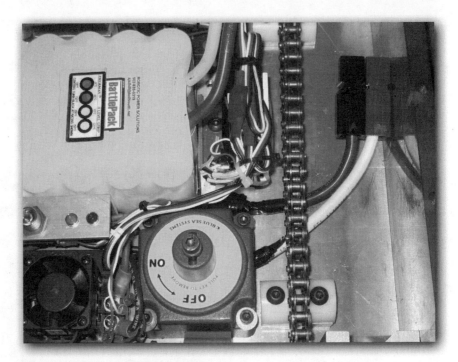

FIGURE **16.19:** Battery connector location.

Connecting the Battery Leads

Now that you've got the Powerpole connector on the battery leads, you need to connect the other ends to the robot's master power switch (for the positive lead) and ground connection (for the negative lead). You will be using crimpers (both the hand-held and hammer style) to make these connections.

1. Route the power leads for the speed control further away from the master power switch around the battery and underneath the chain. They should nestle between the speed control near the master power switch and the secondary power-switch bracket.

2. The positive power leads for both speed controls should be trimmed near the location for the master power switch.

3. Strip the red (positive) speed control power leads and crimp on a 12-10 GA 3/8-inch stud ring terminal to each of the ends.

4. Fasten the white battery lead onto the 3/8-inch terminal of the master power switch that faces the chain, and the speed control power leads should be routed so that the wires enter under the master power switch on the side that faces the battery, as shown in Figure 16.20.

Figure 16.20: Routing the leads under the master power switch.

5. You may have to grind away some of the master power switch brackets at the bottom with a Dremel tool to prevent pinching the wires. Don't worry — it's just plastic down there — the actual switching mechanism is at the top of the switch.

6. Try out the position of the switch and make sure that the leads aren't getting pinched. You may have to tweak the positions of the terminals a little bit. When you're satisfied with the positions, apply liquid electrical tape to all exposed metal underneath the master power switch.

7. Trim the black (negative) power leads just past the secondary switch bracket, on the battery side of the master power switch.

8. Strip the leads and crimp on a 12-10 GA 1/4-inch stud ring terminal to each end.

9. Fasten the black speed control power leads to the 8-gauge battery lead with a 1/4"-20 × 3/8" long button-head cap screw and a nylon-insert 1/4"-20 locknut, forming a star junction, as shown in Figure 16.21.

10. Completely wrap the negative star junction with electrical tape, making tight spirals around the nut.

FIGURE 16.21: Negative power lead star junction.

11. Strip the main power LED leads and crimp a 22-18 GA #8 stud ring terminal onto each lead. Because the wire is smaller than 22guage, you should strip the end twice as long as you normally would, and then fold it back to double the diameter before crimping.

12. Screw the main power LED leads onto the speed control that's closer to the master power switch, along with the fan leads, and the power input (12guage) leads. You may have to shuffle things around and bend the LED's ring terminals a bit to get everything to fit. The fan leads, main power input leads, and main power LED leads should be installed on the "power in" terminals of the speed controls. Make sure to connect red (or white) to +V and black to GND.

Wrapping Up

The theme in this chapter is keeping you out of trouble by developing good wiring techniques. Poor wire and connector choices could limit the performance of your drive system, just as if you'd improperly geared your motors, or underrated your batteries. It only takes one loose connection to lose a match. One bad crimp, or a screw terminal that wasn't tightened down enough, could send you back to the pits in defeat. A single stray aluminum chip could kill an expensive receiver or speed control. By following the techniques I've described in this chapter and in Projects 7 and 8, you can avoid defeating yourself with poor wiring, and make your opponent really work for a victory.

The First Test Drive

Well, the motors are installed, axles and sprockets in place, chains tightened, and wheels mounted. The electronics are in, and the battery is charged. What's left? Try it out! This chapter tells you what to expect during the first test drive of your robot. There are tips on preventing accidents, putting together a preflight checklist, tuning your radio's trims and failsafes, testing the radio's range, and most importantly, improving your driving. You can also find your final project for your robot, Project 9: Testing.

First Test Safety

Resist the urge to put the robot down on the floor and just go for it. Instead, get a pair of scrap 2 × 4s to put under the robot, as shown in Figure 17.1, keeping all the wheels off the ground. This will prevent you from having to patch that robot-sized hole in the garage door.

If your wiring is all done, then you can skip ahead. If you don't have competition-ready wiring, then you need to tidy a few things before you test to prevent such unfortunate consequences as the battery flying off the robot, or the receiver being dragged on the ground.

Cable ties and tape are the order of the day for this situation. If your wiring is still in disarray, put a few cable ties in various strategic places to secure the wiring to the frame. If any long wires are dangling where they might get caught or tangled, tuck them away and hold them there with masking or electrical tape. Make sure that all exposed electrical connectors on the speed controls and battery terminals are covered by electrical tape.

Make sure that your battery is well secured, as described in Chapter 15, "Choosing Batteries." Make sure that your speed controls and receiver (or robot controller) are held down by at least a few pieces of Velcro and some big cable ties. You can go back and properly strain relieve everything later, after you've done a brief road test. If things don't work, you'll have to tear everything apart anyway.

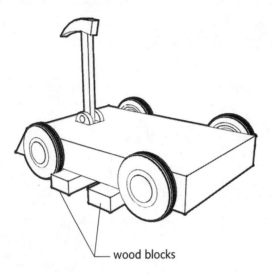

wood blocks

Figure **17.1: Block up the robot.**

The Preflight Checklist

This is a good opportunity to start a step-by-step pre-test-flight checklist. This is a list of all the things you need to do to get the robot running. It should be as detailed as possible, including items you may think are completely obvious. Getting into the habit of using the checklist now will help you later on when it comes to competition. You'll remember everything that needs to be done and the correct order to do it in. Trust me, if you've been madly repairing your robot in the pits and you've got 20 seconds to line up to fight, you'll be glad to have a little guide. Make sure to follow the same routine every time.

1. Put robot up on blocks.

2. Master Power OFF.

3. Receiver Power OFF.

4. Check chain tension (if applicable, see Chapter 11, "Working with Roller Chain and Sprockets").

5. Put new drive batteries in robot and secure.

6. Plug in drive batteries and tape (or cable tie) connector.

7. Put new receiver battery in robot and secure.

8. Plug in new receiver battery and tape (or cable tie) connector.

9. Put old drive batteries on charger.

10. Put old receiver battery on charger.

11. Turn on transmitter (transmitter battery OK?).

12. Receiver Power ON (secondary power LED OK?).

13. Master Power ON (main power LED OK?).

14. Spin the wheels to check drive directions and make sure that polarities and mixing are correct.

15. Put robot on floor.

16. Ready to rumble!

In Chapter 20, "Going to a Competition," I'll discuss the preflight checklist in more detail. During your testing, you should continue to update the list, especially with regard to weapons activation. Your checklist will evolve as your robot evolves. Don't be afraid to revise it. Just make sure you rewrite it so that it's readable even when you're in your most panicked condition.

Tuning Your Radio

Every time that you make modifications to the drive or radio system, you should do your first test up on blocks. This makes it easier to observe your tweaks without having to chase the robot around. Following are some adjustments you may have to do during each first test.

Tweaking the Trims

Most radios or control sticks have *trims*. The trims allow you to make a small adjustment to the center or neutral position of the controller. Sometimes, the combination of a speed control and receiver or control system and joystick will produce a signal at neutral that's actually not stop, but a little in one direction or another. This means that when the robot should be sitting still, it will actually be inching towards the door, like a dog that wants to run away. Using the trims, you can easily correct this "bad dog" behavior. The reason you've got the robot up on blocks is that it's not usually obvious which way to adjust the trim, especially if you're mixed.

Why Don't the Trims Work?

You would think that the trim right next to the stick would be the one you use to correct that stick. That's not always the case with radio-control equipment that's configured for flying airplanes (as opposed to cars or helicopters). The trims for channels 2 (elevator) and 3 (throttle) may be reversed, as shown in Figure 17.2, which works for R/C pilot, but may be a source of confusion for robot builders. If you're having trouble getting the trims to respond, double-check your transmitter manual. If you haven't got the manual, or if it's too confusing, then unplug the speed controls. (Make sure to put a piece of tape on them first to keep track of what channel they were plugged into.) Then, plug a few servos into the channels that aren't responding. By watching the movement of the servos, you can clearly see the effect of the control signal changes.

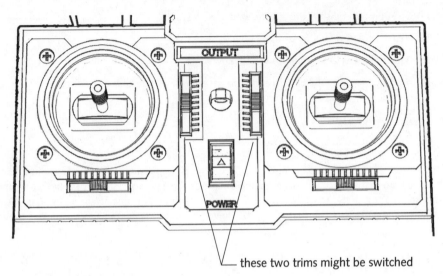

these two trims might be switched

FIGURE **17.2: Standard R/C aircraft controls and trims.**

For example, if you've got elevon active, or you're doing a PWM mixing routine in the IFI program, moving a single trim may cause both motors to turn. In the case of the forward/back trim, both motors will turn forward or backward, and in the case of the left/right trim, one motor will turn forward and the other backward.

After playing with the trims for a bit, you should get a feel for how they are responding and be able to correct it so that all motors are stopped. If you have two motors on a side, each with their own speed control, and both are getting the same PWM signal, but only one motor is turning, then you will have to retune the individual speed controls as per the manufacturer's instructions. Some speed controls (such as the IFI controls) have a button to help you calibrate them.

Cross-Reference

For a review of PWM signals, see Chapter 13, "Choosing Your Control System." Speed controls are discussed in Chapter 14, "Choosing Speed Controls."

Preventing Runaways with the Failsafes

You want to prevent your robot from doing anything when it's not under your control. Some competitions will fail you in the safety/technical inspection for this item. It's relatively easy to take care of this using *failsafes*. In a standard radio-control system, a failsafe takes over when the receiver loses contact with the transmitter. More specifically, when a loss of signal is detected, all PWM channel outputs are set to the failsafe value. These values may be predetermined by you (F/S position), or set to continue with the last known good value received (HOLD or NORM position). The choice is yours, depending on the situation. For speed controls on drive motors, you definitely want the failsafes set to the neutral (centered and stopped) condition. This will help prevent runaway robots.

Figure 17.3 shows the failsafe setup screens for the Futaba FP-7UAPS and 9CAP transmitters. The screen for the FP-7UAPS indicates that channel 1 is set to a failsafe value of 50%, while channels 2 to 7 are set to HLD, and will maintain their last known positions in the event of signal loss. The screen for the 9CAP shows that channel 3 (THR) is set to a failsafe value of 20%, while all other channels are in NOR (normal) mode, and will continue with their last known positions in the event of signal loss.

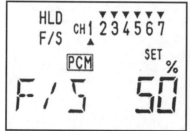

Futaba FP-7UAPS   Futaba T9CAP

FIGURE 17.3: Typical Futaba failsafe setting screen.

Note   If an IFI system loses contact, it will stop, and is usually exempt from the failsafe requirements of most competitions.

Range Testing

An important (and quick) test for your radio system is range. The range test is pretty simple. Put the robot up on blocks and start running the drive back and forth. Have a friend stand by the robot to keep an eye on things, and start walking away as you continue to run the drive. When you're out of range, the drive will stop. If not, check your failsafes as described in the previous section. A minimum safe range that should cover most (if not all) arenas is 100 feet.

Cross-Reference   If you've already got radio signal problems, you might want to skip ahead to Chapter 19, "Troubleshooting," and try some of the radio signal fixes.

Drive It Hard

If you've got your failsafes set and are confident in your radio range, then it's on to actually driving the robot. First, you test to make sure everything works. Then, you test to make sure everything works *well*. You do this by driving the robot hard.

Make It Break

Run the robot at full speed back and forth. Make hard turns. These are the kinds of maneuvers you'll be asking your robot to perform in the arena. Your goal here is to try and break the drive. You don't necessarily have to run the robot into walls at full speed (though it is a pretty good impact test). Just drive it hard and make sure the wheels don't fall off, the drive chains don't go flying, and the sprockets don't skip.

 Tip This is also a great time to take an in-progress picture. Get a camera and take a few pictures for the Web site later on. Also, have a friend take some video while you drive. Reviewing the tape later on may help you solve some problems.

Did It Break?

Did anything break? Check the chain. Is the master link okay? Is the chain slack, or is it still properly tensioned? Are the wheels still mounted where you put them? Are any axles bent? If not, then wait for a while so the motors can cool down, and then send it out again, and if nothing breaks, then wait a bit and do it once again. If everything is still solid after at least three tries, then your drive will probably survive your first match.

 Tip Are your motors hot? Are they too hot to touch? You may have to add small DC fans to cool them.

If something broke during testing, then that's great! This means that it would have failed you for sure in the arena and you've caught it ahead of time. Your next task is to figure out exactly why it broke and fix it. Then go through the same test process and keep repeating it until nothing breaks.

You should continue testing periodically. You want a *bulletproof* drive system. Remember, you can have the most kick-ass (fully functional) weapon in the world, but if your drive is disabled and you can't move, then you *lose*.

Tips and Exercises for Developing Driving Skill

Matches are won and lost by driving skill. In some competitions, there are individual events devoted to driving alone. If you can't drive, you can't deliver your weapon to the opponent, and if you can't cause any damage, then you won't get any points.

 Tip Start out slow when you're just learning to drive your robot. The control stick is pretty sensitive, and you only need a light touch.

The Left/Right Dilemma

When the front of the robot is facing away from you, it's pretty straightforward. Just as if you were sitting in or on the robot, left is left and right is right. The dilemma, as shown in Figure 17.4, happens when you turn, and suddenly the front of the robot is facing toward you. Now, left and right are reversed. You push the stick left and the robot turns to the right! What happened? If you think about it, the robot is still doing exactly what you tell it to do. When you command the robot to turn left, the motor on the right side of the robot spins forward and the motor on the left side spins backward. This process happens regardless of the robot's position relative to you.

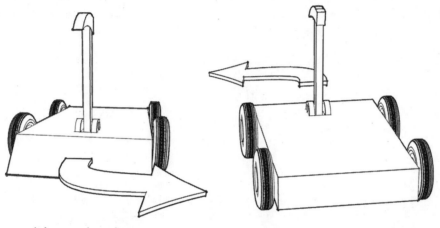

left turn - front facing you   left turn - front away from you

FIGURE 17.4: The left/right dilemma.

Think about it from the robot's perspective. When the robot's front is facing you and you command it to turn left, it spins the right side forward and the left side backward, just as before. The difference is that this time, from where you're standing, it looks like the robot turned the wrong way, but it really didn't. It was just following your orders.

How do you correct this? Well, there's nothing you can do to the robot. The correction must come from your brain, and that's where practice comes in. With enough practice, you will begin to think like the robot.

Tip Pick a point on the robot (with Deadblow I use the hammer head) and associate that one point with the front of the robot. Sometimes I imagine a big arrow floating over the robot pointing towards the front. After a while it will become second nature to think about your controls in terms of what the robot will do.

Drive the Square

Try driving your robot slowly in a square (believe me, it's harder than it sounds). As illustrated in Figure 17.5, your imaginary square should be at least 10 feet to a side. Keep driving the square until you can do it perfectly at a low speed, and *then* do it faster. Then, start over, slowly driving in the *other* direction around the square, increasing your speed as you gain more confidence. Try and make your turns as crisp as possible, or in other words, plan your move ahead of time. For example, if your robot has a lot of torque in the drivetrain, then you probably won't have a problem. Your turns should be pretty crisp to begin with. However, if your robot is prone to overshoot, or is drifting after you've told it to stop, then it will turn too far at each corner. By planning ahead, you can adjust how long you tell the robot to turn (usually shorter for overshoot), so that the overall outcome is a turn that looks crisp. Of course, we know that it's half-powered-turn and half-drift, but the final result is the same: a perfect square.

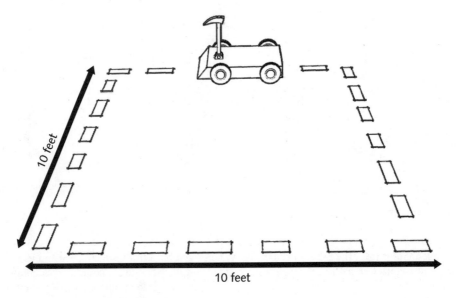

FIGURE 17.5: Drive the square.

Obstacle Course

Next, put some big tape squares on the ground as shown in Figure 17.6 to represent arena hazards and try to avoid them. These tape marks should be at least 4 feet to a side, and you should put down three or four of them. Navigate a zigzag obstacle course driving away from you and then turn around and drive back toward you. Again, start slowly and increase your speed when you get the hang of it. You can also remove the obstacles and just try driving a figure 8 over and over.

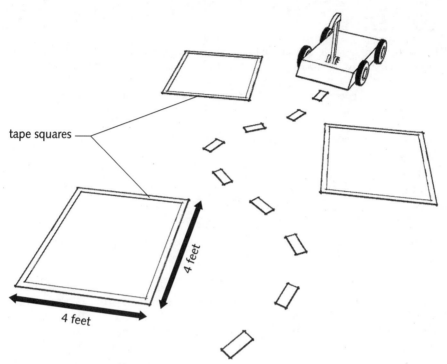

tape squares

4 feet

4 feet

FIGURE 17.6: The obstacle course.

Chase a Box

I throw out a medium-size cardboard box (about the size of a typical opponent) and push it around. Use the tape markers that you put down on the ground to represent the hazards, and push the box into each one of them, being careful not to cross the tape marks yourself. Control the box; don't let it control you.

Live Target Practice

Buy a cheap little R/C car from the drugstore and have a friend drive it around. Chase the car with your robot. I suggest a cheap car (under $10) because they have a habit of getting run over by the much larger 60- to 400-pound robots.

Tip

Take your robot to an empty parking lot to test. Make sure there are no cars or people around. You need the space to make sure you don't hit anything, but more importantly, it will allow you to get used to driving your robot when it's far away from you, which is the same perspective you'll have during competition. You should get used to driving a minimum of 20 feet away from you.

Project 9: Testing

You're in the home stretch. This is the last project. As you've learned in this chapter, testing is only second to safety when building combat robots. Here, you will clean up loose ends such as adding strain relief and managing the antenna. Next, you'll tune the radio and perform your first test run. If all goes well, then you'll finish up the top armor, and tidy up the last few details before going into hardcore survival testing.

Eye protection is required for all drilling and grinding operations in this project.

Review all of the general power-tool safety protocols described in Chapter 5, "Cutting Metal," as well as the sections that correspond to the specific tools used below.

Strain Relief

After installing the chain, you probably realized that the wires running underneath it are a weakness. By adding 1-inch plastic adhesive-backed anchors to the base and using cable ties to hold the wires in place (as shown in Figures 17.7 and 17.8), you can keep the wires away from the chain, and potential disaster. Also, you should add Velcro to the battery rails and brackets, as shown in Figure 17.7, to give you a little tighter hold on the battery.

Figure 17.7: Strain relief for speed control power wires.

Figure 17.8: Strain relief for main power connector.

Routing the Antenna

You need to route the antenna into rows and tape it to the top of the battery plate with electrical tape as shown in Figure 17.9. Make sure to leave enough slack so that you can easily remove the battery plate without pulling the receiver.

 Note It's important to keep the antenna away from speed controls and motors, which may cause interference. Also, since the polycarbonate freely allows radio waves to pass, you won't need to install a Deans antenna, as described in Chapter 19, "Troubleshooting." If this were a completely metal shell, then a Deans antenna would be a necessity.

First Test

This will be the first time that you'll be running the robot under its own power. Before putting it down on the floor and watching it go, you need to make sure your mixing is working correctly, so that when you tell the robot to turn, it will do exactly what you expect. Nothing is more dangerous than attempting to drive a robot that does not respond as expected.

1. Put the robot up on top of some scrap wood blocks, with the wheels off the ground.

2. Put the keys for the wheels in the keyways on the axles.

3. Slide the wheels onto the axles and secure them in place with the shaft collars.

FIGURE **17.9:** Receiver antenna routing.

4. Make sure that the master power switch is in the OFF position, and plug in the main battery.

5. Turn on the transmitter.

6. Verify that you have elevon mixing active, as described in Appendix A, "Advanced R/C Programming."

7. Make sure that all radio channels are in normal (or forward) position and not reversed.

8. Make sure all radio trims are set to zero.

9. Set the failsafes as described earlier in this chapter, so that a loss of signal won't cause the robot to run away.

10. Turn on the secondary power switch. You should see the green receiver light turn on.

You and any visitors to the shop should be wearing eye protection when testing the drive motors. They may throw off little sparks and sometimes debris if you jam them back and forth quickly.

11. Turn on the master power switch. The red master power light should turn on and the speed controls should have solid orange lights. If they're blinking orange, then turn off all power and check the leads going from the speed controls to the receiver. Also check the receiver battery voltage with a voltmeter.

12. Push the stick forward slightly. The wheels on the left and right sides should turn slowly in the same direction. If this is not what you want to be forward, then turn off all power, and swap polarities on both motors where they connect to the speed controls.

13. If forward and backward are working okay, then try a turn. If the robot turns the wrong direction then stop and power down. Swap the channel 1 and channel 2 leads on the receiver.

14. If the wheels are turning slightly, use the trim controls on the transmitter to get the wheels to stop turning, as described earlier in this chapter.

15. Test the failsafes by holding the stick full forward and shutting off the transmitter. The wheels should stop turning within a second or two. If they continue turning, double-check your failsafe settings.

16. If the wheels do not turn while the robot is idle, and the robot passed the failsafe test, then you're ready to give it a road test.

Holes in the Top

Before going into more serious testing, it's necessary to finish off the top armor. You'll take the 1/4-inch thick piece of Lexan you cut way back in Project 2 and drill the perimeter holes for mounting. Then, you'll cut access holes for the master and secondary power switches.

1. Peel off the protective coating on one side of the top armor.

2. Print out and tape together the base pattern. The pattern for the perimeter holes is the same. Apply this pattern to the polycarbonate using spray adhesive.

3. Mark the perimeter hole positions with an automatic center punch.

4. Drill the holes using a 17/64-inch drill bit, which is slightly oversize (1/64 inch larger) from the target 1/4-inch hole size. Make sure to use adequate lubrication.

My Road Test

Everything was working fine until I hit a big bump, and then everything went dead. Walking over, I saw that the main power light was off. The main power switch had not moved, and there was no smoke. Upon further inspection, I found that one of the Powerpole pins was not fully engaged. It was in there enough to get things going, but on the first big jolt, it popped out of the back. That's why testing is so important. This is one of those little things that could have easily killed me in the arena. This also identified the need to grind a small indicator line into the top of the screw head to show the position of the main power switch.

5. Chamfer all the holes, top and bottom, with a countersink.

6. Remove the pattern and get rid of any sticky residue with Goo Gone. Also remove the protective cover on the bottom of the plate.

7. Install the top armor on the robot with one or two screws on each side. Look through the clear top and visually mark the locations of the master power and secondary power switches with a marker. Use an automatic center punch to put a divot at each mark.

8. Clamp the plate to a table with a piece of scrap polycarbonate under each clamp to keep the top armor from getting scratched.

9. Use a 3/8-inch drill bit for the master power switch and a large Uni-bit for the secondary power switch. Make sure to use adequate lubrication.

10. Clean up the plate and put it back on the robot. If you can't get all the screws in, then you can try the trick below:

   Get as many screws as you can going all at once. Make sure the plate is sitting flat on the robot. Chuck a tap (in this case, 1/4"-20) into a cordless drill. Then retap the holes with the lid in place. This will actually thread the through holes a bit where they interfere, but will also allow you to get the screws in much easier. The trick is that the threaded hole will guide the tap. Since it's tapered, it won't start threading the through-hole until it's already started in the right thread.

11. Make sure all the screws fit, but don't tighten down the top cover just yet. There are a few more final touches that you'll need to perform in the next step.

Final Touches

This is the last part of the process. You'll be taking care of the little details that could easily kill you in the arena. High impact forces can cause connectors to pop right out. You'll be taking care of all those next, as well as insulating any remaining exposed electrical terminals, and checking your chain tension.

1. Grind a little line using a Dremel tool with a cutoff wheel (full face protection, please) in the socket-head cap screw on the master power switch (as mentioned in the "My Road Test" sidebar) to help you keep track of the actual position of the switch.

2. Apply hot melt glue on connectors that can pop out, such as the PWM signal driver leads (at both the receiver and speed control sides), as well as on all of the other servo leads going into the receiver. You don't need to smother the connectors in glue. Just apply a dab to the connection point to give them a little extra resistance to being tugged out.

3. Apply the number tape mentioned in Project 7 to the PWM leads on both the receiver side and the speed control side. Pull off the masking tape after you've applied the numbers, so you don't get confused.

4. Stretch some electrical tape over the unused receiver channels. The exposed pins carry power and can short out, causing damage to the receiver.

5. Apply liquid electrical tape to all of the exposed electrical (screw) terminals on both speed controls.

6. Make sure that the chains are still tensioned properly. They will stretch over time, and you may have to add more spacers under the tensioner blocks.

7. Make sure the receiver battery stays plugged in by putting a piece of electrical tape around the plug and receptacle so that half of the tape is on the plug and half is on the receptacle, as shown in Figure 17.10. Also make sure that the main battery power stays connected by slipping an 8-inch long cable tie through both of the roll pin holes and tightening it down (also shown in Figure 17.10). These procedures should be repeated every time you change batteries and are listed in the sample preflight checklist in Chapter 20.

8. Put the top cover on, insert all the screws, and tighten them down.

9. Congratulations. You're done with building. Now test this robot until it breaks, as shown in Figure 17.11. Keep breaking it and fixing it until it doesn't break anymore. If it can survive this kind of abuse, it's going to take one tough robot to kill it in the arena. Now you have all the skills you need to build your own combat robot. Good luck!

FIGURE 17.10: Securing the battery connections.

FIGURE 17.11: Testing the project robot.

Wrapping Up

I know I've made a big deal out of the first test drive. For me it's just a symbolic thing. It's the mystery and anticipation of never having seen something work before. Will it work? What will it be like? Savor the thrill of seeing your creation move for the first time. It's one of the sweetest moments in the whole process.

Sometimes, robot builders are afraid to test their robots for fear of breaking something. Breaking something is actually the point. It's far better for you to break it in your workshop than for someone else to break it in the arena. Yes, I know it will take time to fix, but it will save you the embarrassment of not leaving the starting square for some silly reason that would have been obvious if you had just tested the robot when you had the chance. Go ahead and fire it up. You'll be glad you broke it.

Remember to practice, practice, practice, because good driving is what wins matches.

Choose Your Weapon

So far, I've talked in depth about each of the other robot parts, from the armor, to the drive system, to the control system. This chapter will contain no detailed instructions or step-by-step procedures. Instead, this is your chance to put your own creative stamp on the robot. First, I'll review some weapon safety guidelines, and then I'll describe the basic weapon types so you can see what other competitors have done. Then, it's up to you to choose your weapon.

Basic Weapon Safety

In the early days of combat robots, weapon systems weren't required to have any safety restraints. Then again, in those days, wimpy weapon motors often stalled, high-pressure *pneumatics* (air power) hadn't come into widespread use, and there were only a few spinner robots out there. Since then, things have changed dramatically: The competitors have become more skilled, the community has grown, and the weapons have become very, very dangerous.

If you believe that your weapon is capable of hurting another robot, then you should be easily convinced that it's more than capable of hurting you. Personal injury from a weapon is a real possibility, and if any part of the robot has the most chance of injuring you, it's the weapon. Pay special attention to the safety requirements described in this section, as well as those that are specific to your competition.

Eye Protection

It's difficult to predict what exactly will happen when you test a weapon. Perhaps you forgot to tighten down a bolt on your spinning blade. The tip of your hammer may miss the target block and hit the sidewalk, sending concrete fragments flying upward. A block of wood used as a test target may splinter. I've personally witnessed all of these scenarios happen in real life, and they demonstrate that the use of eye protection during weapons testing is extremely important. There are too many unknowns to simply say, "That'll never happen to me."

Safety Covers

Some robots have sharp or pointed edges as part of their weapon systems. Many competitions now require safety covers to prevent you from accidentally bumping into a sharp edge while moving the robot in and out of the arena or in the pits. Read the rules for your competition thoroughly for specific requirements, such as safety covers that use Velcro or bungee rather than adhesive tape. Make sure that you can remove your safety covers quickly and easily.

Motion Restraints

Several competitions now require a pin, chain, or similar device (called a *motion restraint*) to immobilize a moving weapon when the robot is deactivated. Although you might not think that an unpowered or depressurized weapon isn't dangerous, if it can still move, it represents a safety risk.

Many robots have weapons that fold into the body until they're fired. Others have moving arms or levers. Any joint or place where you can get your hand or other part of the body caught is a *pinch hazard*. By implementing a motion restraint, you can minimize the risk to yourself in the pits and more importantly, during activation before a match, which is one of the most dangerous times with a robot, since everything is powered up, and the robot is primed for combat.

Like the safety covers, motion restraints should be easily and quickly removable so that you don't delay a match.

Cross-Reference

Prematch activation procedures are covered in detail in Chapter 20, "Going to a Competition."

Safety Lessons from My Rookie Year

Though some veterans feel that these rules are a pain compared to the old days, I'm actually really glad that there are so many safety restrictions now. It makes you more aware of your robot's destructive capability. Also, having a motion restraint gives you the added confidence that the weapon won't accidentally deploy while you're standing right there powering up.

In my rookie year, I remember pressurizing my robot's weapon system, which was a 6.5-pound sledgehammer that folded into the body. It was right before a match, and things were moving pretty fast. BattleBots CEO and cofounder Trey Roski gently leaned over and suggested that I move my head out of the *direct path* of the hammer. At that moment, I realized that having been distracted by the impending match, I had put myself in great danger. The restrictions are a good thing because you won't always have Trey to look over your shoulder.

The Wedge

The idea with the wedge (see Figure 18.1) is to get the opponent off their wheels by using an inclined plane. Once they no longer have traction in the arena, you can push them around as you please. The more traction and horsepower you have, the less chance they have. The beauty of this strategy is that it's simple and efficient. No extra weight goes towards a weapon. The *armor* is the weapon.

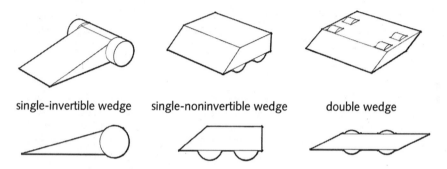

single-invertible wedge   single-noninvertible wedge   double wedge

FIGURE **18.1: Single-invertible, single-noninvertible, and double-wedge robots.**

The Leading Edge

Many wedges have been defeated by having their front edge bent over, propping their wheels off the ground. You've got to make sure that the leading edge is well supported and not likely to bend, dent, or otherwise deform under stress. Remember, this is the single part of your robot that will be making contact with the opponent all the time. Make it robust.

Drive System

Horsepower and traction are what the wedge needs. There aren't any complicated lifting arms or pneumatic systems. All of your weight should be going towards armor, motors, wheels, and of course, batteries.

Shock mounting is important in any robot design, but more so for robots like this that are designed for constant contact. Every time you use the weapon, the whole robot will be subjected to a shock load.

Invertible or Noninvertible?

If you're designing a wedge, then you should plan to have a lot of contact in the arena. Constant, violent contact means that it's much more likely for you to be flipped over. That's why most wedges can drive inverted. As with any robot design, you should avoid any flat sides on the robot where you could land and become immobilized.

Anti-Wedge Prejudice

Most builders scorn the lowly wedge as an uncreative design. They're just jealous of the simplicity and effectiveness. How to avoid being hated: drive well. Drive *really* well. Nothing's worse than a match between two pushers or wedges where they both drive serpentine around the field and don't hit each other or anything else.

Note If you choose to drive inverted, then you should set up an inverted driving control that will switch the left/right steering as described in Appendix A, "Advanced R/C Programming."

Single- or Double-Wedge?

Some single-ended wedges lose their weapon when they're turned upside down. Even if they can run inverted, the leading edge is pointing up in the air, and no longer has any contact with the ground. Double-ended wedges are shaped like a parallelogram when viewed from the side. They will always have a leading edge to work with. However, when this type of wedge is inverted, you've got to reverse not only the left/right steering, but also the forward/back response, which is also described in Appendix A. Otherwise, you'll be leading with an inverted edge, just like the single-ended wedge described earlier.

Driving the Wedge

All of your energy comes from momentum. To build up momentum, a wedge needs speed and space. As a driver, you should get as much of a running start as possible by lining up at the opposite end of the arena from your target. Once you've gained control of your opponent, use the arena as your playground and slam them into a wall at maximum speed.

It's important that you don't get your leading edge stuck in the arena. Since many arenas have a floor that is composed of rectangular metal plates, one of the best strategies is to try to drive at diagonals to the plate seams, so you don't hit the edge of a plate straight on, which could cause you to come to a screeching halt, and possibly damage the wedge tip.

The Rammer

Much like the wedge, the rammer (shown in Figure 18.2) is a very simple strategy. Slam into your opponents at high speed and try to break something loose. Instead of an inclined plane, rammers usually have spikes to inflict damage. Although more spike length may buy you deeper penetration into your opponent, the tradeoff is that they're more likely to bend and break.

Driving a rammer is similar to driving a wedge. Getting a running start from across the arena is the best way to build up momentum. The drive system should be equipped with high horsepower motors and good traction. The rammer should be designed to be inverted and have switchable mixing as described in Appendix A.

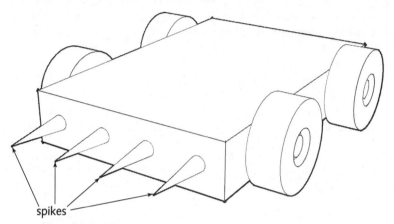

spikes

FIGURE **18.2: Rammer robot.**

The Pusher

The Pusher (see Figure 18.3) is similar to the rammer, but instead of slamming into opponents, the pusher bulldozes them around using a scoop or plow, trying to push them into arena hazards. Unlike the wedge and rammer, you're not as concerned with speed as you are with torque and traction. You've still got to be quick enough to catch your opponent, but you won't be careening across the arena at high speed and slamming into walls.

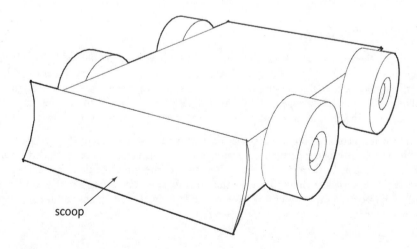

scoop

FIGURE **18.3: Pusher robot.**

A pusher should be driven close to the opponent, trying to outmaneuver them and gain control. No running start is necessary for this strategy. Since the scoop will be your primary weapon, make sure to protect your vulnerable backside.

The Lifter

Unlike most weapons, a lifter (shown in Figure 18.4) seeks not to destroy, but to overturn or control an opponent using a powered arm of some sort. There are a few different ways to implement the lifting arm. Most robots use *linear actuators*, which are electrically driven extending cylinders, although regular gearmotors and pneumatics have also been used. Since the lift can be relatively slow, most competitors avoid the complexities of pneumatic systems, in favor of a slower linear actuator.

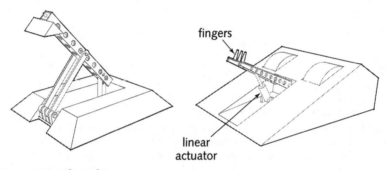

fingers

linear
actuator

FIGURE 18.4: Lifter robot.

Designing the Arm

Arm designs differ dramatically. It can be tricky coming up with a good design. You don't want the target to simply slide off the arm as you lift. Some competitors have little fingers that deploy to grab the underside of their opponents. Others have stationary teeth or barbs to hook the other robots. The four-bar linkage used by Biohazard is particularly effective because it is able to rise up parallel to the ground as well as push forward in a gentle arc, all at the same time.

The lifting arm (in combat robot terms) is a complex system, which means that you pay the price of having more moving parts that are susceptible to damage. However, if the lifting arm can retract into the body, then it will be protected most of the time, and doesn't need to be made incredibly robust. Much like a wedge, this strategy relies on getting a part of your robot under the opponent. Many competitors have benefited from an arm tip shaped like a flat spatula or a small wedge.

The pivot points for the arms vary with design. Some straight arms are pivoted towards the rear of the robot. Some are pivoted near the front. The important thing to consider is what will happen to your robot when you have an extra amount of weight balanced at the tip of your arm. Are you likely to fall forward? If so, then you should balance the robot by putting heavy items like batteries near the rear.

Driving a Lifter

Your drive system has to be accurate in positioning, as well as robust enough to push. Although you've got to be quick enough to maneuver around your opponent, this isn't about speed. It's all about precisely positioning the lifter tip in the most favorable spot on your opponent. Be patient and wait for your moment. It will be far worse to have the arm exposed to damage on multiple lift attempts. If you've got an arm that retracts into the body as described earlier, then it should only be deployed when you're lined up perfectly enough to get a good lift.

Once you've got them up, push into your opponents to complete the flip. Some lifters try and take advantage of the curb at the base of arena walls as a stop, rather than let the opponent simply slide backwards, giving them a pivot point for the flip. Once they're hooked, you can also try to drag your opponents across the arena, and into the arena hazards.

Since this robot has a leading edge like a wedge, you would benefit from attempting to drive at diagonals and avoid hitting the arena seams head-on.

Bidirectional Limit Switches

Linear actuator systems need *limit switches* to tell them when they've reached the end of travel (and to stop applying power). Limit switches are momentary electrical switches. They're spring-loaded so that they return to their original position when you stop pressing them. Although some linear actuators have them built-in, most off-the-shelf items do not. With an IFI control system, you can connect your own limit switches to digital inputs on the robot controller, and connect the linear actuator to a speed control. Then, you can program the system to shut off the power when a limit switch is tripped.

If you don't have an IFI system, you can still add limit switches of your own that will automatically kill the power when your actuator reaches its end stops. It's a bit more complicated than simply hooking the actuator to a speed control. This method involves a double-throw, double-pole relay, as shown in Figure 18.5. You connect power to a relay's coils using a servo switch (described later in this chapter). Energizing the coils switches one set of contacts (the *poles*) between the two sets of contacts (the *throws*).

Continued

Continued

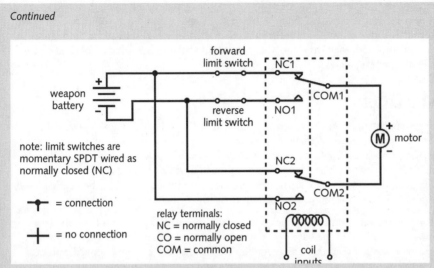

FIGURE **18.5: Double-pole, double-throw switch wiring.**

You should position the limit switches so that they trip in the correct place. During the first test, be prepared to remove power immediately, since confusing the two switches will mean that the actuator won't stop where you want it, or will bind. Some actuators have a built-in clutch that protects them and keeps them from running past their endpoints. In this case, the activator will just use up power rather than burn the motor, and a limit switch system isn't critical to protecting the actuator. However, it's better to add the switch system so that you don't have to remember to turn off the power in the middle of a battle.

The Clamper

Control. That's what this strategy is about. The clamper (see Figure 18.6) is a variation of the lifter with a twist. It has a similar lifting arm, but it adds a finger-like mechanism that firmly grabs the opponent so that its whole body can be lifted off the ground. Most robots become helpless when their wheels no longer have contact with the arena, which allows the clamper to do as it pleases with its opponent. It can then carry the other robot around the arena (great for you, but highly demoralizing for them) and use the hazards to inflict damage.

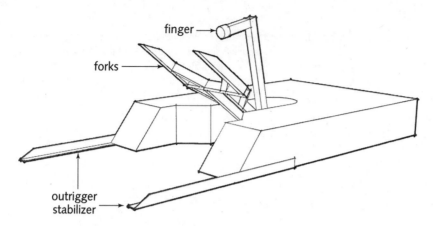

FIGURE **18.6: Clamper robot.**

Weight Distribution

Clampers have to guard against tipping over once the opponent is captured and raised, since their center of gravity changes drastically, making them unstable. To minimize this effect, the clamper should be designed with a low profile and extra stabilizers (skids) or a long, wide body. You can also compensate by keeping the pivot point for the arm near the center or farther back in the robot. Heavy items should be placed in the rear to help keep the back from tipping up during a lift. Also, the lift will place a great strain on the robot's frame. Make sure that you have adequate reinforcement to prevent the frame from flexing too much.

Arm Design

The difference between a lifter and a clamper (in terms of the arm) is that the lifter tips the opponent up, leaving some part still supported by the arena. A clamper needs to be able to lift the opponent completely off the ground. This means that the actuator that raises the arm must be stronger than one for a lifter robot. Electric motors with significant geardown are usually used for this job, since the lift needs a lot of power, but doesn't need to be very fast. The finger should be designed so that it can accommodate a variety of sizes of robots. (If you can't get the finger around them, you can't clamp.) It also needs to be pretty quick to catch your opponents. Pneumatics are usually used to activate the finger because they're fast and powerful enough to keep the robot from slipping.

See Appendix C, "Pneumatics," for safety and implementation details.

Driving a Clamper

Clampers can drive with many different strategies. When fighting a wedge, it's best to go out to the center of the arena and wait. Your drive should be quick enough to swivel in place, keeping the arm always facing the opponent. When fighting other types of robots, be patient, and wait for your chance to catch them. Getting good positioning on an opponent is the hardest part, but once you've got hold of them, then it's off to the arena hazards.

Make sure that your drive system is quick enough to maneuver and catch an opponent, but also strong enough to carry not only your own weight, but also the overhung weight of your hoisted opponent.

Making a High-Power Servo

If you want fully proportional positioning, then you need a servo (as described in Chapter 13, "Choosing Your Control System"). The problem is that most off-the-shelf servos have a torque rating in the inch-ounce range. What if you need something in the foot-pound range? No problem. You can make your own.

Servos consist of three main systems: a motor and gearbox, a servo amplifier, and a feedback potentiometer. The motor and gearbox provide the torque. The servo amp is an electronic device that controls the motor. The feedback pot is a variable resistor that tells the amplifier the position of the output shaft from the gearbox. In a hobby servo, all of these parts are included in a single plastic case. To make your own high-power servo, you need to increase the size of the motor and the capacity of the servo amplifier, and add your own potentiometer.

Keep in mind, though, that this is a pretty expensive proposition. If you don't really need the ability to stop in the middle of travel, then the simple bidirectional limit switch system described earlier would be a significantly cheaper way to go. If you're intent on your own proportional system, then there are a few choices.

One of the few off-the-shelf servo choices is the high-power Tonegawa Seiko servo (SSPS series) from Japan. They are carried by many U.S. vendors, including Vantec. These servos take a standard radio-control PWM signal in and a separate supply of 12 or 24 volts. They have a few options regarding angle of travel (in 90-degree increments up to 360 degrees). These are well-proven units, but you've got to make sure that your application won't exceed the torque specifications, or you may damage them. (An example might be a hammer head that continues past the servo's stop with its own momentum.)

If you have your own motor in mind, you can use a third-party external servo amplifier as long as the amp can handle the current requirements of the motor. The controllers I use most often for custom servos are Advanced Motion Controls (AMC) amplifiers. They cost about $300 each, and handle 12- to 180-volt motors. Their current rating is a respectable 25 amps. These are fairly bulletproof amplifiers that have seen a lot of film work. The three models (12A8, 25A8, and 20A14) have slightly different current ratings and prices, and are available from Servo

Systems. Much like the 4QD speed control (described in Chapter 14, "Choosing Speed Controls"), these amplifiers require an R/C interface to translate the PWM signal from the radio into a varying DC control voltage. This can be accomplished simply by using a servo connected to a potentiometer. The amplifier provides the control voltage, and the servo turns the pot to vary the voltage up or down.

I've used the Vantec Bully (RBSA) amplifiers on a few occasions, which work very well, although they cost almost twice as much as the AMC controllers. They do, however, take a direct PWM input.

A newer amplifier that shows great promise is the Roboteq AX2500 controller (also described in Chapter 14). This is a versatile two-channel unit that accepts a number of different types of input (PWM, RS232, and analog pot) and has an excellent output current rating of 120 amps per channel.

The Crusher

Much like the clamper, the crusher (as shown in Figure 18.7) has jaws that grab the opponent, but instead of lifting them, the crusher tries to pierce the armor or otherwise bend or compress the frame. This sort of damage might be achieved with pneumatics, but is more suited to the use of a *hydraulic* system, which can run at much higher pressures, and uses oil instead of air. Crushers have been introduced in both horizontal and vertical configurations.

FIGURE **18.7: Crusher robot.**

Jaw Design

The jaws have to be strong enough to endure forces that would crush other robots. This means that you'll probably be using some sort of steel, which can become quite heavy in large quantities. Hydraulic systems are incredibly powerful, but are usually quite slow. Unfortunately, it can be a bit frustrating trying to catch an opponent with a slow-moving jaw. Some competitors use variable-displacement pumps for the system, which allow you to switch between a low pressure, high-rate response (to catch the opponent) and a high-pressure, low-rate response (to crush them), sort of like a gearshift.

 Hydraulic systems are very dangerous because of the exceptional pressures involved. A detailed discussion of these systems and safety requirements is beyond the scope of this book (as well as my personal experience). If you wish to pursue hydraulic weapons, I suggest that you consult a hydraulics manufacturer for advice.

The Hammer

The point of the hammer robot (Figure 18.8) is to pound the opponent into submission, knocking loose connectors or bending their frames. Since the top armor has gotten better in recent years, it's become pretty difficult for a hammer to knock out an opponent. The hammer remains a crowd favorite, however, and can score points with rapid-fire pounding.

FIGURE 18.8: Hammer robot.

Speed and Power

Because it's difficult to line up a shot in the first place, hammers should be designed with the ability to fire many, many shots. In the past, some robots have used an electric winch and ratchet system to prime a large spring for quick release.

Although some hammers are electric, these weapons are usually pneumatic because you need the speed to inflict damage. In fact, the speed of the tip has more of an effect on damage than the mass of the head. With pneumatic systems, you have the choice of carbon dioxide (CO_2) or high-pressure air (HPA)/nitrogen (N_2). When firing multiple shots at a high rate, CO_2 tanks tend to freeze over, and with the drop in temperature comes a drop in pressure.

Straight pneumatic actuators are usually chosen over stock rotary pneumatic actuators because they offer an advantage in speed and power that outweighs the convenience of already having the rotary motion. Besides, rotary pneumatic actuators can be quite expensive. You can convert the linear motion to a rotary swing by a lever arm, but this limits the hammer's swing angle. If you desire more than 180 degrees of swing with a straight cylinder, then you will have to build your own rack-and-pinion system.

Although a short, fat cylinder and a long, skinny cylinder may have the same internal volume, and thus use the same amount of air, the fatter cylinder will have much more power because the force is proportional to the area of the piston. Cylinder diameters of 3 to 4 inches are not unheard of among top competitors in higher weight classes.

Cross-Reference

For the complete detailed description of pneumatic systems, see Appendix C. It contains all of the information that I've learned over the years in building my pneumatic hammer robot Deadblow.

Arm Design

The arm should be as tough and light as possible. Some robots use aluminum or titanium. Most of the weight should be concentrated in the head. The point should be harder than anything that it comes into contact with. Usually tool steel is a good choice. The length of the arm is also important in your strategy. An arm that is too long is difficult to aim and will continually miss the target.

In most hammer designs, the hammer is mounted in the middle of the robot or near the front to give the most striking range. The internal weight should be distributed so that the rest of the robot doesn't pop up when the tip impacts the opponent. The addition of a wedge ramp in front can help keep your front down (underneath the opponent) while you hammer. It also provides a useful secondary weapon.

Driving a Hammer

Since the hammer is a unidirectional weapon, it requires great driver skill to line up the weapon in a favorable position. Precise maneuvering ability will greatly increase your chances for scoring hits. An aggressive rapid-fire approach yields good results.

Servo Switches

In order to trigger a pneumatic valve or apply voltage to a relay, you're going to need a *servo switch*. This is an on/off switch that's controlled by your radio. If you have an IFI system, they have a line of relay modules (called Spikes) that you can plug directly into the robot controller. If you've got a standard radio-control system, then you can purchase an aftermarket servo switch similar to the IFI Spike that will plug into your receiver from Team Delta and other vendors. You can also make a servo switch with a hobby servo and a single-pole, single-throw (that is, single connection) momentary switch.

Protect your vulnerable backside. Although it is possible to design a hammer so that it may attack both ways, it's usually too confusing for the operator to try and attack from both front and back.

The Launcher

While spinner robots (described later in this chapter) have the edge on damage, the launcher (see Figure 18.9) wins on sheer dynamic performance by literally *throwing* its opponents through the air. Imagine dropping your robot from a height of 5 or 6 feet. For many builders, this isn't really the sort of abuse you think of immediately. Dropped from that height, something usually breaks loose or bends out of position. This is what the launcher robot strategy counts on.

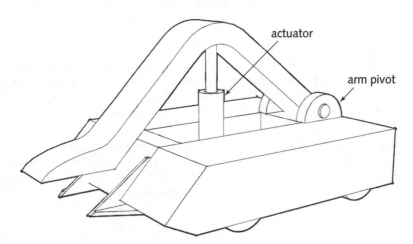

actuator

arm pivot

FIGURE **18.9**: Launcher robot.

What separates a launcher from a lifter is speed and power. You don't need any fancy fingers to catch the edge of the armor. A flat spatula tip will do. It's usually connected to a lever arm that's fired by a pneumatic cylinder. The speed and power requirements are much higher for a launcher than for a lifter because you must toss your opponent, making pneumatics the unanimous choice. This comes at a price, however. The number of shots you'll have may be less than a dozen.

See Appendix C for safety and implementation details.

The highest-powered launchers come from Team Inertia Labs. They use unregulated CO_2, which can reach up to 1,200 psi, and custom modified cylinders to handle this pressure. Using this system requires special preapproval from competition organizers and a demonstrated knowledge of techniques for managing unregulated high-pressure gases and the dangers associated with them. Rookies and even most experienced veterans should avoid doing this until they've had extensive experience with standard regulated systems.

Arm Design

Since getting a high firing speed is more important than a lot of leverage, the cylinder usually fires up in the middle of the arm, between the pivot and the load. Consequently, in order to get a favorable middle position for the pneumatic cylinder, most launchers end up being fairly tall.

Most launchers position the pivot point for the arm near the rear of the robot. Like lifters, most launcher arms fold into the body for protection when they're not being deployed. It's also important to find a way to stop the arm with some sort of tether at the end of its travel to prevent the momentum from damaging the cylinder.

Driving a Launcher

It requires a lot of air to move a large cylinder. You may not have many shots, so you'll have to choose your moments carefully. Your first attempt is the most critical for upsetting your opponent's game. Be patient and hunt for the right time. This strategy is all about maneuvering, so your drive system should be geared for torque and precise positioning, not speed. Make sure to protect your vulnerable backside.

The Spinner

One of the most deadly weapons to emerge in robot combat is the spinning mass. These weapons take advantage of *angular momentum*, which means that when you get the mass up to speed, it will accumulate a large amount of energy and store it, resisting sudden changes in motion. When something causes the mass to stop (like hitting your opponent), a huge amount of energy is released.

This is a knockout robot, usually inflicting enough damage to incapacitate its opponents, sending parts flying in all directions. When you've got a spinner robot, your most dangerous opponent is usually yourself. Avoiding self-destruction is a very big challenge for a spinner.

Usually these robots are driven by powerful electric motors (see the "Switching Big Electric Motors with Solenoids" sidebar) or internal combustion gasoline engines.

The Shape of the Mass

The shape for the spinning mass can take one of four main forms: bar, disc, full-body (dome), or drum. The distribution of the mass establishes the *rotational inertia*, which ultimately determines (along with the speed) how much energy is released into the target. The important thing (regardless of the overall shape) is that the mass is concentrated at the edges of the shape, which increases the inertia, and thus the ultimate hitting power. Unfortunately, this also increases the time it takes to get up to full speed.

For *full-body* spinners that have domes (as shown in Figure 18.10), any additional weights or blades should be mounted low to keep the center of mass close to the ground, which will give you more stability. Hanging weights can produce unpredictable results, because balance is a major consideration with such a large spinning mass.

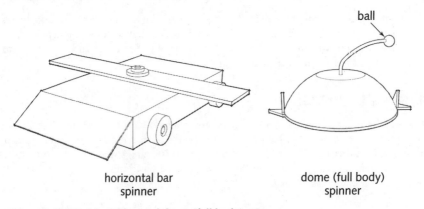

horizontal bar
spinner

dome (full body)
spinner

ball

Figure **18.10: Horizontal bar and dome (full body) spinners.**

For bar and disc horizontal spinners, where the whole body doesn't spin, the height of the spinning mass is an issue. You want to get it down as low as possible so that you're guaranteed to impact the other robot instead of harmlessly passing overhead. However, because of motor placement and other design issues, it's usually difficult to get the bar or disc down low while still keeping the spin axis at the center of the robot. One solution is to offset the center of rotation so that the bar or disc spins in front of the robot, as shown in Figure 18.11. Unfortunately, this leaves a blind spot in the rear, and means that your weapon becomes directional. Another variation is to angle the bar or disc so that it tips close to the ground in front of the robot. Note that placement of the spinning mass relative to the center of the robot will also affect driveability, as discussed later.

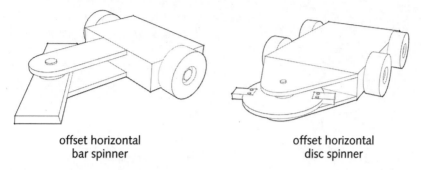

offset horizontal
bar spinner

offset horizontal
disc spinner

FIGURE **18.11: Offset horizontal bar and disc spinners.**

For drum robots (see Figure 18.12), the spinning mass is concentrated along the outside of a wide drum, usually with horizontal bars or little teeth to catch opponents and pop them up into the air.

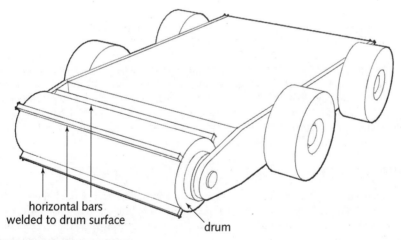

horizontal bars
welded to drum surface

drum

FIGURE **18.12: Drum robots.**

Horizontal or Vertical?

Another major choice in your design is how you want the mass to spin: horizontal or vertical. Although they share the same theme of delivering a large hit to the opponent, the horizontal spinner seeks only to bludgeon, while the vertical spinner (shown in Figure 18.13) can bludgeon as well as flip the opponent. Usually, dome spinners are horizontal, while bar and disc spinners have been quite successful in either configuration. Drum spinners have a horizontal spin axis, running the drum so that the teeth spin upward.

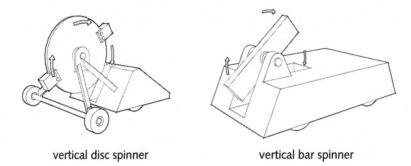

vertical disc spinner   vertical bar spinner

FIGURE **18.13: Vertical disc and bar spinners.**

Special Bearing Considerations

Because spinners need to rotate with as little energy wasted in friction as possible, you're going to need a bearing. However, since the impact energy can be so high, it needs to be better than an average bearing.

Special bearings called *tapered roller bearings* can handle both radial and axial loads, as well as high shock loads. (Standard bearings are only really good at handling radial loads.)

Note Using bearings greatly decreases the friction that the spinning mass experiences. As a result, a mass rotating at maximum speed could take a long time to spin down at the end of a match. Therefore, you should pay close attention to the required spindown time for your competition. This is closely regulated in some competitions, and could be as short as 30 seconds. This may require an additional circuit to actively reduce the speed when the robot shuts down.

Limiting Torque

As I mentioned earlier, self-destruction is a big issue for spinners because the huge amount of energy that's released into the opponent is also experienced by the spinner. That's why it's a good idea to build in a way to try and minimize the amount of force that's transmitted back to the robot. You can use a *torque limiter* or *slip clutch*, which transmits power up to a point before releasing and spinning free. You want to set the release torque just high enough to make sure that you can get the mass going quickly without slipping.

You can also take advantage of the characteristics of V-belts, which can slip a little in a large impact because they use friction to transmit power. This slippage prevents the system from absorbing all of the energy. Note that belts require *sheaves*, shown in Figure 18.14, which are the equivalent of sprockets in a roller chain system, to transfer power. Also, you will have to provide some sort of means to tension the belt, just as with sprockets and chain.

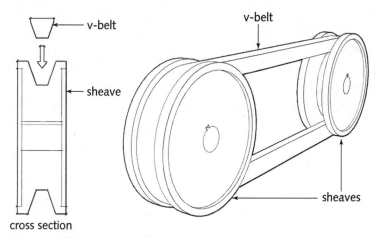

FIGURE 18.14: V-belt and sheave.

Gyroscopic Effects on Steering

Consider the gyroscopic effect of a spinning top. The gyroscopic force wants to hold the top upright against gravity. This is because of a side effect of angular momentum called *precession*. Spinner robots experience the same sort of gyroscopic forces, which have an effect on the way the robot steers. It isn't so much of a problem for horizontal spinners, because when you move in the plane of rotation, the effect of precession isn't nearly as pronounced—they will tend to drift in the direction of rotation (see the following note). Vertical spinners are the ones that experience the full impact of this phenomenon, because they turn perpendicular to the spin axis, which causes them to lean as they turn, as shown in Figure 18.15. One way that they can compensate is to have a really wide base, or a pair of outrigger wheels. Although drum robots have the same spin direction and turn axis as vertical spinners, they don't experience the effect as much because the drum rides much closer to the ground, which lowers the center of the rotating mass. A lower center of mass means that the twisting (tipping) gyroscopic force has less mechanical advantage on the robot's frame, and is less likely to cause it to topple over.

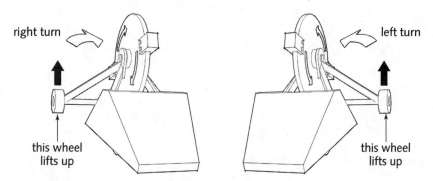

FIGURE 18.15: The effect of precession on a vertical spinner.

Switching Big Electric Motors with Solenoids

To drive a big spinning mass weapon, you need a big motor. Speed controls are great for running moderate-sized electric motors in a proportional way. What if you just need to turn a really big motor on and off? That's where a *solenoid* comes in. A solenoid (shown in Figure 18.16) is a relay that's capable of handling an enormous amount of current for an extended amount of time (at least as long as a match). Usually, they're used in cars and trucks to engage the starter motor, a high-torque, high-current DC motor.

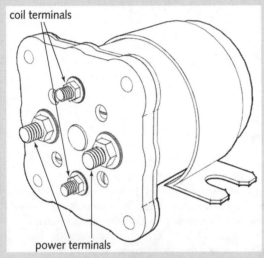
coil terminals

power terminals

FIGURE **18.16: Solenoid for switching high currents.**

Solenoids usually have a pair of terminals for the high-current connections, and a pair of terminals for the low-current *coil* connections. The coil is an electromechanical device that when energized, actually makes the high-current connection by physically moving a conductor into contact with another conductor. The coil has an activation voltage of 12 or 24 volts that tells you what you need to apply to charge the coil and activate the solenoid. Usually, you use a servo switch (as described earlier) to switch the voltage to the solenoid.

You should make sure that the solenoid that you select is battle-proven, and not subject to contact bounce, which may cause the connection to make or break during a large shock load.

Note Some competitors use radio-control helicopter *gyros* to help compensate for the drift in the direction of rotation with horizontal spinners. These go in between the receiver and the speed controls, and help to keep the robot moving straight. Be warned, however, that some gyros do not operate when upside down, and that you should trim your controls (as discussed in Chapter 19, "Troubleshooting") after the addition of the gyro.

Driving a Spinner

Effectively driving a horizontal spinner doesn't require as much timing and accuracy as many other weapons that need to be triggered, such as a lifter, clamper, or hammer. Since the weapon is always active, all you have to do is make sure that it's pointed at the target. In fact, horizontal dome spinners can simply wait for the opponent to come to them. All spinners should be equipped with a fast enough drive so that they can evade an opponent and buy themselves more time to spin up to full speed. You will also need a reliable way of telling which direction you're facing with dome spinners, since the whole body is moving. Some competitors use flags or bright lights to indicate which direction is forward.

Get out into the middle of the arena, since you will probably end up flying a fair distance upon impact, and you don't want to be thrown up against any walls.

Note There is a series of proportional speed controls designed by IFIrobotics specifically to handle spinning weapons. Based on the popular Victor line, the Victor SC (Spin Controller) dedicates all of its circuitry to spinning in a single direction, while still maintaining the same size and weight as a regular Victor. It also features a soft startup (so that you don't waste a lot of battery power getting up to speed) and an automatic ramp-down in the absence of a valid receiver signal, which is handy in meeting competition spin-down requirements. Since you have proportional control of the speed, you don't have to run at full speed all the time (as with a solenoid switch), which is useful if your robot becomes unstable at very high speeds. The Victor 883 SC handles 90 amps and costs $170, while the 885 handles 150 amps and costs $220.

Internal Combustion Engines

Another viable option for powering spinner weapons is the internal combustion engine (ICE). Like all other combat robot choices, ICEs have their pros and cons. Knowing the differences between power sources will let you choose the appropriate one for your application.

ICEs really shine when it comes to power-to-weight ratio. An ICE and an electric motor of the same weight can have comparable horsepower ratings, but the fuel to feed each for 3-5 minutes is vastly different. For the same amount of run time, the ICE uses a few ounces of gas, while the electric motor requires pounds of batteries. In addition, as those batteries drain, the motor loses peak power. An ICE runs just as strong on an almost empty tank as it does on a full one.

Available torque is another consideration. An ICE's torque increases with RPM (to a point). An electric motor has max torque at 0 rpm and loses torque as RPM increases. So for low-end grunt, you may want the electric motor; while for high-end power, the engine would be a better choice.

Continued

Continued

An ICE's biggest drawback is complexity. An electric motor only needs a solenoid switch or speed control to operate it. An ICE has more moving parts and requires more controls to operate it. The electric motor can be controlled using only one radio channel, but an ICE may require four: You need to control the throttle (1), choke (2), a kill switch (3), and I highly recommend having an onboard starter (4).

Servos can handle the throttle and choke. Each servo moves the appropriate butterfly flap in the carburetor, thus controlling engine speed. A servo switch can act as a kill switch by cutting off the ignition system. The kill switch is necessary to stop the engine in an emergency or at the end of the match, because the throttle will only be able to bring the engine down to idle speed.

An onboard starter is necessary for reliability in my mind. Some competitors use the "pull-and-pray" method of starting their engine; either as a pull start, or with an external starter, and then hope it keeps running. I personally feel this is too risky. An onboard starter does add weight and complexity, but it makes the match startup procedure easier to do as well as allows you to restart the engine in case it konks out in the middle of a match. After all, once the match has started, you can't run out into the arena to restart your robot! I must note that I built four starter setups before I made one that worked consistently!

A robot must be motionless at the start of a match, yet an ICE runs constantly; therefore a centrifugal clutch becomes necessary. A centrifugal clutch allows the engine to idle freely, and doesn't engage the clutch until a certain RPM is reached. Don't worry though, centrifugal clutches work automatically and don't require a separate radio channel.

A few other issues for ICEs are vibration, electromagnetic (EM) interference (from the spark plug), and sound limits. Vibration can be dealt with using dampeners and shock isolators. For EM interference, the use of resistor-type spark plugs and positioning electronics away from the spark system solves most problems. Sound limits are determined by the event coordinators and can usually be met by using the stock muffler. Crowds love the noise of an engine—event coordinators worry about hearing loss.

 Caution Remember that handling fuel is a potentially dangerous fire hazard! The last thing you want is for your pride and joy to go up in flames, or worse yet, a person to get burned. So, it is in your best interest to secure and armor your fuel tank and lines.

As you can see, there are a lot of factors to consider when it comes to power sources. A drive system using an internal combustion engine could prove too complicated for most builders, but an ICE offers tremendous power for your robot's weapon. So, for your power needs, don't be afraid to consider the internal combustion engine.

—John Duncan, builder of the ICE-powered spinner robot Claymore.

The Thwack Bot

Thwack bots (see Figure 18.17) are generally two-wheeled robots that have a long arm with a mass on the end that serves as the weapon. As the robot rotates in a circle, the mass accelerates like a spinner. While driving, the arm drags either in front or back, but when they spin, the arm usually lifts up due to centrifugal force. Because of the nature of this design, most thwack bots are completely invertible.

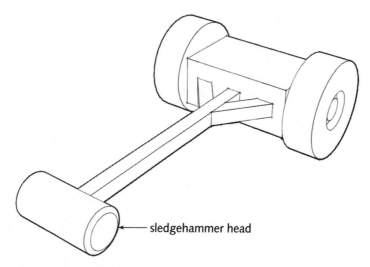

sledgehammer head

FIGURE **18.17: Standard thwack bot.**

Horizontal or Overhead?

Most thwack bots spin horizontally to build rotational momentum. Some, however, use an overhead approach to the attack (see Figure 18.18). Much like a hammer, they bring the weapon crashing down onto their opponents. The actual mechanism is unique because they build linear momentum by rushing at the opponent, but then stop just before impact, and slam their drives into reverse. The momentum continues through the arm, while the wheels run in reverse, and swings the entire body around on the wheel axis.

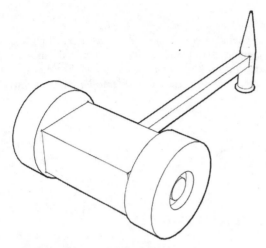

FIGURE **18.18: Overhead thwack bot.**

Head Types

Your choice of what type of head to put at the end of the arm will depend upon what type of damage you want to inflict. A sledgehammer will give you bludgeoning power, for shaking something loose internally or bending frame components. This is a robust solution and should hold up well in battle, provided that the head is well mounted. A spike will give you the possibility of penetration and damaging internal components. A longer spike buys you the possibility of deeper penetration, but will also be more subject to bending. Finally, a wedge tip can give you the ability to pop opponents up and possibly onto their backs when used horizontally. Make sure that the wedge is double-sided (a parallelogram, as seen in Figure 18.19), so that it can be used even if the robot is inverted.

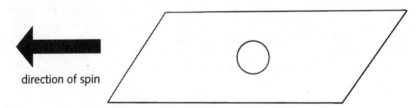

direction of spin

FIGURE **18.19: Parallelogram head profile.**

Driving a Thwack Bot

Most thwack bots operate in either driving/positioning mode, or stationary spinning mode. It's impossible to do both at the same time without some help. Some competitors have been developing controllers that will allow the robot to move in a definite direction while spinning.

Generally, a thwack bot's strategy should be to get out to the center of the arena, away from the walls, so that they have enough space to spin up and swing the arm. From then on, it's generally a game of move, spin, move, spin.

Note   If a thwack bot becomes inverted, it can benefit from an inverted drive control as described in Appendix A.

Cutting Weapons

Traditionally, cutting weapons have produced sparks, but don't really carve into opponents to cause any real damage. It's a difficult task because a saw (or drill for that matter) needs to stay in one place to really get anywhere. In the arena, things happen too fast, and unless you can latch on to an opponent, all you'll end up doing is causing cosmetic damage on the outside. However, using a saw blade with the teeth spinning in the *up* direction actually works as a variation of the vertical spinner. The saw blade usually grabs the opponent and tosses them up. If you want to use this strategy, then a saw blade with fewer, larger teeth should be used, which will make it easier to catch an opponent.

Wrapping Up

This is a snapshot of what's been tried so far. By no means is it a road map or a blueprint. Some of the most successful robots out there are variations of an original idea. Also, some favorites have combined old ideas to form something new and fresh. Don't be afraid to mix and match as you see fit.

The important thing to keep in mind is your weight budget. Most would-be robot designers call out for "a flipper on every side" for example, without considering the amount of weight that each weapon requires. While this would be possible, it would divide the amount of weight that you can spend on making that one weapon more effective, which is more valuable to me as a competitor.

You never know what's going to work in the arena. You've just got to take a guess. Competitors change all the time. That's part of what makes this sport so fun: seeing what different people will come up with next, and how your idea works with theirs.

Troubleshooting

Y ou've got the robot basically together. All the parts are in place, and it all seems like it should work, but something's not quite right. Discussed in this chapter are some common problems, including being overweight, having a drive response that's difficult (or impossible) to control, and various electrical/radio problems. I'll list some techniques for putting your robot on a quick weight-loss program, which includes thinning armor, changing armor types, and possibly even changing battery types. Here, you'll find fixes for uncontrollable drive such as adding an exponential response to your steering and changing gear ratios. Finally, I explore electrical/radio problems that frequently occur, and suggest solutions like switching to a Dean's antenna and reducing electrical noise within the robot.

Some of these solutions may seem a bit drastic, but your choices become limited at this late stage of the game. Throughout the rest of the book, I've tried to stress techniques to avoid these problems in the first place, but sometimes in the design process, you have to make compromises and adjustments that cause other unanticipated problems, such as the ones addressed in this chapter. All robot builders have experienced at least one of these problems at some time or another, so don't worry, it's normal.

Weight Problems

Almost everyone experiences the horror of having a robot that's just over its target weight. The scales don't lie, and in most competitions, there's no grace on weight limits. What do you do? Here are some of my tips for weight loss:

 Caution — Before making any drastic decisions, try calculating the amount of weight you stand to lose by each procedure. Then, make your choice and go for it.

Make the Armor or Baseplate Thinner

In losing robot weight, you've got to think in terms of volume. Look at the square footage on your robot. You can get the most bang for your buck by making a large plate thinner. Following are two different strategies applied to the same piece of armor for comparison. First, let's say that you have a piece of 1/4-inch aluminum for top armor that measures 24" × 20".

$Volume= L \times W \times D = 24 \ in. \times 20 \ in. \times 0.25 \ in. = 120 \ in.^{3}$

$Weight= volume \times density= 120 \ in.^{3} \times 0.1 \ lbs/in.^{3} = 12 \ lbs.$

So, the overall weight of this piece is 12 pounds. The first thing to try out is a thinner plate. Shave off 1/16 inch make the thickness 3/16 inch and check the result.

$Volume= L \times W \times D = 24 \ in. \times 20 \ in. \times 0.1875 \ in. = 90 \ in.^{3}$

$Weight= volume \times density= 90 \ in.^{3} \times 0.1 \ lbs/in.^{3} = 9 \ lbs.$

So, just by switching to a slightly thinner armor, you can save 3 pounds. That's great. Next, try switching materials from aluminum to Lexan for the original thickness and check this result.

$Volume= L \times W \times D = 24 \ in. \times 20 \ in. \times 0.25 \ in. = 120 \ in.^{3}$

$Weight= volume \times density= 120 \ in.^{3} \times 0.043 \ lbs/in.^{3} = 5.16 \ lbs.$

Switching to Lexan saved over half the weight. Of course, the tensile strength isn't as high, and you should also factor in to your decision what weight class you're competing in, and how strong the weapons are.

Drilling Lightening Holes

Just for kicks, compare the good old lightening hole method, using a 1-inch diameter hole saw. This is popular in the pits, because it *seems* like a good idea for aluminum, and it's pretty quick to drill a big hole, but you'll see why it's likely to take the most effort.

$Volume= \pi r^{2} L = \dfrac{\pi d^{2}}{4} L = \dfrac{(3.1416)(1 \ in.)^{2}}{4} = (0.25 \ in.) = 0.196 \ in.^{3}$

$Weight= volume \times density= 0.196 \ in.^{3} \times 0.1 \ lbs/in.^{3} = 0.02 \ lbs.$

So, in order to lose 3 pounds as in the first example, you would have to drill 150 holes in your 24" × 20" × 1/4" top armor plate. You can see the futility of turning this item into swiss cheese. Your chances of success with the lightening hole method rise as the material's thickness and density get larger (as in a thick steel plate), but always do a little calculation before diving in, since you could end up weakening your robot a lot more than if you'd just replaced the plate with something slightly thinner.

Attack Steel

Steel is the heaviest material you've got and it can make an immediate difference. How far are the axles extending into the wheel hubs? Are they hanging over the bearing blocks? If your axles are fixed and you don't need keyed shaft, you may consider going to hollow 4130 axles. (I would only recommend doing this for double-supported shafts — no overhung loads.) Are there any other items made out of steel that could be replaced with aluminum?

Tip

A hole saw (see Figure 19.1) works surprisingly well on mild steel. First, however, you need two things: slow speeds (under 100 RPM) and lubrication. Automotive motor oil will work fine. You should make sure that you immobilize whatever you're cutting, and that the part is rigid (as little vibration as possible) and secure. You should cut slowly, but with a lot of downward pressure into the part. It will smoke a bit, but that's okay. It will take awhile, but that's to be expected. Double-check Chapter 7, "Drilling and Tapping Holes," for tips on safe setups for drilling.

FIGURE **19.1: Hole saw for drilling steel.**

Change Battery Types

By changing battery technologies, you may be able to maintain performance while losing precious weight. I know this may sound like a pretty drastic measure, but it resulted in quite a bit of weight savings for my robot Deadblow. Generally, you will gain the most benefit by changing from SLA (sealed lead acid) batteries to nickel cadmium (NiCad), but you can also change from NiCads to nickel metal hydride (NiMH).

Of course, there are a few things that you have to consider. First of all, you can't just copy the number of amp-hours when you switch to NiCads because of the difference in de-rating factors (as described in Chapter 15, "Choosing Batteries"). Remember that the actual amp-hour capacity of an SLA battery is 30 to 40 percent of the manufacturer's rating, as opposed to 85 to 90 percent for NiCad and NiMH packs. The best thing to do is to go back to your initial calculation of the required current and re-select the NiCad packs based on those calculations. Also bear in mind that the NiCad packs don't put out as much instantaneous current as SLA batteries, and are generally available in much smaller amp-hour ratings, so you may have to put several NiCad packs in parallel to make sure that you have enough battery capacity. The NiMH packs are a little lighter than NiCads, and have the same de-rating factor, but are not capable of the same instantaneous current output. That's why your best performance tradeoff is switching from SLA batteries to NiCad packs.

By far, the biggest limiting factor in this upgrade is cost. The best NiCad packs available for robot combat are custom made, and usually cost upwards of $100 each. You could use packs made for radio-controlled cars, but they aren't available in convenient 12-volt packs. You could make your own packs out of individual cells, but that is a tremendous pain in the ass. In addition to the expense of the battery packs themselves, you will also incur the cost of new chargers, since they're not interchangeable with those made for charging SLA batteries.

For a review of battery terminology and technologies, refer to Chapter 15.

Your Drive Is Uncontrollable

This is usually a symptom of two-wheeled robots, and lightweights in particular. I've seen countless matches where two small robots are zooming around the arena at high speed, but *never even touch*. Besides being incredibly boring to watch, it's really difficult for the judges to score such a match. The whole point of having a weapon is to bring it into contact with your opponent.

Add Exponential Response to Your Stick

If you've got a computer radio (one that has programming capability and, usually, an LCD screen), then you're in luck. There are some easy things you can do to make the left-right turning response less sensitive on your stick.

Normally, the control stick for an R/C channel has a proportional linear response. This means that each movement of the stick produces an equivalent change in the output. This is fine for most things, but if you're having trouble driving straight, you can change the side-to-side (turning) response so that it's less sensitive ("touchy"). Oh, you'll still turn, but what you're doing is adjusting things so that turning requires more of a move from the stick. This makes it easier for you to make minor adjustments to your path without swerving all over the arena. Figure 19.2 shows the difference between linear and exponential responses.

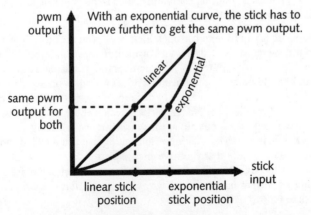

FIGURE 19.2: Linear and exponential response curves compared.

Note This is only applicable to mixed channels (whether mixed in the radio with elevon or PMIX, or with an external mixing unit) where one stick controls both sides of the robot's drive.

Cross-Reference For more radio-control programming tweaks and techniques, see Appendix A, "Advanced R/C Programming."

Change Your Gear Ratio

Your drivetrain may be geared for too much speed and not enough torque. You may also experience the robot coasting to a stop and drifting after you've told it to stop. Gearing your drivetrain so that the output RPM is lower may help you a lot in this situation. Of course, you've got to pay a price, and that is top speed. But consider this: If speed is a big part of your weapon, then you've got to line up on a target before you can hit it. If you can't accurately aim for your opponent, you'll be careening across the arena at top speed right into a wall. Don't feel bad about sacrificing the speed; you'll earn it back in maneuverability.

Cross-Reference For a review of gear ratios, check out Chapter 9, "Selecting Drive Motors."

Adjust Your Weight Distribution

Try moving the battery forward or backwards in the frame, as demonstrated in Figure 19.3. After trying a few different positions, you may find that there is a spot where the balance helps out your steering significantly. This has to do with how the majority of the weight is carried by the wheels.

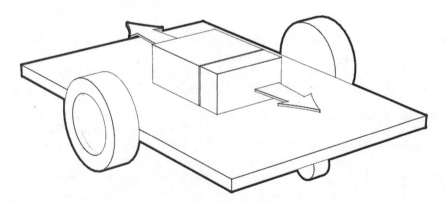

FIGURE **19.3: Shifting the battery forward and back.**

Exponential Programming Example for Futaba

Here's how to program exponential rates on a Futaba 9C series radio. Check your manual to see how exponential is programmed on your radio. It should be similar, but not exactly the same.

1. Press and hold the MODE button to activate programming mode. Most radios make it harder to activate the programming mode so that you don't accidentally activate it while flying (or driving, or whatever).

2. Use the dial to highlight D/R, EXP and press it to select this option. *D/R* stands for *dual rate*. By activating the dual rate function, you can assign a different response (rate) to a channel, and exponential is one of the options.

3. Use the cursor keys to move up to the CH line. This will allow you to select the channel to assign the exponential response to. The channel that you want to adjust is channel 1 (aileron). Make sure that channel 1 is highlighted.

4. Use the cursor keys to move to the SW line. This will allow you to select the switch that you use to activate the exponential response. Rotate the dial until switch D is highlighted. When switch D is highlighted, you can flip the switch up and down, and the display will say AILE (UP) or AILE (DN) according to the position. For our purposes, let's assume that UP = ON, and set switch D to the UP position. The display should read AILE (UP).

5. Use the cursor keys to move to the D/R line. You will notice that there are two columns. One is under an arrow pointing left and the other is under an arrow pointing right. The reason is that you can set a different rate for each side of neutral. For example, if you put a dual rate on the elevator channel, which controls the forward/back response on the robot in a mixed situation, you could make the robot drive slower in reverse. (I'm not sure why you would want to do this—it's just an example of what you could do with different rates on either side of the stick.) In the current case, you want to make sure that both rates are 100%, or else your robot will turn more slowly in one direction that the other. You can ensure this by moving the joystick to the left (you should see a line in the display move also) and letting it come to rest, which will highlight the left column. Turn the dial until it says 100%, if it doesn't already. Move the joystick to the right and repeat the procedure.

6. Now that you've got all the initial items set up, it's on to the actual exponential programming. Just like the dual rate, you should make sure to set both sides of the exponential to the same value. That way, the robot will handle the same way whether you're turning right or left. Use the cursor keys to move down to the EXP line. Movie the joystick to the left and let it come to rest, highlighting the left column. Turn the dial counterclockwise until it says –100%. You should see the response line in the graph start to bow more like Figure 19.2 as you adjust the EXP rate. Move the joystick to the right and repeat the procedure.

Note You want a *negative* exponential response because you want to decrease the sensitivity near neutral.

7. Now, if you flip switch D down, you can see the response pop back down to linear (0% exponential). Flipping the switch back up should reactivate the exponential. Hit the END key twice to exit from programming mode.

8. Try driving with the exponential OFF (switch D down) for a little while and then flip switch D up to activate the exponential response. Setting the level of sensitivity at neutral is a matter of personal preference, so feel free to experiment with the amount of exponential that you apply.

Change a Caster to a Drag Wheel

This trick has helped out a few builders before. Try replacing your regular caster with a fixed (nonswivel type) caster, as shown in Figure 19.4. You won't lose any speed due to friction when you're traveling straight, since a fixed caster will roll just fine forward and back. Since the wheel doesn't swing around when you turn like a swivel caster, it has increased side-to-side resistance that will help keep you going straight.

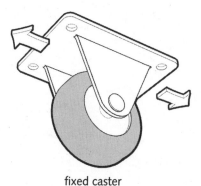
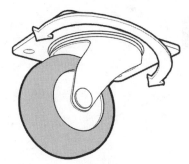

fixed caster

swivel caster

FIGURE **19.4: Fixed and Swivel casters.**

Practice

It seems like an obvious one, but I have to mention it. If you've done all that you can with the above steps and you've still got a robot that's snaking all over the place, then all you've got left is to accept that and practice. You've got to drive your robot and then try driving it some more. Nothing can replace time when you're learning to drive. The key is to keep at it. Eventually, it will become second nature.

Refer to Chapter 17, "The First Test Drive," for some great tips and exercises for developing your driving skills.

Electrical Problems

Well, the robot works great until you put the lid on. You've got no response, or the robot is twitching, and barely controllable. What went wrong? I have to admit that I suck at the actual physics of antenna theory (not one of my favorite subjects in engineering). But here are some proven techniques for the robot builder.

Managing (not Mangling) Your Antenna

If your outer armor is all metal, as many combat robots are, then you need to make sure the antenna is off this metal surface by a minimum of 1/4 inch, as shown in Figure 19.5. If you've got Lexan armor, then you needn't worry about this. You can prop the antenna up with a non-conductive material or make a little holder. It doesn't matter. All that matters is that you keep the antenna away from the armor. Also, make sure that the antenna doesn't cross itself. You can make a tight zig-zag pattern, but no loop should cross another loop.

Don't change the length of the antenna wire. This means *do not* cut it to make it shorter or add wire to extend it. The length of the wire is related to the frequency that it's trying to pick up, and changing that length makes the antenna less efficient at doing its job.

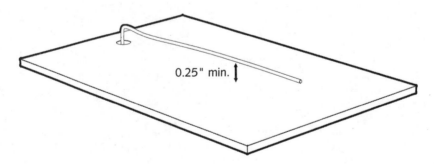

0.25" min.

FIGURE **19.5: Body clearance for antenna.**

Switch to a Deans Antenna (R/C only)

The *Deans antenna* (see Figure 19.6) became the miracle solution for the fighting robot community. It's an antenna that is added to the end of your receiver antenna. It's relatively cheap, durable though easily replaceable, and it works great. You actually cut your receiver antenna (scary, I know, but it works). For best results, you should follow the instructions exactly.

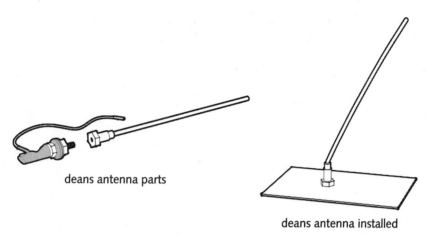

deans antenna parts

deans antenna installed

FIGURE 19.6: Deans antenna.

Separate Power and Signal Leads

Your motors and the wires that supply them are capable of spewing out a lot of electronic noise that can interfere with your radio. All standard radio-control leads are unshielded, meaning they have no metal jacket to help reduce noise. This means that they are big targets for picking up nasty noise signals and confusing your radio, servos, and speed controls.

Tip

Motor noise suppression capacitors can help alleviate the amount of noise that the motors put out. The caps should be 0.1 µF (usually has 105 printed on the cap) nonpolarized and rated for at least two to three times the voltage that you're running the motors at. A ceramic disc cap is small and works great for this application. Figure 19.7 shows a typical configuration, but you may be able to get away with just putting a cap across the motor leads alone. You'll have to try it out and see what works.

Also make sure that all motor and power leads that connect to the speed controls are twisted together. Specifically, you should twist the positive and negative wires of the motor around each other, to form a tight spiral wrap, as shown in Figure 19.8. The same goes for the power leads.

Cross-Reference

Chapter 13, "Choosing Your Control System," has some radio tweaks that might bear reviewing if you've still got radio problems. You can also check out Appendix A for some more tips.

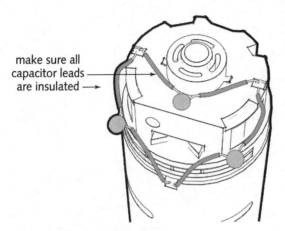

make sure all
capacitor leads
are insulated →

FIGURE **19.7: Motor noise suppression capacitors.**

FIGURE **19.8: Wrapping the leads.**

The Drive Is Cutting In and Out

When you purchase an IFI speed control, it usually comes with a complementary 20A thermal self-resetting breaker. If you've put it inline with the speed control input, then you should remove it from the system. In this sport, it's all or nothing.

Replace your drive battery with a known good fresh battery. Sometimes, if the battery voltage is too low, it will drop below the voltage threshold to keep the electronics inside of the speed controls running, and they'll shut down momentarily until the voltage comes back up.

Replace the receiver battery with a known good fresh battery. You may have been using the same receiver battery for quite a while. It's easy to forget about this. If the battery's really low, the signal may be intermittent, which will cause the speed controls to lose track of their commands.

Wrapping Up

The most common problems combat robot builders have are outlined in this chapter. If what you've got isn't listed here, check out the Web sites listed in Appendix D, "Online Resources." The competitors' Web sites are especially helpful here.

Going to a Competition

This is it! All those weeks (and possibly months) of work have come down to this event. Are you ready? Your tasks related to a competition can be divided roughly into before, during, and after the event. Before the event, you've got to do a little research to find an event you want to enter and make sure to fill out all the registration paperwork to get yourself into the competition. You can't just show up unannounced at the door with a robot under your arm — organizers like to know who's coming well ahead of time. By the time competition rolls around, you'll probably be really, really tired. Don't worry, it's normal. You'll have to make a checklist of tools and supplies so that you don't forget anything.

When you first get to the event, it will probably be pretty chaotic, and you'll have to get your pit table set up, make sure your robot passes safety, and find out when your fights are. During the fight, your mind will probably be racing. I'll try and explain some of the things that are important to focus on, like your driving, and give you a few tips that will help you get prepared, like setting up an all-important preflight checklist.

Before the Competition

There are a lot of things you've got to do in order to get the chance to compete, and building a robot is just *one* of them. By taking care of a few preparations ahead of time, you can make things a lot easier on yourself at the competition.

Finding a Competition

BattleBots, Robot Wars, and Robotica have hosted the major televised robot combat events in the past. Check the respective Web sites for rules and event information. Check Appendix D, "Online Resources," for the addresses of these Web sites and others.

A national organization called the RFL (Robot Fighting League) was recently formed to help bring together all the regional competitions and put forth a unified, but *extensible* rule set. (Extensible means that the rules remain basically the same, but certain parts can be adapted to the specific arena's safety capabilities, since not every arena is the same.) This is the best source of information for local fighting. Events are constantly being added, and you may find one nearby.

Getting Your Paperwork Done

Every competition has paperwork. It's inevitable. Applications, tax forms, and safety releases are just a few of the documents you may be required to fill out. Different competitions require different documents, so you should make sure that you're fully aware of what you need to turn in. For example, you may have to sign a release form to have you and your robot appear on TV, and where royalties are involved, there may be tax forms that you've got to fill out. You may also have to provide proof (in the form of a picture) that you have more than just a *vapor-bot* (a robot that exists only on paper or in the computer). There are deadlines for the submission of all these materials, so make sure to get your entry fees and paperwork in early. You've spent all this time and money getting your robot together; don't let a minor thing like submitting your forms on time be the deciding factor that prevents you from playing.

Finishing Your Robot

The best-case scenario: The robot is pretty much done before you go to competition. You've tested the weapons system on several different targets and had a few days of driving practice. You have a short list of things to tidy up. You followed the safety requirements to the letter, and you're a few pounds underweight. You're pretty well rested, though a little nervous.

The worst-case scenario: You're on your third all-nighter. There is a long list of things to take care of and instead of getting shorter as you complete tasks, it just seems to keep getting longer. You've long abandoned trying to make fancy machined parts and are doing most things with the jigsaw or bandsaw. You're pretty sure you can make safety, though there are a few things that might be a little questionable. You weighed the robot a week ago, but a lot has happened between then and now. Nothing has been tested, short of your first "yeah, it works" test weeks ago. Your first time driving the robot will be in the arena tomorrow afternoon.

The reality: You should be somewhere in between, hopefully a lot closer to the best case than the worst case. Let it be said: Try your best to finish the robot ahead of time. Give yourself artificial deadlines. Don't leave it to the last minute. Trust me. I know.

Make a Checklist of Tools and Supplies

You can't bring the shop with you. Some have tried, but unless you've got an 18-wheeler with a machine shop already built in, it's not going to happen. Instead, make a list of all the *essential* tools and supplies that you need. These will be the items that are difficult or impossible to get on the road.

You should pack as lightly as possible, since you won't have much room in the pits (see the description of pit conditions in the next section). For instance, don't pack the bandsaw, but do pack the jigsaw. It's the tool of choice in the pits because you can change blades easily, it's compact, and the precision is good enough for a pit repair.

Cross-Reference
Appendix F, "Tables and Charts," has a sample list of tools and supplies for your travel kit.

Special equipment may not be available to you, so you'll have to bring it or find a local source ahead of time. Unless it's a big competition, you most likely will not have access to a welder.

Finishing the Robot (My Rookie Year)

The Deadblow that you may have seen on TV was actually the second generation of that robot. The first generation Deadblow had a 6.5-pound steel sledgehammer head that swung 180 degrees and retracted into the body. It fought in the first BattleBots competition, and there were no network cameras, no announcer Mark Biero, and a pretty small audience.

I wasn't done. I wasn't even close. Nobody was, except for Carlo Bertocchini (of Biohazard fame). I was up until 3 A.M. the night before trying to drill "quietly" in my hotel room. I'd stuffed towels under the door to muffle the sound, and left a trail of aluminum chips all over the room. I'm not sure what housekeeping might have thought I was doing in there, but I didn't have time to care. It was my third or fourth night of being up until 3 or 4 A.M., but I had so much adrenaline flowing through my system that I just kept working.

When I arrived at the competition, Deadblow had no main power switch, and I didn't have any place for the batteries inside the frame, so I had to hang them on the outside. Some friends at work pitched in and cut the aluminum for me, which I brought to the competition, intending to assemble and mount the external battery boxes in the pits. The only problem was that I made a measurement error, and the boxes were 1/32 inch too small on the sides. Running out of time, I resorted to pounding the batteries into the boxes. I just left them in there the whole competition, because I would have had to disassemble the boxes to get them out.

I finally made safety and weight on the second day, after having added the master power switch. I was exhausted and elated all at the same time. It didn't really matter if I won or lost. At least I could play! Luckily, Deadblow went on to win the middleweight rumble that year, but I won't forget how many little things could have gone wrong (and *did* go wrong) at the last minute. After that, I vowed to always finish ahead of time, whatever it takes.

Note

Fortunately, you can count on at least a local hardware store, if not a big chain somewhere in the area.

Welding supply shops have ready supplies of N_2 and CO_2. Hopefully, the event coordinators will have a listing of local shops, or you can ask around in the pits. Scuba supply shops can fill scuba tanks, but usually only for those who have dive cards.

Loading all your stuff into the pits can be a real pain. Some competitions offer free dollies that you can borrow. I like to bring my own hand truck. It doubles as the carrier for the robot, since the distance from the pits to the arena can be quite far, and you can't drive the robot outside of the arena.

Tip

What you really want is a convertible hand truck with big, fat pneumatic tires (see Figure 20.1). Big fat tires help you easily get over the numerous cables and bumps between you and the arena. Dollies have tiny little caster wheels that tend to get stuck in cracks and have a low load capacity, which means that they get damaged easily, and don't swivel very well after that. You'll be fighting the cheap dolly throughout the competition (instead of other robots). Spend a little more for the big wheels. It's worth it.

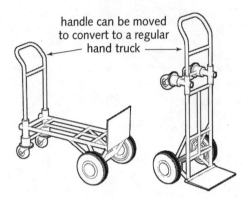

handle can be moved to convert to a regular hand truck

FIGURE 20.1: Convertible hand truck.

At the Competition

The first day is going to be incredibly stressful. You'll have to register, find your pit table, unpack your robot, and get your safety/technical inspection. And that all assumes that you've already finished the robot.

What Will the Pits Be Like?

It's going to be crowded and noisy. You never have enough space, so you've got to be economical. There will be people all around you who are probably just as busy as you are. Keep the aisles clear for traffic. There will be robots moving to and from pit tables all the time. Your stuff should be on or under your pit table, not blocking the aisle.

Environmental conditions may be difficult in the pits. You may not have enough light, and you can't always count on being inside, which means it may be either very hot or very cold, and sometimes both.

Set up Your Pit Table

How big will it be? Not big enough. Usually 36" × 60" is all that you get. (Then again, if you've been building the robot on the kitchen table, this may be an improvement.) Will you have electricity? Yes, but you will have to bring a power strip and an extension cord or two. Plug the extension cord into the outlet, and then run it to your pit table, where you plug in the power strip. (The second extension cord is so that you can run power tools, such as a jigsaw or a grinder, a little bit farther away from the power strip and a little bit closer to the robot.) This way, you can use a single outlet, which is sometimes all you'll get, but should be all you need. You will be responsible for your own valuables, so keep them out of sight.

Tip

Set your battery chargers up first. Sometimes batteries will take a while to charge, and you always want fresh batteries charged up and ready to go. If you need to take fully charged batteries off to make room for charging others, first take off the charged batteries and check their voltage with a digital multimeter (DMM). Sometimes chargers will lie. Next, mark the full batteries with a piece of tape over the terminals that says FULL or NEW. That way, when you're in a hurry, and you have to change batteries fast, you won't get confused. Only remove the tape from the terminal once the battery has been installed into the robot and secured, and you're ready to plug it in. Make sure to do it the same way every time.

Stow the tools and supplies that you're not using under the table, and try and keep your work surface clear for the robot. If you brought tools that you find you're not really using, then relegate them to the trunk of your car. Usually, your car isn't too far away, and is close enough to run to if you find you need something after all.

Tip

If you only use a few sizes of Allen wrenches and sockets, separate those out from the rest of the set. If you're in a hurry, the other sizes just get in the way. I like to keep a small tray or box on the table with the most-often-used tools.

Tip

A lazy susan (flat ring bearing), as shown in Figure 20.2, can make working on your robot in the pits much easier. You can spin the robot around to work on one side or another.

Safety Inspection and Weigh-In

The safety inspection is a make-or-break situation for every robot (and robot builder). You must pass in order to compete. If you fail, then you must scramble to fix everything and then get retested. It's not a pretty sight if you fail. There's no use arguing the logic or goodness of a rule with the safety inspector. You had the rules and technical regulations ahead of time. Your reward for following rules is flying through the safety inspection. Your penalty for not following rules is either not playing, or suffering through whatever modifications it will take to bring you up to code.

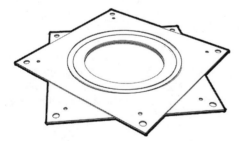

FIGURE 20.2: Lazy susan (ring) bearing.

Note   You should have all of the information and data sheets for all of the pneumatic or hydraulic components used in the robot ready for presentation to the safety inspectors. (It's also useful to have a system diagram similar to the one in Appendix C, "Pneumatics.") They will be checking to make sure that the components you used conform to psi requirements. Don't try to fool them. Many inspectors have been chosen because they can spot an underrated component on sight. If they say you can't compete until you change it out, then it's back to the pits for a lot of scrambling.

The weigh-in is the other stressful part of the process. If you don't make weight, then it's back to the pits to hack away parts of your armor, frame, or weapon. The scales don't lie, and if you're over, then you can't fight.

Cross-Reference   See Chapter 19, "Troubleshooting," for tips on robot weight loss.

Transmitter Impound

Since the radio-control frequency spectrum is divided into a small number of channels (50 for aircraft and only 30 for surface models), competitions have instituted a transmitter impound. Why? With so many competitors, it's guaranteed that two people will have the same frequency. If both people have their transmitters turned on at the same time, they can interfere with each other because they're sending out signals on the same channel, and the robots won't know who to listen to. Out of control robots at an event (or anywhere else, for that matter) are a big no-no. By keeping all the transmitters in a special room under tight control, event coordinators can minimize the risk of having runaway robots. What this means to you is that you've got to relinquish your transmitter (your "baby") as soon as you set foot at the event. You can only see it during certain designated times (such as when you're going to fight, obviously) and not at any other times. I personally found this to be a source of great anxiety at an event. If you want to check something, then you've got to go to transmitter impound and wait in line to get your transmitter back. You may have an hour (or much less, depending on the size of the event) to make your adjustments, and then they send someone after you to get it back.

Smaller events may use a frequency-clip system. The idea is that there is a big board somewhere that has all the available channels listed on it. Next to each channel, there is a little clip (usually a clothespin) that has the channel numbers on it. You check out the clip for testing or fights. You get to keep your transmitter (yay!) but you must not turn it on without the clip. You must also return the clip immediately after you're done.

Tip   If you're using an IFIrobotics system, then most competitions will allow you to keep your control system. This keeps you out of transmitter impound, which is great. (And did I mention that transmitter impound is a real drag?) This allows you to test whenever you feel like it, tweak your code, or just stare at your controller. The price that you pay is that you must plug in the tether so that you don't transmit to the arena.

Know Your Fight Times

You should check the event schedule and know your fight times. Even the best-run events get slightly behind schedule, but don't count on it. You will usually be called early to line up, so be prepared. You want to have everything in order when they call. In addition, if you need to fill your air system, you should take care of this *before* they call you, since things like this take time and are dangerous, and you *don't* want to rush it. You may also need time to check your transmitter out of impound or grab your frequency clip. Panic is not a good prefight emotion.

Tip
Don't scream at the runners who come to ask you to line up. It's not *their* fault if *you're* not ready. They're just doing their job.

You should also copy down the event bracket to see who your potential opponents are. The bracket should be available at all times, although the actual fight times will probably not be posted until the night before, or the morning of each day.

Talk to Other Builders

One of the great secrets of this sport is that fighting is only part of the fun. Meeting other builders and hanging out with them is sometimes just as much fun as fighting. In past interviews, I've described a combat robot event as "a really cool party, and your robot is your ticket to enter." Where else can I sit down with just about anyone in the pits and delve right into a conversation about the merits of neoprene shock absorption for armor, or brushless motor controls? These are people whose lives have *also* revolved around these combat robots for the past number of months or years.

Tip
Talking to other builders and making friends will be handy later on if you have to borrow tools or the odd screw. Personally, I'm not really all that good at approaching people I've never met before and asking for stuff. If I've chatted with them about where to find the doughnuts or how they like their Colson caster wheels, or whatever, then I'm not so intimidated to ask for help when I need it.

This is your best chance to see other robots up close and personal. It's your opportunity to gather ideas and see how other people solved their problems. Don't try to remember everything. Take pictures instead, but make sure to ask the builder *before* you take a picture of his or her robot. If you see something of interest, ask. Usually, builders are proud of the items that make their robots cool and unique.

Tip
It's important to be sensitive to the builder's situation. If they're getting ready for a match or in the middle of fixing a serious problem, then it's not a good time to interrupt. If they're just sitting around, then go ahead. I'm sure they'd love the company. I always do.

Also, if you're approached by a fellow builder, be prepared to talk about your robot and its capabilities. Take the top armor off and show them around the inside, even if you're going to fight them. Doesn't this sound like I'm giving away an advantage? Not really. Consider this: My robot Deadblow has an extremely fast hammer arm, and I've been practicing with it for several years now. If I could aim Deadblow's hammer during a match to hit *exactly* where I wanted it to every time, I'd be a great shot. No, I would be a *phenomenal* shot. But in reality, my opponent's robot will be moving, my robot will be moving, there will be lights and arena hazards, and a million other distractions. Not to mention the amount of adrenaline coursing through my veins causing me to oversteer. Even if it was just the two of us out in the parking lot, and my opponent were standing completely still, it would still be challenging hitting a specific target.

What I'm saying is that the likelihood of an opponent identifying a target and hitting it is more *luck* than it is *skill*. You have nothing to lose. If you choose to keep your armor closed and your robot a secret, then you'll look like a jerk — and losing the respect of your fellow builders is far worse than losing a match.

Leave a Contact Number

Although I'm sure you'd like to hang out in the pits all day and night, eventually, you'll have to leave to get something to eat and perhaps some sleep. You should leave a little sign with your name and contact information. Occasionally, event coordinators will need to get in contact with you and if you're not there, having a contact number clearly displayed enables them to find you.

Access to HPA/N_2 and CO_2

Some competitions have tanks on hand and a qualified professional to help you fill up your robot. Most do not, and you're pretty much left on your own. In this situation, personal safety will be up to you. There should be a "safe" area for filling air tanks. You should find the exact location and then secure your tanks there. Make sure that you put your name and pit number on the tanks.

You should bring eye protection (and full face protection, if you've got it) with you every time you fill your tanks. You should clearly label the controls to your air-fill tanks (turning them on and off) and practice with your equipment. If there's a problem and you need to shut off the tank immediately, having the controls labeled will help out a lot. The more practice you have, the better you'll be under pressure (no pun intended).

Access to Machine Tools

You can't count on having access to a machine shop or welding facilities. For insurance reasons alone, it's simply too much to ask a host to provide those items to the competitors. This means that every repair you'll be making in the pits will be with simple hand tools.

The Spirit of Sportsmanship

It may not look like it on TV, but behind the scenes, sportsmanship plays a big part at these events. It dates back to the days (not so long ago) when this sport was much, much smaller, and builders had to pull together and help each other in order to get their robots working.

Realize that your fellow competitors have suffered and sacrificed just like you, and gone sleepless nights to get to the competition. We all share a common bond, and there exists an unwritten set of rules for pit etiquette based on this bond. I'll attempt to list as many as I can think of, but the basic idea is one of mutual respect between builders and their tools and robots.

- Lend tools to those in need. If you borrow a tool, it's your responsibility to get it back to its owner. Usually it's best to write down the name of the owner's robot somewhere, so that the tool and the owner can be reunited as soon as possible.

- Do the dirty work outside. Some things are a lot noisier and messier than other things. If it's going to be really loud or throw a shower of sparks that might hit someone, do it outside. There's nothing worse than having 100 people simultaneously staring at you in disgust.

- As I mentioned before, talk about your robot and share information freely among competitors. It's still a small enough community that there aren't many "trade secrets." In fact, I've poured all of what might be considered my trade secrets into the pneumatics section of this book.

- You can look, but don't touch any robots, tools, or other equipment without permission. If you want to check out a robot, find its owner and ask if it's okay to take pictures and see the inside, or find out about a unique feature of his or her creation.

The rest of the list pertaining to the actual fight is contained in the next section. Please consider these suggestions, and I think it will enhance your experience at these events.

I do my best to keep this spirit alive by helping rookie teams and anyone else in need. It's important to me, as it should be to you, to earn the respect of your fellow builders.

During the Fight

When you really think about the time that an average combat robot actually spends in combat, it's very, very short compared to the amount of time it takes to build one. You've got to maximize your time in the arena by being prepared.

What Will the Arena Be Like?

Usually, the arena will be an enclosed box, or an area that is behind a Lexan wall. There will be an announcer and an audience. It can be extremely loud and difficult to concentrate with all the music and cheering. You'll be called ahead of time to line up in a "fight line," where contestants will be waiting with their robots to go into the arena. You might be able to carry your lightweight robot to the line, but larger robots should bring their own wheels, as described in the "Tools and Supplies" section of this chapter.

Preflight Checklist

Hopefully, you've been updating your preflight checklist throughout the build. It evolves over time as you begin to see things that you need to check every time you go up for a match (think chain tension). Don't be afraid to revise it. Just make sure you rewrite it so that it's readable even when you're in your most panicked condition. Try not to skip steps and follow the same routine every time. It helps if you say each step out loud as you perform it. (Okay, you can *mumble* it under your breath so the competitor in the pits next to you doesn't think you're crazy.)

If you haven't composed your checklist yet, then try and take a spare (calm) minute to sit down and list all the procedures you need to perform to get ready for a match. Remember, it should include even things you think are *ridiculously* obvious. When you're under the gun, things won't be quite so obvious (or ridiculous) to you.

A sample preflight checklist:

1. Get CO_2 tank filled.
2. Get transmitter from impound.
3. Put robot up on blocks.
4. CO_2 tank valve CLOSED/SAFE.
5. Air purge valve OPEN/SAFE.
6. Check chain tension.
7. Master Power OFF.
8. Receiver Power OFF.
9. Put new drive batteries in robot and secure mounting screws.
10. Plug in drive batteries and tape (or cable tie) connector.
11. Put new receiver battery in robot and secure mounting screws.
12. Plug in new receiver battery and tape connector.
13. Put old drive batteries on charger.
14. Put old receiver battery on charger.
15. Quick check receiver ON/OFF (battery OK?).
16. Quick check drive battery ON/OFF (battery OK?).

17. Check for activation tool (example: 5/16-inch Allen wrench).

18. Check for activation tool spare.

19. Put robot on the cart.

20. Go line up.

21. Relax!

Tip

I actually derive great comfort from the preflight checklist. It eliminates the horrible feeling that you've forgotten something important, and allows you to relax a bit before your match.

Final items like turning on the receiver and master power switches and activating the air system will be handled in the arena, and should be written on the robot's outside armor right around the switch, as described in the next section.

Marking the Case

In addition to your preflight checklist, you should mark on the outside case the sequence of things that need to be turned and/or flipped to activate the robot. You won't have much time in the arena to get things going—maybe less than a minute, so you should do your best to make it quick and easy. Holding up a match because you forgot what to do is bad form. Losing a match because you forgot to activate the air system is even worse.

I use red to activate and green to deactivate. I draw the arrows with red and green markers to help indicate which way to turn something, as shown in Figure 20.3, or to indicate the on and off sides of a switch. You could also use a label printer to clearly indicate ON and OFF.

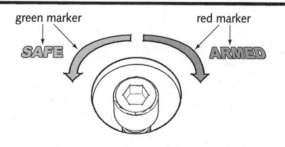

FIGURE **20.3: Marking the case.**

You should be carrying with you two of each tool that you need to activate the robot. This could be an Allen wrench or some other tool. Some competitors carry them around their necks.

Plan Your Strategy

As I mentioned before, you should be prepared to talk about your robot with your opponent. After weeks of laboring on your creation, you should be quite proud of your accomplishment. It's also a nice gesture to go over and say, "Hi." Don't be intimidated. This is how I've met some of the nicest people at these competitions. Really. They're probably just as intimidated as you are. I show them the inside of my robot, and we check out each other's systems. It won't harm you unless you've got something ridiculously mounted or badly designed, in which case, it's only a matter of time before someone breaks it.

What's more important for you as a competitor is to watch your competitors' matches. Observe how *they* drive and what *their* strategies are. See how they've beat their previous opponents, then come up with your own plan on how to fight them. The event bracket, which shows the tree of which robots will be fighting in each round, will give you a clue as to whose matches you should be scouting.

Relax While You Wait

Right before the match, as you load your robot into the arena, your heart will probably be pounding pretty hard. There will be people telling you to wait here, and then go there. There will be lights and music and the sound of the crowd. It's an incredible, overwhelming experience.

Sometimes I chew gum to help calm my nerves before a fight. No matter how many times I've fought, I always get a little nervous before a match. Try to relax. Chat with your opponents in line. You'll be waiting there together for a little while at least—you might as well make the best of it.

Do yourself a favor and don't try to psyche-out your opponent. It just makes you look like an ass, and word will spread in the pits. I haven't seen many people do this, but when you do, it stands out like a sore thumb.

If your opponent requests a postponement of your match, grant it. It doesn't hurt you—you came here to fight, right? It's common good sportsmanship. Besides, I can only think of one instance where a competitor didn't grant a postponement. That kid was booed so badly by the audience, I'm not sure if he'll ever recover. It was pretty ugly.

Focusing on Your Driving

Driving is what wins and loses matches. Even if you've got the most awesome weapon in the world, it won't make a difference unless you can bring that weapon into contact with your opponent. There may be arena hazards and house robots to contend with that are far more destructive than your opponent. Avoiding them with driving skill can make all the difference.

Cross-Reference See Chapter 17, "The First Test Drive," for a review of techniques to improve your driving skill.

Things happen. I have a favorite saying. It goes, "You never know what's going to happen in the BattleBox." What it means is that unexpected things happen all the time, and you can never, ever be certain of the outcome of a match. Luck plays a big part. All you can do is drive your best.

When It's Over

When the match is over, smile and shake hands with your opponent, *no matter who won*. I smile after every match, even if I'm really, really disappointed. If you lose, it's natural to feel upset and disappointed. You may feel that the arena hazards all had your name on them, or the house robots didn't like your looks, or the judges were scoring some other match. That's fine, but whatever you may be feeling, don't get mad at your opponent. He or she was trying to do exactly what you were trying to do: win. The best that you can hope for is to have fought a close and exciting match.

After you're safely back in the pits, share your damage with your opponent. I usually go over and visit my opponent, and check out his or her damage. Then we come back and check out my damage. It's a fun way to relive the excitement, since you really only had about 3 minutes of *actual* combat. Occasionally, if your opponent is feeling generous, you will be given a great honor: a piece of their robot that was severed during battle, a souvenir of your fight. Likewise, if you find a good chunk of your robot that came off, sign and date it, and give it to your opponent. It does wonders for the spirits.

Above all, remember that you're there to have fun. If you're doing this to make money, then you've chosen the wrong sport.

Honor Your Fans

If you happen to be lucky enough to be asked for an autograph from a fan, make time and give him or her a little of yourself, no matter how busy you might be. I remember when a child recognized my robot, saying, "Wow, Deadblow" and then looked up, wide-eyed, and said, "Grant Imahara!" I was floored. At that moment, I was on top of the world. I felt like Michael freakin' Jordan.

Where to Go from Here

Just because you're done with the competition doesn't mean you're done with the project. Gathering and organizing your information now will help you a lot when it comes to the next round of improvements.

Making a List of Upgrades

Right after the competition, you'll be bursting with ideas on how to improve your robot. You may have seen something cool or gotten a tip from a fellow robot builder. Write these ideas down as soon as you get home, while they're still fresh. You probably won't be able to make all your upgrades at once, so it's helpful to have something to refer back to later.

Tip Write all your ideas down, no matter how ridiculous or expensive or weird. You can edit the list later for reality. Who knows? A ridiculous idea now may not seem so strange later on. Remember, people used to think putting drill motors in a robot was a silly idea, and now that's one of the most popular methods for locomotion.

You should also keep very clear notes on what worked and what didn't work, although I'm sure the things that failed will haunt you for a while (as they do me). You want to be able to look back on your notes for specific cases where a certain part failed when you go to redesign. Since you may not be getting back into your robot immediately, it's an excellent idea to record this right after the competition.

Putting Together the Binder

I like to put together a three-ring binder that has all of my notes for a particular season. I organize all of the notes by system: drive, weapons, frame, electrical, and so on, and put them into a binder with a tab for each category. Invariably, for this robot or another one down the road, you'll want to use the same part or vendor. ("Now where did I get that custom battery pack or special bracket?") Also, the invoices usually have all the part numbers neatly printed on them, making ordering easier and more accurate.

Hopefully, you followed my advice earlier and put dates on all of your diagrams. This way, you should be able to track changes chronologically. When you start to repair or upgrade, you'll find it handy to have a little note about why this spacer is here, or why you moved that hole.

Note Personally, I find it therapeutic to put things in order when it's over. At the end of a season, I have a mountain of papers with different bits of information scribbled on them. I sort it all out, and when I'm done, I can clearly see all of the work that went into my robot piece by piece. Also, (and this is probably the most important part), information is one of the most valuable commodities in robot building. It loses value significantly, however, if you can't find it.

Building a Web Page

Setting up a Web page for your robot gives you a chance to show off all those pictures that you've taken during the process. (You did take some pictures like I mentioned in Chapter 1, "Getting Started," right?) Also, get some of that video captured and uploaded.

Some of the competition Web sites may provide links to your Web site from theirs. This will give you more of an identity on the Web. In addition, you can point potential sponsors to your Web site. This is your chance to showcase your accomplishments in whatever way you want. Take the time to write about the making of a part, for example. Was it easy or hard? Were there things about the part that you wish you could or plan to change? You could write about the competition. If you thought to bring a journal, then create a trip report. Talk about the people you met and the matches you've fought.

Derek Young (www.automatum.com) and Jim Smentowski (www.robotcombat.com) both have excellent trip reports that you can check out and share their experiences. See Appendix D for the addresses of other notable builder Web sites.

Volunteering Your Time

If you're aching for some robot action during the off-season, then consider volunteering for a high-school event, such as BattleBots IQ or FIRST. These events are national robotic competitions for high schools. They represent a way for you to use your hard-earned skills to give a little back to the community. For more information, check the BattleBots IQ Web site at: www.battlebotsiq.com and the FIRST Web site at: www.usfirst.org.

Wrapping Up

My hope is that this chapter has painted a picture of what an actual competition might look and sound like, but it's very hard to fully appreciate the experience without actually having been to one. My first competition was an incredible experience, as I suddenly realized that there were hundreds (possibly even thousands) of people who were just as nuts about these robots as I was. The amount of energy and enthusiasm buzzing around through the air was infectious. I was so tired, yet so pumped up at the time.

This is one of the most exciting experiences you're likely to have in life, if you're not already a star athlete or a stage actor, because you'll be performing live in head-to-head combat in front of hundreds, maybe even thousands, of people. It will probably make a lasting impression on you, as it did on me. When it's all over, you can take a look back at where you started, and then gaze upon your creation (whatever state of repair it's in) and feel an immense sense of accomplishment. You've done it. You've created your own combat robot. Congratulations. You should be proud of yourself.

Advanced R/C Programming

The basics of radio control were introduced in Chapter 13, "Choosing Your Control System." The discussion now turns to some more advanced topics. Like a scientific calculator, your radio-control transmitter is capable of so many more complex and powerful functions than just the basic add, subtract, multiply, and divide. In this chapter, you'll find out how to unlock those powerful mixing capabilities to get the output signals that you need. I'll show you how to tweak the response of your radio-control system to maximize performance using end point adjust, and how to implement your own custom mixing programs. All of the variations of tank turn mixing will be explored, and you will learn the quick and direct way to debug problems that you may be experiencing with the tank-turn mixing on your own robot.

Note While all of these techniques will apply directly to the Futaba brand of radio controls, most families of radio-control transmitters have similar features and functions with slightly different names.

Using EPA to Control Your Servos

EPA stands for *end point adjust*. It's also known as ATV, or *adjustable travel volume*. It allows you to set the end position (maximum travel) in each direction for a servo. Normally, when you command a servo to go to a position, it will go there and do whatever it takes to stay there until you tell it to move somewhere else. This means that even if the servo encounters an obstacle in its path, it will continue to fight to get to its commanded position, wasting power and eventually burning itself out. You can prevent this from happening by limiting the servo's travel using EPA, so that it stops before bumping into anything at the end of travel. Using the following procedure, you can reduce the EPA in the desired direction until the servo no longer bumps into the obstruction.

Tip Sometimes, it's easier to adjust the EPA by first setting the value to zero and then slowly increasing it until the servo travels far enough to do the job, but not touch the object. It's much quicker this way because you can watch the servo move as you adjust it. Make sure to have the control stick in the correct direction when you make your adjustments.

Checking and Setting EPA Limits

The following procedure applies to the Futaba 9C series of transmitters. Other models may have slightly different controls.

1. Turn on the transmitter.

2. Press and hold the MODE/PAGE button to activate programming mode.

3. Use the dial to highlight END POINT, and press it to select this option.

4. Use the cursor keys to select the different channels along the right side of the screen.

5. To increase an EPA value, turn the dial clockwise. To decrease the value, turn the dial counterclockwise.

6. Every channel has an EPA value for each direction. In order to choose the other direction, move the control stick associated with that channel until the value that you want is highlighted. (Usually, there is a corresponding arrow on the screen.) You can use the dial to adjust the value as described in Step 5.

7. Press END to accept the changes and return to the previous menu.

8. Press END again to deactivate programming mode.

Turn It Up to Maximum

When the recipient of a control signal is a servo, limiting the travel to a safe value makes sense. If the signal is going to a speed control, then you usually want the maximum range that you can possibly get. Otherwise, your motors, batteries, and speed controls will be capable of putting out higher RPM and more power, but you're artificially limiting them with the control signal.

With most things, 100% is the maximum that you can get. In the radio-control world, 100% is just the default setting, not the maximum travel. In fact, you can increase the EPA in each direction of a channel above 100% to a value of 110% to 140%, depending on the brand and model of your transmitter. For the Futaba 9C series, you can use the procedure described earlier to set all of your EPA values.

Tank-Turn Steering with Elevon Mixing

As mentioned in Chapter 13, the most direct way to implement single-stick mixing is to activate the built-in elevon mode. This section will help you figure out how to implement tank-turn steering with elevon mixing and make sure everything's working well.

What Elevon Actually Does

Before we delve too deeply into the usage of elevon mixing, let me explain what it actually does. When elevon is enabled, the transmitter reads the control sticks for the aileron and elevator channels. Instead of passing them on to the receiver outputs as usual, it performs scaling and addition (which is what mixing really is) to form new output signals. This scaling and addition is controlled on the elevon mixing screen. The overall scale factor and the polarity are changed to actually get subtraction for part of the signal. The signal path and the resulting sum that is sent to the aileron and elevator output channels are shown in Figure A.1.

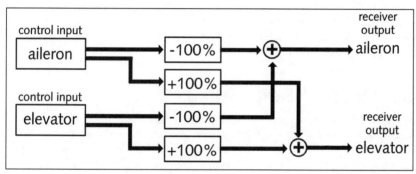

FIGURE A.1: Typical elevon mixing scheme.

If you activate any additional *programmable mixing* (see the "What Is Programmable Mixing" sidebar), it acts like an algebraic sum on top of the elevon mixing. For example, in order to invert a channel that's mixed at +100% (that is, change it to −100%), you've got to add −100%, which makes the algebraic sum zero, and then add another −100% to get it to the correct −100% level, as shown in Figure A.2. This double mixing is usually implemented as two separate programmable mixes with exactly the same settings, and is the basis for creating your own switchable *inverted driving mode*. An inverted driving mode is useful if your robot has the capability to run upside-down. You can set up a switch to invert the left/right steering controls, which will allow you to drive inverted with the same steering sense as when the robot is right side up.

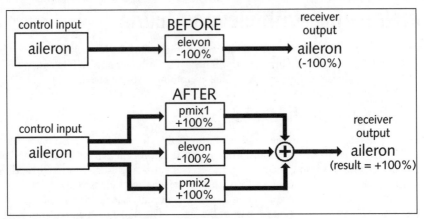

FIGURE A.2: Adding programmable mixing to invert channels.

Note   A simpler way to invert the channel would be to reverse it using the *servo reverse* function. However, this method does not allow you to control the reversing on the fly with a switch, which is what is necessary for an inverted driving control.

Checking the Stock Elevon Settings

In some radios, enabling elevon mode will bring up a default value of 50% mixing all around. This means that you won't be getting the maximum amount of travel from the signal, much like EPA described earlier. To correct this, make sure that all mixing values are set to 100% (maximum for the mode) and that all EPA values are set to maximum (usually 140%).

The stock settings for elevon should be as follows:

AIL1 (AIL to AIL) = −100%

AIL2 (AIL to ELE) = +100%

ELE2 (ELE to AIL) = +100%

ELE1 (ELE to ELE) = +100%

What Is Programmable Mixing?

In addition to the preprogrammed built-in mixing (such as elevon) discussed previously, most computer radios also have at least two fully programmable mixing positions, which will allow you to take input signals and combine them as you wish to form new outputs. You can choose whatever channels you want to be mixed. The *master* serves as the source of the control signal, while the *slave* is the destination channel where the new mixed signal will be added.

Elevon Programming Examples

In the following examples, elevon mode is used to perform the necessary mixing for tank-turn steering. Normally, the output channels used with elevon mixing are channel 1 (aileron) for the left-side speed control, and channel 2 (elevator) for the right-side speed control.

Example: Activating Elevon Mode

Following are the steps required to activate elevon mode for the Futaba 9C series transmitters. Other models may have slightly different controls.

1. Turn on the transmitter.

2. Press and hold the MODE/PAGE button to activate programming mode.

3. Press the MODE/PAGE button again briefly to bring up the next page of options. Use the dial to highlight ELEVON and press it to select this mode.

4. Use the cursor keys to move up to the MIX line. Turn the dial to the left to activate this mix mode. The screen should say ACT next to MIX.

5. Use the cursor keys to return to the previous screen and make sure that the default mixing settings are identical to those listed previously.

6. Press END to accept the changes and return to the previous menu.

7. Press END again to deactivate programming mode.

Example: Standard Inverted Driving

To set up an inverted driving mode along with the standard elevon mixing, you have to first enable elevon mixing as described earlier, and then create four programmable mixes (two doubles) as previously discussed to invert the aileron mixing. Basically, what you're doing is adding aileron to itself at +100% twice and aileron to elevator at –100% twice, which will invert only the left/right steering. Since the aileron control input is the left/right stick, you need to flop those control signals to have the correct response.

Listed next are the steps required to create the programmable mixes for the Futaba 9C series transmitters. Other models may have slightly different controls.

1. Turn on the transmitter.

2. Press and hold the MODE/PAGE button to activate programming mode.

3. Press the MODE/PAGE button again briefly to bring up the next page of options. Use the dial to highlight PROG.MIX1, and press it to select this mode.

4. Use the cursor keys to move down to the MIX line. Turn the dial to the left to activate this mix mode. It should switch from INH to ON or OFF.

5. Use the cursor keys to move down to the MAS line. This is the master channel select. Turn the dial until it says AILE.

6. Use the cursor keys to move down to the SLV line. This is the slave channel select. Turn the dial until it says AILE also.

7. Use the cursor keys to move down to the SW line. This will select what switch activates the mixing. Turn the dial until it reads D, which is the upper-right-corner switch. Although switch D is used in this example, you can select whatever switch you feel comfortable with.

8. Use the cursor keys to move down to the POSI line. This will select which position of the switch will actually activate the mixing. Turn the dial until it says UP. Flipping the switch UP will engage the inverted driving mode. Although the UP direction is used in this example, you can use whatever direction you feel comfortable with.

9. Use the cursor keys to move back to the RATE line. Turn the dial until the number under the right arrow is +100%. Move the right control stick slightly to the left, and turn the dial until the number under the left arrow is also +100%.

10. Press END to accept the changes and return to the previous menu.

11. Use the dial to highlight PROG.MIX2, and press it to select this mode.

12. Repeat Steps 4 to 10, making sure that you maintain the same switch choice and activation direction. (This is the second of the double program required for a correct algebraic sum on this channel.)

13. Use the dial to highlight PROG.MIX3, and press it to select this mode.

14. Repeat Steps 4 to 10 with MAS = AILE, SLV = ELEV, and RATE = −100% in both directions. Make sure that you maintain the same switch choice and activation direction.

15. Use the dial to highlight PROG.MIX4, and press it to select this mode.

16. Repeat Steps 4 to 10 with MAS = AILE, SLV = ELEV, and RATE = −100% in both directions. Make sure that you maintain the same switch choice and activation direction. (This is the second of the double program required for a correct algebraic sum on this channel.)

17. Press END again to deactivate programming mode.

 Note The RobotLogic IMX-1 external mixer has the ability to correct for inverted driving. It has a separate "inverted" input that connects directly to one of the receiver channels. You can use this channel to switch to inverted mode yourself, just as if you had programmed the radio as previously.

Tank-Turn Steering without Elevon Mixing

Perhaps you don't have a built-in elevon program. Can you still implement tank-turn steering inside of the radio? Yes. You can use manual programmable mixing as described later in this chapter.

Programmable Mixing Examples

In the following examples, individual programmable mixing positions are used to approximate the type of mixing performed by the elevon program.

Example: Manual Tank-Turn Mixing

If you don't care about running upside-down (if you have self-righting capability, for example), then manual mixing is pretty straightforward and requires only two programmable mixing positions. If you were to copy the elevon mixing exactly, you would need four channels total, but in this situation you only need two, because each channel has an *implied* mixing from its own control stick to the corresponding output (aileron to aileron and elevon to elevon). The only catch is that you have to swap speed-control leads between aileron and elevator receiver outputs. (The right-side speed control should be plugged into channel 1 and the left-side speed control should be plugged into channel 2.)

Following are the steps required to create manual tank-turn mixing for the Futaba 9C series transmitters. Other models may have slightly different controls.

1. Turn on the transmitter.

2. Press and hold the MODE/PAGE button to activate programming mode.

3. Press the MODE/PAGE button again briefly to bring up the next page of options. Use the dial to highlight PROG.MIX1, and press it to select this mode.

4. Use the cursor keys to move down to the MIX line. Turn the dial to the left to activate this mix mode. It should switch from INH to ON or OFF.

5. Use the cursor keys to move down to the MAS line. This is the master channel select. Turn the dial until it says AILE.

6. Use the cursor keys to move down to the SLV line. This is the slave channel select. Turn the dial until it says ELEV.

7. Use the cursor keys to move down to the SW line. Turn the dial until it reads D, although it doesn't really matter because the next step will force mixing ON all the time.

8. Use the cursor keys to move down to the POSI line. Turn the dial until it says NULL, which will force the mixing to be active all the time.

9. Use the cursor keys to move back to the RATE line. Turn the dial until the number under the right arrow is −100%. Move the right control stick slightly to the left, and turn the dial until the number under the left arrow is also −100%.

10. Press END to accept the changes and return to the previous menu.

11. Use the dial to highlight PROG.MIX2, and press it to select this mode.

12. Repeat Steps 4 to 10 with MAS = ELEV, SLV = AILE, and RATE = +100% in both directions. Make sure that POSI = NULL, so that the mixing will be active all the time.

13. Press END again to deactivate programming mode.

14. Make sure that the right-side speed control is plugged into channel 1 (aileron) and the left-side speed control is plugged into channel 2 (elevator) on the receiver.

Example: Manual Tank-Turn Mixing (Invertible)

If you have the ability to run upside down and want a switch to swap the left/right steering so that it's correct when you're inverted, then you'll need six channels total to perform this task without elevon. Two are used to set up the constant elevator (forward/back) mixing while the other four are used to invert the aileron (left/right) mixing. You will also be plugging the speed controls into different channels. This is required because you have to intercept the implied mixing mentioned earlier. In this example, the left-side speed control is plugged into channel 6 (flap), and the right-side speed control is plugged into channel 7 (aux1). Flipping switch D to the UP position will engage inverted mode, although you can choose any switch and activation position as long as you remain consistent in all of the mix programs.

The steps required to create invertible tank-turn mixing for the Futaba 9C series transmitters are listed next. Other models may have slightly different controls.

1. Turn on the transmitter.

2. Press and hold the MODE/PAGE button to activate programming mode.

3. Press the MODE/PAGE button again briefly to bring up the next page of options. Use the dial to highlight PROG.MIX1, and press it to select this mode.

4. Use the cursor keys to move down to the MIX line. Turn the dial to the left to activate this mix mode. It should switch from INH to ON or OFF.

5. Use the cursor keys to move down to the MAS line. This is the master channel select. Turn the dial until it says ELEV.

6. Use the cursor keys to move down to the SLV line. This is the slave channel select. Turn the dial until it says FLAP.

7. Use the cursor keys to move down to the SW line. Turn the dial until it reads D, although it doesn't really matter because the next step will force mixing ON all the time.

8. Use the cursor keys to move down to the POSI line. Turn the dial until it says NULL, which will force the mixing to be active all the time.

9. Use the cursor keys to move back to the RATE line. Turn the dial until the number under the right arrow is +100%. Move the right control stick slightly to the left, and turn the dial until the number under the left arrow is also +100%.

10. Press END to accept the changes and return to the previous menu.

11. Use the dial to highlight PROG.MIX2, and press it to select this mode.

12. Repeat Steps 4 to 10 with MAS = ELEV, SLV = AUX1, and RATE = +100% in both directions. You should also set SW = D and POSI = NULL.

13. Use the dial to highlight PROG.MIX3, and press it to select this mode.

14. Repeat Steps 4 to 10 with MAS = AILE, SLV = FLAP, and RATE = –100% in both directions. You should also set SW = D and POSI = UP.

15. Use the dial to highlight PROG.MIX4, and press it to select this mode.

16. Repeat steps 4 to 10 with MAS = AILE, SLV = FLAP, and RATE = +100% in both directions. You should also set SW = D and POSI = DOWN.

17. Use the dial to highlight PROG.MIX5, and press it to select this mode.

18. Repeat Steps 4 to 10 with MAS = AILE, SLV = AUX1, and RATE = +100% in both directions. You should also set SW = D and POSI = UP.

19. Use the dial to highlight PROG.MIX6, and press it to select this mode.

20. Repeat Steps 4 to 10 with MAS = AILE, SLV = AUX1, and RATE = −100% in both directions. You should also set SW = D and POSI = DOWN.

21. Press END again to deactivate programming mode.

22. Make sure that the left-side speed control is plugged into channel 6 (flap), and the right-side speed control is plugged into channel 7 (aux1).

There is a crude and cheap way to implement inverted mixing. You can face the transmitter away from you with the antenna still pointing up. This will correctly invert the left/right steering. Unfortunately, you will have to reach around to the front to drive, and the control stick will be on the other side of the transmitter. You will lose valuable time while you turn the transmitter, and more importantly, you run the risk of dropping your transmitter on the floor.

Double Inverted Driving

Certain robots, such as double-ended wedges, require that both steering and forward/back control be switched, because when they are inverted, not only will the steering be flipped, but they will also be attacking using the *rear* of the robot as described in Chapter 18, "Choose Your Weapon." This is called *double inverted driving*. Unfortunately, this would require a total of eight programmable mixing positions in addition to elevon, which is one more mixing position than the 9C transmitter has available, ruling it out as a possibility. However, this type of mix is easily implemented by setting up a *condition* in a Futaba 9Z transmitter. A condition allows you to change mixing ratios as well as trims and ATV settings, all controlled by a switch that you choose. This feature allows you to simply invert the mixing in the elevon screen for an inverted driving mode, instead of using double negative programmable mixing described previously. Unfortunately, this little-documented, but very powerful feature is only available on the 9Z transmitters and not the 9C series.

There is a crude and cheap way to double invert the mixing. You can simply turn the transmitter upside-down, so that the antenna is pointing down. This will correctly invert both the left/right and forward/back controls. Unfortunately, the control stick will be on the other side of the transmitter, and you lose valuable time while you turn the transmitter, not to mention the possibility of dropping your transmitter on the floor.

Debugging Tank-Turn Mixing

In Chapter 13, I made it sound pretty easy to get everything all hooked up and running. The reality is that sometimes, things get a little confusing, and the robot doesn't behave as you expect it. No problem. The following list will help you figure out what's wrong and fix it.

- The left-side motor polarity should be set up so that a positive voltage will cause the robot to move forward.

- The right-side motor polarity should be set up so that a positive voltage will cause the robot to move forward. (This should be the opposite polarity of the left-side motor.)

- All channels should be set to normal direction in the transmitter. You can use the *servo reverse* function to verify this.

- If using elevon mixing, make sure that the left-side speed control is connected to channel 1 (aileron) on the receiver, and the right side is connected to channel 2 (elevator). You should also verify that the mixing is active and all mix values and polarities are correctly set up as described earlier in the chapter.

- If you're using individual programmable mixing (noninvertible), then make sure that the left-side speed control is connected to channel 2 (elevator) on the receiver, and the right side is connected to channel 1 (aileron). You should also verify that the mixing is active and all mix values and polarities are correctly set up as described earlier in the chapter.

- If you're using individual programmable mixing (invertible), then make sure that the left-side speed control is connected to channel 6 (flap) on the receiver, and the right side is connected to channel 7 (aux1). You should also verify that the mixing is active and all mix values and polarities are correctly set up as described earlier in the chapter.

- If you're using an external mixer, make sure that all programmable mixing and built-in mix modes (like elevon) are inhibited. Connect the mixer as specified by the manufacturer.

If moving the stick forward causes the robot to turn, then probably one of the motors has its leads reversed. You should make sure that the polarities are as described in the previous list. If the polarities are okay, then one of the channels is reversed. If reversing one of the channels causes all of the functions to work in an inverted fashion (forward is back and left is right), then change the channel that you modified back to normal and reverse the other channel.

If forward and back are working as expected, but the robot turns opposite of what you expect (left is right and right is left), then the left and right channels are switched at the receiver outputs. Swap the two PWM connectors at the receiver, and the robot should turn as expected.

IFI System Programming and Troubleshooting

Originally, I was intimidated by the sheer volume of code in the IFI Isaac system default program. I had heard a lot of good things about the IFI system, and the prospect of avoiding radio impound was highly attractive. But would I be able to learn all those commands in time for competition? Well, it turned out that you don't have to learn all that code. As I mentioned in Chapter 13, "Choosing Your Control System," the IFI system comes preprogrammed and ready to go with the most popular features already implemented. However, if you need to add some custom code to your IFI system, this appendix will give you the ins and outs of the default program. Once you learn a few simple things, then programming becomes a snap.

System Review

The IFIrobotics system consists of an operator interface (OI), which is what you plug joysticks into for controlling the robot, and a robot controller (RC), which rides inside the robot and receives, interprets, and implements your commands. The OI and RC transmit information back and forth wirelessly using radio modems.

How the robot responds to various commands is determined by a program that you load into the RC. IFIrobotics provides you with a *default program*, which is the basic framework that you can customize to create your own programs. The whole setup is extremely flexible because you can quickly and easily change the program and try it out. You can keep modifying the code until it works for your needs.

Inputs and Outputs

The discussion begins with a review of the system's inputs and outputs because the data that they provide is what you will be manipulating in your own programs. Both the OI and the RC have analog and digital inputs. How they're used by the overall system varies a bit, as described in the following sections. The outputs on the RC are for driving relays and PWM-compatible devices, while the outputs on the OI are informational only.

Analog versus Digital Inputs on the OI

Both the analog and digital inputs on the OI are accessed through the four joystick input ports. They are the primary interface for controlling the robot. Each of the four input ports has four analog inputs (two of which correspond to the X-axis and Y-axis of the joystick), and four digital inputs (two of which correspond to the thumb and trigger switches). What you get out of each analog input is an 8-bit byte of digitized information that ranges from 0 to 254, depending on the position of the joystick. Each digital input gives you a single-bit ON or OFF signal (0 or 1).

Note   In programming jargon, a *bit* is a single character memory location, while a *byte* consists of many bits. Depending on the processor, the number of bits in a full byte differs. For the Basic IIsx chip that powers the IFI system, 8 bits constitutes a full data byte.

Note   You don't necessarily have to plug joysticks into the input ports. You can make your own custom input box, using switches and potentiometers that you choose. The *potentiometers* (variable resistors) are what are connected to each axis of the joystick. You must use 100k-ohm pots, and wire them correctly, as described in the *Operator Interface Reference Guide*, provided on the IFIrobotics Web site.

Analog versus Digital Inputs on the RC

The RC also has both analog and digital input ports. They are for gathering onboard information about the robot and its immediate surroundings. The typical use for a digital input on the robot would be a limit switch that tells the robot that an arm or actuator has reached the end of travel, and it should stop (limit) or else damage may occur. An analog input might be used to read a potentiometer (variable resistor) that tells the robot the exact position (rotation) of an arm, or a sensor that provides a voltage level based on the pressure in an air tank. The Isaac 16 system provides access to 8 digital and 4 analog inputs, while the Isaac 32 system has 16 digital and 8 analog inputs. Just as with the input data on the OI, you get an 8-bit byte for analog inputs while the digital inputs are a single bit (0 or 1). It should be noted that the digital inputs on the RC have *active low logic*, which means that the switch input line needs to be grounded (zero volts) to be *activated*, which registers a logic 1 for that switch input. For more information, you can refer to the *Full-Size Robot Controller Reference Guide* and the *Robot Control System/Mini RC Users Manual*.

Note In the default code, the digital inputs on the robot controller are referred to as *switches*, while the analog inputs are referred to as *sensors*.

Outputs on the RC

The PWM outputs on the RC are for interfacing with devices such as servos and speed controls that use the proportional radio-control PWM standard (see the sidebar). There are 8 PWM outputs on the Isaac 16 system and 16 PWM outputs on the Isaac 32 system. Although the PWM outputs default to reading analog inputs on the OI (joystick X- and Y-axes), you can perform whatever math or processing you need on your input data and send that out to any PWM output. As you'll see later in this appendix, that's exactly how the preprogrammed mixing is implemented.

While the PWM outputs are used for proportional control, the relay outputs are used for turning things on and off. Each relay output has a variable name that can be set to 1 or 0 to turn it on or off. The relay outputs normally read digital inputs on the OI (switches on the joystick), but you can choose whatever data conditions you want to trigger them. For example, you could also read the value of an analog input for an automatic gear change at higher speeds, or a switch input on the RC to automatically fire a weapon when an opponent hits a bump switch on the robot. You can also set up an internal timing loop (discussed later in this appendix) to set up delays for time sequence events, such as firing pneumatic valves.

The relay outputs are designed to be used with IFI Spike Relay Modules for bidirectional relay control. This arrangement gives you three possible output states for the relay: positive polarity voltage, negative polarity voltage, and *no connect* (no voltage). There are 4 relay outputs on the Isaac 16 system and 8 relay outputs on the Isaac 32 system. If you don't require bidirectional (reversing) control, you can connect double the number of single-ended solid-state relays (also sold by IFIrobotics) to the same outputs.

PWM Values versus Output Pulses

Each PWM output can assume an 8-bit value from 0 to 254, with 127 as the midpoint of the range. With a speed control, for example, a value of 254 would be full forward, while a value of 127 would be stop, and 0 would correspond to full reverse. As described in Chapter 13, the radio-control PWM specification is based on pulse lengths in milliseconds, rather than digital values. How does the IFI system resolve this? Well, writing a value of 254 to any PWM output actually produces a 2.0 msec pulse at the output (full forward), while a value of 127 produces a 1.5 msec pulse (neutral), and a value of 0 produces a 1.0 msec pulse (full reverse) to maintain compatibility with radio-control equipment.

Outputs on the OI

As mentioned earlier, the outputs on the OI are for informational purposes only. They provide useful visual feedback about the current status of PWM outputs 1 and 2, as well as relays 1 and 2. The most important thing to realize here is that if the PWM and relay lights are changing along with your control actions, then the RC and OI are communicating back and forth, because the LEDs on the OI are actually controlled by commands from the RC. During actual combat situations, the LEDs on the OI are of limited use. If you're focusing on the robot, then you don't really have time to look down and check the LEDs.

Note   The dashboard port of the OI can provide you with a direct feed of the data stream that the radio modem receives from the RC. Although software is available to read that information, all you really get is a graphical display of the same status LEDs that are already on the OI. Specifications are available to decode the full data stream in more detail, but you're pretty much on your own, because IFI doesn't have the resources to fully support this feature.

Programming

What does all this code mean? At first glance, the default code may seem horribly complicated and pretty daunting, but really, it's pretty easy. The thing to keep in mind is that there are only a few lines you need to change to do your operations. Everything else remains the same. In this section, I'll give you the hit-and-run, guerilla programmer's version of what you need to know to get in, insert your code, and get out. To follow this section in any coherent manner, you're going to need to have a copy of the default code in front of you, which can easily be downloaded from the IFI Web site for free.

Note   IFI has created an excellent *Programming Reference Guide* that is also available from its Web site for free. In addition, PBasic (the programming language used in the IFI system) is wildly popular among electronics hobbyists, and there are several excellent books on this subject. *Programming and Customizing the Basic Stamp Computer* by Scott Edwards (McGraw-Hill, 1998) and the *Basic Stamp Programming Manual*, published by Parallax (I have version 1.9) are the two best books that I can recommend. They are both easy to read and follow. If you have any questions about the command set or syntax, these books can provide quick answers. Also check Appendix D, "Online Resources," for a listing of Web sites that offer free downloadable support materials and tutorials for PBasic.

Program Flow

First of all, how does the program work? Well, there's a lot of setup stuff and subroutines defined in the beginning. The actual program is at the very end. It's very short and runs in a continuous *loop*, which starts by taking a snapshot of all the inputs, including the remote input from the OI and the onboard switches and sensors. Next, it processes the information, and performs any custom code that you've added. Finally, it sends the results to the appropriate outputs. As long as stable power is applied to the unit, it will continue running the loop.

Parts of the Default Program

Next is a section-by-section description of the default program, with an emphasis on what's important to you. The names of each of the numbered sections that follow are the same as those used in the default program, so you can follow along sequentially. To help illustrate issues that might come up when trying to implement your own operations, I'll use the single-stick tank-turn mixing code. Although it's built into the default program, it behaves very much like a piece of code that you might write for yourself.

Note By placing an apostrophe in front of any statement in the program, that statement effectively becomes a *comment*, and is ignored by the compiler. You only need one apostrophe—you don't have to enclose the text. If the apostrophe appears in the middle of a line, the text before it is interpreted as valid code, while the text after it becomes the comment. It's useful for temporarily removing statements without deleting them, and for leaving yourself bits of information. You'll see its usage throughout the default program to help explain various statements.

Section 1: Declare Variables

The program begins by creating *variables* (storage locations in memory) for the analog joystick and sensor inputs. These are the locations where the snapshot values will be stored. When manipulating them in your own custom programs, you will use the names declared in this section. Each of these locations is a full 8-bit byte. Also, if you want to perform any math of your own, you'll need to create your own variables as storage containers for the results.

You should take note of the naming conventions for the analog joystick inputs. The first part is the input port, while the second part is the axis. For example, p1_x is the X-axis of a joystick connected to input port 1.

In the single-stick tank-turn programming example, they've created two memory locations called PWM1 and PWM2:

```
PWM1 VAR byte
PWM2 VAR byte
```

Section 2: Define Aliases

All of the full 8-bit variables were declared in the previous section. However, many of the variables that you will be dealing with (like relay outputs or switch inputs) are only a single bit. This section further subdivides the previously specified bytes into bits with various names so that you can keep track of them separately.

Take note of the naming format. It will help you decode things later. For example, p1_sw_trig is the trigger button on the joystick plugged into port 1, and rc_sw1 is digital switch input 1 on the robot controller.

There are no changes to this section for the single-stick mixing program because we're not renaming anything.

Section 3: Define Constants for Initialization

This section allows you to select which pieces of data you want to read in the initial snapshot. It takes precious processing power and time to read all the variables that are available to the processor. It's best to conserve your resources by only enabling the data that you need. Variables that have a 1 next to them will be read, while those with a 0 will be ignored. The total number of variables that you can read in is limited to 26.

Note that in the default code, the processor reads the X-axis and Y-axis of each joystick input port, as well as the wheel, but not the aux input or the RC sensors. If you want to read a sensor, for example, then you have to change the value next to that sensor name to 1 to enable it.

Modifying this section can be tricky because you've also got to change the serin command in the *main loop* section later in the program to exactly match what you have in this section. If you're not careful, you may end up with a nonresponsive output or the wrong data because this section was incorrectly configured or you didn't enable the right input stream.

There are no changes in this section for the single-stick mixing program because it uses the joystick inputs that are already enabled in the default program.

Section 4: Define Constants (Do Not Change)

The constants defined in this section are used for system configuration. Nothing should be changed here for any custom program.

Section 5: Main Program

This section defines inputs and outputs for the processor and sets initial values for analog inputs. If a joystick is mistakenly unplugged, or otherwise absent, the program will default to a neutral (safe) value of 127, or whatever has been specified in this section. When adding your own sensors, you should define an initial condition that's safe.

There are no changes in this section for the single-stick mixing program because it uses the joystick inputs that already have safe initial values in the default program.

Section 6: PBasic — Master uP Initialization Routine

This section is used to send configuration commands to the master microprocessor. Do not change anything in this section, or move it around.

Section 7: Main Loop

The most important thing here is the serin command. As mentioned above in the *define constants for initialization* section, you have to alter this command so that it matches any changes you've made in that section. There is a specific order for variables, and you either add or remove variables from the statement without changing the basic order.

Also, don't remove the toggle 7 statement. This blinks the Basic Run LED on the RC, which will help you troubleshoot looping and program branching problems later.

There are no changes in this section for the single-stick mixing program because it uses the joystick inputs that are already enabled in the default program.

Section 8: Perform Operations

This is the section where you'll add your own custom code to perform whatever operations you need. There are also a lot of other things to see and do here, including specifying triggers for relays, PWM limit switches, and LED outputs.

Using Relays

The relays in the IFI system are bidirectional, which means that when a relay output is connected to an IFI Spike Relay Module, you can reverse the polarity. However, you don't always need bidirectional relays. For example, when using a relay to activate a pneumatic solenoid valve, polarity doesn't matter — only whether there is voltage or not. In this case, you have two choices. You can either connect a bidirectional relay module or a solid-state relay. Both devices would use *only* the forward and off conditions, ignoring the reverse condition. In fact, you can connect two solid-state relays to *each* relay output on the RC for independent control of each, using the forward condition to activate one and the reverse condition to activate the other. This doubles the number of available relay outputs.

If you want to activate a relay in a program, all you have to do is set its variable name to 1. This will turn the relay on until you actively turn it off. Conversely, to turn a relay off, set its value to 0. The example below will turn relay 1 on in the forward direction.

```
relay1_fwd = 1
```

If you want to use a switch to control a relay, then you set its name equal to the relay, as shown next. This is actually the normal case in the default program, where the relays are controlled by the joystick buttons. If you recall the naming conventions for relay outputs and OI switch inputs, you should be able to decode what's going to where. For example, the following statement will turn relay 1 on in the forward direction when the trigger switch of the joystick connected to input port 1 is activated, and while the limit switch connected to RC digital input 1 is not activated.

```
relay1_fwd = p1_sw_trig &~ rc_sw1
```

The &~ symbol used here is a *logical AND NOT* operation that will only produce a logic 1 when the condition before it is logic 1 and the condition after it is logic 0. In the case of the RC switch input described previously, if you don't have a switch connected, then the latter condition will *always* be logic 0 and the relay will be controlled solely by the joystick trigger. If a limit switch is connected and activated, then it will produce a logic 1 condition, which will prevent the relay from firing. Note that if you're programming your own relay control, you don't need the limit switch part of the above statement. It's just a part of the default program.

Bump Switch Example

If you were going to add the bump switch mentioned earlier in the appendix, this would be the section where you insert your code. The idea is that you have some momentary switch on the robot's body that causes the weapon to fire when the opponent touches it. Ideally, the switch would be placed so that your weapon is perfectly aimed when it fires. Because this would be a dangerous thing to have active all the time, this example will include a manual enable for the bump switch, so that it will only be active when the joystick trigger button is held down. You can feel free to substitute whatever switch, port, or relay you want.

```
'relay3_fwd = p3_sw_trig
relay3_fwd = p3_sw_trig & rc_sw1
```

The first statement comments out the default control for relay 3, so that it's ignored by the compiler, but you've still got it there if you want to go back. You connect a joystick to port 3 and wire the momentary (normally-open) bump switch to digital input 1 on the RC. The second statement fires relay 3 only when the trigger button is held down (manual enable) and the bump switch is activated. The & symbol used here is a *logical AND* operation that will *only* produce a result of logic 1 when *both* of the conditions are also logic 1. Note that if you just wanted to use the trigger button to fire the relay (no bump switch), you could have used the original statement.

Feedback LEDs

The feedback LED outputs on the OI are defined here. As mentioned before, they're useful for diagnostic purposes, but not much else. These statements are disabled in the single-stick mixing program because they rely on the raw joystick inputs, and not the mixed output. This would be confusing to attempt to use in a mixed situation, because the LEDs would not correspond to the actual PWM outputs and speed controls. You'll see that later in the program, after the mixing math has been completed, these LEDs are added back in as a diagnostic tool.

PWM Outputs and Limit Switches

As mentioned earlier in this appendix, a *limit switch* is used to tell the speed control to stop running a motor, or else damage may occur. The default program only includes code for limiting on a few of the inputs, but you can assign whatever switch and input combinations that you want, as long as you make sure to comment out other references to those switches and inputs that might interfere with your limiting operation.

Linear Actuator Example

The default program gives us an opportunity to explore an example of using a limit switch. In the following code, you see that RC limit switches 5 and 6 are being used with the Y-axis of a joystick connected to input port 3.

```
if rc_sw5 = 0 then next1:
    p3_y = p3_y MAX 127
next1:
if rc_sw6 = 0 then next2:
    p3_y = p3_y MIN 127
next2:
```

Later on in the default program, you will see that the p3_y input is connected to the PWM3 output. If you connect a speed control to PWM3, then you can use this preprogrammed code to control a linear actuator, as well as stop it when it reaches a limit switch. Wire each switch input to the normally-open terminals of a momentary limit switch, and you're ready to go. Just be sure that you've got the correct switch on the corresponding side, or the actuator won't stop when it hits the switch.

How does it work? First, we assume that the speed control is wired with positive polarity so that a value of 254 will produce full-forward motion, while a value of 0 will produce full reverse, and a value of 127 will be a full stop (as described in the sidebar on PWM outputs). If you set a max limit of 127, then it will allow any value up to, but not exceeding 127 (full stop). This means that any *reverse* value will be accepted. Thus, the statement p3_y = p3_y MAX 127 corresponds to a limit switch for the maximum forward travel. It will not allow further forward travel, but allows the actuator to back away from the limit. The p3_y = p3_y MIN 127 statement works in a similar manner, but limits reverse travel, while accepting any *forward* value.

Single-Stick Mixing Code

Finally, at the end of this section, we get to the actual two statements that perform the mixing math, as shown next:

```
PWM1 = (((2000 + p1_y - p1_x + 127) Min 2000 Max 2254) -2000)
PWM2 = (((2000 + p1_y + p1_x - 127) Min 2000 Max 2254) -2000)
```

The extra 2000 term is to keep all the values in the calculation positive, and guarantee that the result will be between 0 and 254.

Now that all the necessary math has been performed and you have the correct mixed signals, the output LEDs (Out 8 to Out 11) on the OI are reenabled, so that they will reflect the actual PWM signals going to the speed controls as a diagnostic aid.

Tip Check to make sure that you're the only one changing your variables. It's entirely possible for you to change a value, and then have some other part of the program change it later on to something else, which might make it appear that your software isn't doing anything, though it's actually performing correctly.

Section 9: Output Data

As mentioned before in this appendix under the heading *Program Flow*, the basic flow of the program is that the processor reads the inputs, does the math, and then sends the results to the outputs. This section determines exactly what data bytes are sent to the PWM outputs. Like the serin command, the serout command in this section has a specific format and order. However, instead of adding or removing what you want, you will be plugging names into the spaces, or setting them to the neutral value of 127. This is another tricky statement that can cause unpredictable results if you make a mistake. Do not remove any of the terms in the serout command. Just rename them or set them to a value.

Specifically, each space in the serout statement corresponds to a PWM or relay output. For example, the seventh value in the serout command corresponds to the PWM3 output. In this space in the default code, we have p3_y, which is the variable that contains the analog input value of the Y-axis of the joystick connected to input port 3. You can substitute one of your own variable names for the name that is currently in any of these spaces, and the processor will write your data to the corresponding output. Just make sure not to move or replace the relay outputs. They are a different situation from the PWM outputs and were defined earlier in the *perform operations* section. Note that in the default program, outputs PWM9 to PWM16 are set to the neutral value of 127. Although it's not necessary, you can use this technique to disable any outputs you're not using if you want.

In the single-stick tank-turn programming example, the third and fifth spaces are filled with PWM1 and PWM2, which are the user variables that contain the result of the mixing program math. (If you were not mixing, these spaces would normally contain p1_y and p2_y, which contain the raw joystick input values.) You could have called those variables anything you wanted, as long as you used the same name throughout the program. For example, PWM1 and PWM2 could have been called LEFT and RIGHT. If they were changed in every location in the program, then processor would execute it exactly the same way as before.

Make sure not to delete the Goto MainLoop: and Stop statements at the end of the program, or you will cause a basic run error. These statements are needed to return the program to the start of the loop for the next run.

Program Examples

The following examples will help illustrate an event triggered by a position of the joystick, and a fully timed sequence, which may be of use to you in implementing your weapon or drive systems.

Example: Active Relay Triggering

As mentioned in Chapter 12, "Let's Get Rolling!," one of the drawbacks of a four-wheeled slip-steer drive system is the amount of friction that you experience when you turn, because all the wheels have to simultaneously slip sideways. In this example, from one of my students' FIRST robots, we used a pneumatic cylinder to pop a swivel caster down *only* when we turned, which lifted the two back wheels off the ground, and reduced turning friction dramatically. At all other times, the caster wheel was up and out of the way, and we enjoyed the benefits of added traction and straight tracking that are inherent to the four-wheeled system. This conserved battery power and prevented our circuit breakers from popping. The bit of code that controlled the caster is listed next:

```
if p2_x > 137 then caster:
    if p2_x < 117 then caster:
            relay3_fwd = 0        'caster UP
            goto exit_caster:
    caster:
            relay3_fwd = 1        'caster DOWN

    exit_caster:
```

Here's how the code works. The first two statements look at the X-axis (the steering axis) of the joystick connected to port 2, and check to see if it is either greater than 137 or less than 117. If either of these conditions is met, that indicates a turn is taking place, and the pneumatic solenoid connected to relay 3 will extend the cylinder, which pushes the caster down and into contact with the ground. If the X-axis is between 117 and 137 (with dead center being 127), then the robot is pretty much going straight, and the solenoid will keep the pneumatic cylinder retracted, lifting the caster out of the way.

Example: Adding a Timed Sequence

When triggering some weapons, you need a series of events to happen in a specific order with accurate timing (pneumatic weapons with a sequence of valves opening and closing, for example). Since you sometimes need timing in seconds between each step, using a conventional delay or wait statement could cause the program to go into a holding pattern for a lengthy amount of time (a second is a long time to your processor). While the processor is waiting, you may miss important incoming data from the OI. Since the same program is responsible for processing commands to control your drive system, the delay could result in an extremely sluggish or nonresponsive drive system, which should be avoided at all costs. The correct solution is to define a *loop counter* and a *step counter*. The loop counter keeps track of how long (specifically, how many loops) you've been in a particular step, while the step counter keeps track of what step you're on. Each timed event should be considered a new step.

Why is this different and much better than a conventional delay? Because the loop and step counters hold on to the current status of the sequence without delaying the rest of the loop. Your drive and other parts of the program will go on processing as normal, while the timed sequence can go on for as long as you want. Check out the sample code that follows (which is based loosely on some great code published by FIRST Team #45, the TechnoKats) for an example of implementing a timed sequence with loop and step counters to fire a sequence of valves.

```
'Declare Variables
stepnum VAR byte
counter VAR byte

'Main Program
stepnum = 0
counter = 0

'Perform Operations
If p1_sw_trig = 0 then SkipIt:        'Keep stepnum = 0 until
                                      'the trigger is pulled.
stepnum = 1                           'Set stepnum = 1 to start
                                      'the firing sequence.

SkipIt:
If stepnum = 0 then EndFire:          'If we haven't triggered,
                                      'then skip over the
                                      'entire sequence.


counter = counter +1                  'Increment the counter
                                      'on every loop.

If stepnum = 1 then Step1:
If stepnum = 2 then Step2:
If stepnum = 3 then Step3:
If stepnum = 4 then Step4:

Step1:
If counter = 15 Then EndStep:
relay1_fwd = 1                        'Activate valve 1
Goto EndFire:                         'and wait 15 cycles.

Step2:
If counter = 100 Then EndStep:
relay1_fwd = 0                        'Deactivate valve 1.
relay2_fwd = 1                        'Activate valve 2
Goto EndFire:                         'and wait 100 cycles.
```

```
Step3:
If counter = 55 Then EndStep:
relay2_fwd = 0                        'Deactivate valve 2.
relay3_fwd = 1                        'Activate valve 3
Goto EndFire:                         'and wait 55 cycles.

Step4:
relay3_fwd = 0                        'Deactivate valve 3
Goto NewFire:                         'and reset the sequence.

EndStep:                              'When you complete a step,
stepnum = stepnum + 1                 'go to the next step number
counter = 0                           'and reset the loop counter.
Goto EndFire:

NewFire:                              'Resets the program for the next
stepnum = 0                           'firing sequence
counter = 0

EndFire:                              'Pauses the firing sequence for
                                      'now where it is and lets you
                                      'continue with the rest of the
                                      'program until the next loop.
```

The trick with this code is getting the sequence to start with a momentary trigger, and ignoring any other weapon triggers until all of the steps are finished. You'll probably be firing the weapon (triggering the sequence) manually from a joystick button, and in the heat of battle, you may generate (accidental) multiple triggers. If the program doesn't ignore these, you might end up stopping the sequence in the middle and restarting, which could be dangerous, especially with weapons that have critical valve timing. That's handled by the first few statements and the SkipIt loop. Note how even when triggered, each run through the loop will only execute a few commands, as opposed to using a wait statement that would bring the whole program to a grinding halt. You'll have to experiment to see how many loops it will take to get the timing you need for each of your steps.

Thanks to good program structure, you can add as many steps as you want, as long as you move the Goto NewFire: statement to the last step. Also, you don't have to turn off each valve in the step that immediately follows its activation. That's just how I set up this example. You can keep them on as long as you want.

Helpful Tools

Next, I'll discuss the various diagnostic and debugging tools, including the download program, the debug command, and the tether, which will help you make sure that your program is working the way that you want it to.

The Download Program

This is the program that you run on the PC to edit and compile your program for loading into the RC. The download program can help you catch errors in your custom programs. It also provides the interface for the debug command described next. You must make sure to have the most recent version of the download program from the IFIrobotics Web site.

Using the Debug Command

A really handy feature of PBasic is that you can use the debug command, which allows you to interactively view any variable or group of variables in real time. It uses your PC screen to display the results of the debug process. You must be connected from the program port on the RC to your PC's serial port using a standard DB9 cable, and running the download program.

Debug Command Syntax

Any bit or byte can be monitored using the debug command. Since it is updated with every loop, the value changes in real time. Following are a few examples of typical implementations of this command in a program:

```
DEBUG dec p1_y
```

The previous command will display the value of p1_y (which is the Y-axis of the joystick connected to input port 1) as a decimal number. For example, if the value of p1_y were 175, then the screen would display 175.

```
DEBUG dec ? p1_y
```

The previous command will display the value of p1_y (which is the Y-axis of the joystick connected to input port 1) as a decimal value along with a label that corresponds to the name. In this case, if the value of p1_y was 175, then the screen would display p1_y = 175.

```
DEBUG bin rc_sw2
```

This command will display the binary value of rc_sw2 (which is digital switch input 2 on the RC). For example, if rc_sw2 were ON, then the computer screen would display 1. You can also use the ? option to get the command to display the variable label as described previously.

You can use multiple debug statements in a program. The only thing to keep in mind is that the extra overhead required to manage all of this information causes the program to execute more slowly. In most cases, it won't make much of a difference, but in programs that have to execute a long series of commands in a timed sequence, you may notice some slowing down.

Using the Tether

The tether is a standard DB9 cable that directly connects the OI and RC without transmitting any radio signals. It also provides power for the OI from the robot's battery. This is useful for checking things in the pits without transmitting any radio signals that might cause interference.

With everything off, you should connect the DB9 cable between the tether port on the OI and the tether port on the RC. Since power is provided to the RC through the tether, you only need to turn on the robot. It will work exactly the same as normal, but you won't be able to drive too far away because of the tether's physical connection. Make sure to observe all the safety precautions for testing, as described in Chapter 17, "The First Test Drive."

If it doesn't connect, make sure that you're connected to the tether ports on both the OI and RC. The program port takes the same size connector, and is often confused with the tether port.

After you're done checking things out, make sure to power the robot down. Using the tether locks the radio modem to channel 40, the default channel, where it will stay until you power down or reset. If your radio modem is locked to channel 40, you won't be able to lock to any of the competition frequencies used in the arena before your match, and your robot will rudely ignore you until you cycle power or hit the reset button on the robot controller.

Common Failure Points

The IFI system is pretty bulletproof, with built-in protection fuses and current limiting, so it's highly unlikely that you'll have a failure due to the equipment itself. However, problems caused by inadequate power, programming errors, or bad connecting cables are quite common.

Bad Power

On an Isaac 16 system, although the low battery light will blink when the voltage drops below 8 volts, the radio modem will work until it drops to 6.4 volts. With the Isaac 32 system, the low battery light will begin to blink at 9.5 volts, and the radio modem will stop transmitting at 8 volts.

Low voltage is only one of the problems you can experience with power, though. One of the more common problems is sharing the power for the RC with the drive system. Because of the fluctuating demands of the drive, and the possibility of a stall (high current) situation, the power provided to the RC can momentarily dip below the minimum voltage required. Also, intermittent failure can happen due to high shock loads. As mentioned in Chapter 13, you've got to be sure that your secondary power switch is up to the task of handling high shock loads.

Basic Run LED

If the Basic Run LED goes out during operation, the cause can usually be traced to a software bug. Make sure that all of your custom programming has correctly terminated loops and that you have not removed any necessary statements. It's also possible that if you removed the `toggle 7` statement from the program, that the Basic Run LED will never turn on. You should keep this statement to help you diagnose software problems.

Another source of errors is deleting the `Goto MainLoop:` statement at the end of the program. Without this statement, the program won't be able to loop back to the beginning to start a new cycle.

Although the program runs in a loop as mentioned previously, you still need the Stop statement at the end of the program. Otherwise, you'll get an execution error, and the Basic Run LED will go out.

DB9 Cables

The cables are the most susceptible part of the system. Intermittent operation and bad reception can often be traced to a faulty DB9 cable. Both the Isaac 16 and 32 systems have a DB9 cable that connects the radio modem to the OI, while only the Isaac 32 has a full-size RC that has a corresponding DB9 cable and radio modem. The integrated motherboard, RC, and radio modem on the Isaac 16 has no DB9 cable.

Processing Time

Some programs end up taking so long to process their loops that the RC misses the next packet of information. This may cause you to experience sluggish response from the robot, and even missed events. This should only be the case if you've modified the default program with your own loops or branching statements. You may also make use of the DELTA_T command to monitor how long your commands are taking to execute. See the IFI *Programming Reference Guide* for details on implementing the DELTA_T command in your program.

Troubleshooting Sequence

The following numbered lists will take you through typical troubleshooting sequences for both constant and intermittent failures. Think of the following list of steps as a checklist to perform in sequence. If everything on the list has been verified, and the robot's still not working, then the last step is to contact IFIrobotics.

Constant Failure/No Response

If the robot just sits there with no discernible drive response, then begin with the following "Establishing Communication" list. If your robot is already working, but only on an intermittent basis, then skip to the "Intermittent Failures" list.

Establishing Communication

1. Is the radio modem firmly connected to the RS-422 RADIO port on the OI through a DB9 M-F pin-to-pin cable of less than 6 feet length?

2. Is the radio modem firmly connected to the RS-422 RADIO port on the RC through a DB9 M-F pin-to-pin cable of less than 6 feet length? (Isaac 32 only)

3. Is the motherboard firmly seated on the robot controller? (Isaac 16 only)

4. Is the correct team number set on the OI?

5. Is the exact same team number set on the RC?

6. Do you have input power to the OI? Is the Power On light on?

7. Do you have input power to the RC? Is the Power On light on?

8. Is the Low Battery light blinking on the RC? Replace the RC battery and try again.

9. Are any of the fuse lights (F1-F4) on? Power down and replace the fuses.

10. Is the Internal Fault light on? (Isaac 16 only) Power down, remove all connections to the RC except for power, and try again. If the light is still on, contact IFI.

11. Is the Servo Fault light on? (Isaac 16 only) Make sure that all PWM devices are plugged in with the correct polarity. You may also have a PWM device that is drawing more power than the current limiting will allow. Disconnect devices until you identify the device. (If this failure is caused by an intermittent servo that becomes jammed during its travel, try to limit the travel with software.) Try to find an equivalent device with less current draw.

12. Is the No Data/Radio light on the OI solid? Check the radio modem connection again, and replace the cable.

13. Is the No Data/Radio light on the RC solid on? (Isaac 32 only) Check the radio modem connection again, and replace the cable.

14. Is the Search/No Data light on the RC solid on? (Isaac 16 only) Check the motherboard connection again.

15. If the No Data/Radio light is blinking on the RC or OI, make sure that the radio modem is not fully enclosed in metal (as described in Chapter 13), replace the cable, and try again.

16. Are both the Search/TX and Valid RX lights are blinking? Move on to "Checking the Program."

Checking the Program

17. Is the Basic Error light on the OI on solid? There is a programming error. Check the program and reload it into the RC.

18. Is the Basic Init Err light on the RC on solid? There is a programming error and the PBasic code did not initialize the main microprocessor correctly. Check the user program against an unmodified default program and make sure that the Master uP Initialization Routine and communications constants have not been modified or moved.

19. Is the Basic Run Err light on the RC on solid? There is a programming error. Upload an unmodified copy of the default code to RC. If it does not cause the Basic Run Err light to turn on, then there is an error with the custom program. Make sure that the `Goto MainLoop:` and `Stop` statements have not been deleted. Make sure that the program isn't caught in an infinite loop.

20. Is the correct custom program loaded?

21. Are the Search/TX, Valid RX, and Basic Run lights blinking? Move on to the "Checking the Speed Controls" list.

Checking the Speed Controls

22. Do the speed controls have power? Verify battery voltage with a voltmeter.

23. Are the LEDs on the speed controls blinking orange? Check to make sure that the speed controls are connected to PWM outputs (not RELAY outputs) and that they are plugged in correctly at both ends (colors match black to black).

24. Are the LEDs on the speed controls solid orange, with no change as you move the joystick? Make sure that the speed controls are connected to the right PWM outputs according to the program. Also make sure that the joystick is plugged into the correct port on the OI. Use the debug command to verify the output of the PWM channels that the speed controls are connected to. Make sure that you have modified the serout command to accept the correct data.

25. Do the LEDs on the speed controls change from green to red as you anticipate, but there is still no movement? Make sure that the motors are firmly connected to the speed controls.

26. Are the motors permanent-magnet DC motors? Have they been damaged?

27. Are any of the RC digital inputs shorted to ground? This can limit the output of certain PWM channels. Also make sure that any limit switches are not activated.

28. Contact IFI.

Intermittent Failures

If the Search/TX and Valid RX lights are blinking and the Basic Run light is blinking, and the robot behaves normally, but only on an intermittent basis, consult the following lists.

Failures from Basic Motion

1. Does the RC have its own separate battery supply? High current demand from the drive system may cause a battery to drop below the minimum voltage for a short amount of time. Add a separate battery for the RC.

2. Is the RC located near any motors or speed controls? Electrical noise from motors and speed controls may interfere with the radio signal.

3. Have the 20-amp inline resettable thermal circuit breakers been removed from the circuit? IFI ships thermal circuit breakers with all Victor speed controls. Large current surges can cause these thermal breakers to trip. They take a brief moment to reset. They should not be used in any combat robot.

Failures from Shock Loads

4. Is the power on/off switch capable of sustaining high shock loads? Replace the switch with one that can handle high shock loads (as described in Chapter 16, "Wiring the Electrical System").

5. Is the RC mounted with adequate shock protection? Add neoprene or natural rubber sandwich mounts to isolate the RC from shock loads.

6. Are the connectors on the power input terminals tightened down?

7. Are the terminals well crimped? (Tug on them — do they pop out?)

8. Are the battery connectors well crimped?

9. Are the on/off switch connectors well crimped?

10. Contact IFI.

Pneumatics

appendix

C

If you've been impressed by what pneumatics can do for robot weapons and have considered a pneumatic weapon, then this is the place for you. If you're already familiar with pneumatic systems, or have been doing a lot of research on your own, then most of this will be fairly familiar. This information has been relegated to an appendix because it's considered a more advanced subject, and unfortunately, there's going to be a little math.

Pneumatic System Safety

Safety is a major concern with pneumatic systems, and the restrictions and regulations imposed by various competitions are tight. This comes with good reason. Pneumatic systems are more dangerous than any of the power tools in this book because they are potentially explosive. These systems store a lot of energy. As with all high-energy systems, when things go wrong, they can go *really* wrong, and you will have very little time to react.

Below is a list of safety guidelines for working with high-pressure gases. Remember that you are the ultimate judge of safety in your shop, and common sense is your best defense.

➤ The maximum pressure that your system can handle is based on the *lowest* maximum pressure rating of any component in the system.

➤ Always wear safety glasses when testing the system, handling pneumatic components, or filling tanks. A full-face shield is preferable, if available.

➤ Always fill your air tanks at a slow rate, as described in the tank-filling procedure at the end of this appendix.

➤ When testing a new pneumatic system, start with the regulator set to a low pressure to verify its operation. If the low-pressure test is successful, then increase the pressure gradually on subsequent tests, working your way up to the desired system pressure.

➤ Even with major flow restrictions in your system, the speed of pneumatics can be surprising to the builder.

Now That's Loud!

I once asked Geoff Herron, the pyrotechnics expert at Industrial Light and Magic, what the loudest explosion he had ever heard was. Geoff's job is to blow things up for the movies on a daily basis. He and his crew use black powder, gasoline, and all sorts of exotic combinations of explosive compounds. Let it suffice to say that he's an expert on explosions, and he's heard quite a few. The loudest explosion he had ever heard was not an explosive, but a pneumatic pressure vessel that had exceeded its ratings.

- Never put yourself in the way of your pneumatic actuator when the system is charged, even with a motion restraint, as discussed previously in Chapter 18, "Choose Your Weapon." Your motion restraint should be designed so that it can be removed without placing your body in harm's way.

- In high-pressure air systems, petroleum-based solvents and lubricants can contaminate the system and potentially cause an explosion (a condition called *dieseling*). To avoid this, make sure to use silicone lubricant such as Dow Corning 111 silicone compound.

Compressible Fluid Power

Air, like all gases, is a *compressible fluid*, which means that it can be pressurized and manipulated in order to do useful work. These systems have become popular for robot weapons because they can store a lot of energy and produce a large amount of force without heavy motors and batteries. They're also capable of extremely quick response.

PSI

Pounds per square inch, or *psi*, is the unit of measure for the pressure of a compressed fluid. In some formulas included in this appendix, you will also see psia, which is the *absolute scale* for psi. This scale includes atmospheric pressure. For most system applications, however, you will use psig, which is the *gauge scale* for psi. It measures only the amount of pressure applied to the system above atmospheric pressure. All system gauges will read zero pressure (0 psig) at atmospheric pressure, which is around 14.7 psia. For convenience, psi is used for psig in all system applications. Any references to absolute pressure will be specifically designated by psia.

Note In other countries, they use Bar to measure pressure. In this unit of measurement, 1 Bar = 14.5 psi. For example, a 200-Bar tank can handle 2,900 psi.

Gas Types

The two major gases used in robot combat are carbon dioxide, or CO_2, and high-pressure air, or HPA. Since nitrogen (N_2) is very similar to HPA, it's often used in place of HPA in robot systems. The use of these gases has its roots in the paintball community. In the early days of robot combat, we had to adapt whatever available systems were out there to get the job done. Paintball systems were ideal because they provided a lightweight, portable, reliable air source. Also, they were in widespread use at the time (and still are), so they were easy to find and relatively cheap.

Carbon Dioxide (CO_2)

Carbon dioxide has been a standard in paintball for many years. Its main advantage is that when compressed, the gas goes into a liquid state, which allows you to fit a large volume of gas in a small space. CO_2 has a pressure of about 850 to 1,000 psi, depending on the ambient temperature. Since CO_2 systems have been around for a while, their components are usually quite affordable (in paintball terms).

There are two major drawbacks to this system. The first is that the liquid CO_2 has a higher density than the gas, and increases your weight. (You have to weigh in the robot with *full* tanks.) In fact, CO_2 tanks are filled by weight, while HPA/N_2 tanks are filled by pressure. The second drawback is that changing state from liquid back to gas is an endothermic process, which means it draws heat from its surroundings. This causes CO_2 tanks to freeze over when you demand a lot of gas very quickly, and as the temperature drops, so does the CO_2 pressure.

High-Pressure Air/Nitrogen (HPA/N_2)

High-pressure air is actually breathing air, and emerged because the scuba (self-contained underwater breathing apparatus) tanks used by divers were relatively easy to get for paintball enthusiasts, and the filling equipment already existed, allowing paintball shops to provide equipment fills. Nitrogen is readily available from welding supply stores, and has become the alternative of choice for builders who are not certified divers, since many scuba shops will only fill tanks for certified divers.

HPA/N_2 can be compressed to very high pressures, but scuba tanks are usually filled to 3,000 psi. The advantages of HPA/N_2 over CO_2 are that you can have a rapid-fire weapon with no freezing over, and lighter system weight (no adding 12 to 20 ounces every time you fill the tanks). Unfortunately, the higher pressure ratings of HPA/N_2 systems mean that all of the components need higher pressure ratings, which translates into greater cost and, often, heavier components. This is coupled with the relatively recent introduction of HPA/N_2 to the paintball world, which causes these systems to be 3 to 4 times as expensive as CO_2 systems, depending on the part.

Note Different competitions have different pressure limits for HPA/N_2. They are lower than the 3,000 psi rating of most of the components. Common limits are 1,000 psi and 2,500 psi. In your tank-capacity calculations, make sure to use the pressure limit for your specific competition.

System Overview

The following configuration (see Figure C.1) is a by-the-book pneumatic system, incorporating all the necessary safety equipment. Think of the system as a linear diagram. You can trace the flow of air starting in the tank, then through the regulator, past the relief and purge valves, into the accumulator, and through the solenoid valve into the cylinder. The discussion in this appendix will describe each of these components in detail.

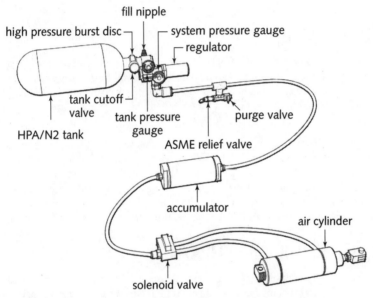

FIGURE C.1: Typical onboard pneumatic system for a combat robot.

Tip

It's a good idea to make a diagram similar to the one in Figure C.1 for your own system. This will help during your safety inspection, because you will probably be required to take the inspector through each component in the system. Also, you should bring all of the data sheets and documentation for each part of the system. The inspector may question you on the psi rating of any component. If it's not printed on the side, you need to be able to produce supporting documents to prove the psi rating. Failure to do so may result in your being barred from competition until you provide the appropriate documents or switch out the component for one whose acceptability can be proven to the inspector. Remember, if the inspectors say you can't compete, then you're out of luck until you satisfy them.

Cylinders

Pneumatic *cylinders*, also called *rams*, are the most common way to execute air power in a weapon system. The cylinder is basically a piston, as shown in Figure C.2. The piston in the middle separates the two sides with an airtight seal, and is connected to the piston rod. Standard pneumatic cylinders have a thin-wall, Type 304 stainless-steel body. The *air valve* (discussed a bit later) routes air into one side of the piston and lets it flow out of the other side. The air pressure coming in pushes on the piston, which moves the rod. To change the direction of travel, you change the direction of air flow, which means that input and exhaust are switched, pushing the piston in the other direction.

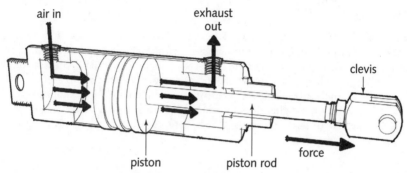

FIGURE C.2: Pneumatic cylinder parts.

Cylinder Force

The force with which the rod is pushed (in either direction) is based on two factors: the surface area of the piston and the air pressure (in psi), as shown in the following formula. Basically, larger-diameter cylinders will produce more force. In fact, each time you double the diameter, you quadruple the force, because the force is proportional to the *square* of the radius.

Cylinder force (pounds) = piston area (in.^{2}) × psi = πr^2 × psi

Air Usage

The air that you use each time you extend or retract is based on the volume of the chamber inside the cylinder, which is based on the surface area of the piston and the stroke of the cylinder, as shown in the following formula. The *stroke* is the change in position from fully retracted to fully extended. Larger-diameter cylinders will require more air, as will longer-stroke cylinders.

Cylinder volume (in.^{3}) = piston area (in.^{2}) × stroke = πr^2 × stroke

Example: Large Bore versus Small Bore Cylinders

To get a feel of what's more important, compare the amount of force versus air usage in the following two cylinders with a similar volume (air consumption):

Cylinder 1 has a 4-inch bore and 1-inch stroke (ridiculously short for comparison):

$$V_1 = \pi r^2 \times stroke = \pi \left(\frac{bore}{2}\right)^2 \times stroke = (3.14)\left(\frac{4\ in.}{2}\right)^2 \times (1\ in.) = 12.56\ in.^3$$

Cylinder 2 has a 2-inch bore and 4-inch stroke:

$$V_2 = \pi r^2 \times stroke = \pi \left(\frac{bore}{2}\right)^2 \times stroke = (3.14)\left(\frac{2\ in.}{2}\right)^2 \times (4\ in.) = 12.56\ in.^3$$

The same volume of air (12.56 cubic inches) will be expended when firing either cylinder. Next, we compare the amount of force produced by each:

$$F_1 = \pi r^2 \times psi = \pi \left(\frac{bore}{2}\right)^2 \times psi = (3.14)\left(\frac{4\ in.}{2}\right)^2 \times (150\ psi) = 1185\ lbs.\ of\ force$$

$$F2 = \pi r^2 \times psi = \pi \left(\frac{bore}{2}\right)^2 \times psi = (3.14)\left(\frac{2\ in.}{2}\right)^2 \times (150\ psi) = 296\ lbs.\ of\ force$$

That 12.56 cubic inches of air will produce 1,185 pounds of force in cylinder 1, but only 296 pounds of force in cylinder 2. The tradeoff here is the stroke length, which means that cylinder 1 will need to be closer to a pivot to get the same swing angle as cylinder 2. Still, as you can see from this comparison, a large-diameter, short-stroke cylinder is preferable to a small-diameter, long-stroke cylinder. As a side benefit, larger-diameter cylinders are also equipped with larger port sizes, which will allow a higher flow. Of course, you can get a large-diameter, long-stroke cylinder to maximize both force and stroke length, but this results in a larger volume, which leads to more air consumption.

Single-Acting and Double-Acting Cylinders

You can select either single-acting or double-acting cylinders. The difference is that *single-acting cylinders* have a single pressure port for a power stroke only in *one* direction. *Double-acting cylinders* have an accessible port on either side of the piston for powered travel in *both* directions. In practice, however, single-acting cylinders should be avoided for high-flow systems because they have a main pressure port with a large orifice, and a tiny exhaust orifice. High-flow systems need as large an exhaust orifice as possible. To remedy this, the builder should use a double-acting cylinder for all applications. Because you're using a double-acting cylinder doesn't mean that you must use *both* ports. You can leave the other port open as a high-flow exhaust, and the cylinder will function just like a single-acting cylinder. Single-acting systems (whether or not they use single-acting cylinders) will need a spring or gravity return to reset the system for the next firing cycle.

Cylinder Mounting Options

There are two main styles of mounting air cylinders: nose mount and pivot mount. As shown in Figure C.3, nose-mount cylinders use a foot bracket for fastening, while pivot-mount cylinders use pivot brackets. A third type of mounting offered by many manufacturers (such as Bimba) is called universal mounting, which combines nose and pivot mounts along with a rear mount.

Cylinder Manufacturers

Bimba, Clippard, Humphrey, Norgren, Parker, and SMC are some of the major cylinder manufacturers in the United States. Models are quite similar among manufacturers to accommodate the possibility of substitution for the repair of a preexisting system (called a drop-in replacement).

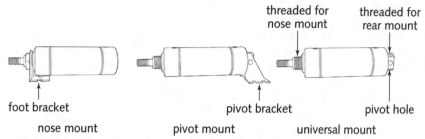

threaded for nose mount

threaded for rear mount

foot bracket

pivot bracket

pivot hole

nose mount

pivot mount

universal mount

FIGURE C.3: Nose, pivot, and universal cylinder mounting styles.

In high-energy, high-flow systems, you are often firing a cylinder at incredible speeds and, usually, there is a great deal of momentum associated with whatever mass you're accelerating that quickly. You must have an end stop or cushion, or nylon webbing, or some other device to limit the travel at the end of the stroke. If you don't, the entire amount of the momentum is likely to blow the end cap off of the cylinder and/or bend the piston rod.

Air Tanks

The *air tank* is the primary air storage vessel that's carried on the robot. It's a sealed, self-contained unit. A larger external supply tank is used to fill the smaller tank in the robot before a match. As mentioned before, one of the great advantages of air power is that a gas stored at a higher air pressure can expand to a higher volume at a lower pressure.

Why not use an onboard compressor? Because of weight considerations, and because they're too slow to produce enough compressed air to power even moderate-flow systems.

Hydraulic Cylinders and Unregulated CO_2

Many builders have expressed interest in developing pneumatic systems for launcher robots similar to those developed by Team Inertia Labs. It's been a great debate in the combat robot community whether the use of hydraulic cylinders with unregulated CO_2 is a wise and/or possible thing. Although empirical evidence suggests that the pressure ratings of hydraulic cylinders hold true when used as pneumatic components, what they don't account for is the incredibly dangerous (and potentially self-destructive) increase in speed and inertia resulting from the use of air instead of liquid. Some British builders have made their own hydraulic cylinders, but the general consensus on both sides of the pond is that if you have to ask, then you shouldn't try. It's very, very dangerous.

Calculating Tank Size

Having already selected your cylinder, you need to make sure that your air tanks have enough capacity to give you the desired number of shots. To get the number of shots, you're going to have to do a few calculations. First, we start out with the *equation of state of an ideal gas*, or the *Ideal Gas Law* for short. It was derived from the work of Sir Robert Boyle, Jacques Charles, and Joseph Gay-Lussac (all *physicists*, not robot builders).

$$PV = nRT$$

In this equation, P is pressure, V is volume, n is the number of moles of gas, R is a constant, called the *universal gas constant*, and T is temperature. The number of molecules (in moles) is used in this calculation instead of mass so that the universal gas constant will work with many different gases. By converting to moles (a dimensionless unit), you can account for the fact that each gas has a different molecular weight. You can then use the Ideal Gas Law to determine how the gas will change based solely on changes in pressure, volume, and temperature, independent of the type of gas. This allows us to rewrite the equation as:

$$\frac{PV}{nT} = R, \text{ which leads to } \frac{P_1 V_1}{nT_1} = \frac{P_2 V_2}{nT_2}$$

If the number of molecules remains constant, which we assume because we are converting, not consuming the gas, then the equation becomes the more familiar:

$$\frac{P_1 V_1}{T_1} = \frac{P_2 V_2}{T_2}$$

If temperature remains constant, then the law reverts to Boyle's Law, which is:

$$P_1 V_1 = P_2 V_2$$

Using the last two equations, we can calculate the gas volume at working pressure for different types of tanks, as shown in the following examples.

Example 1: Gas Volume of a 16-Ounce CO_2 Tank

Given a CO_2 tank filled with 16 ounces of CO_2, we need to find the equivalent volume of gas that this liquid would produce. The molecular weight of CO_2 is 44 grams per mole.

n = 16 oz of CO_2 = 453.59 grams $\times$ (1 mole/44 grams of CO_2) = 10.29 moles

R = universal gas constant = 0.08206 (liters $\times$ ATM)/(moles $\times$ Kelvin)

T = 20° C (room temperature) = 293 Kelvin

P = 850 psi = 57.8 ATM

Plugging these values in to the Ideal Gas Law, we get:
$$V = \frac{nRT}{P} = \frac{(10.29)(0.08206)(293)}{(57.8)} = 4.28 \text{ } liters$$
So, the equivalent gas volume is 4.28 liters, which is 261 cubic inches.

Next, using Boyle's Law, we want to find what this volume of gas at 850 psi will become if converted into the working pressure of 150 psi.

$$P_1 V_1 = P_2 V_2 \text{ can be written as } V_2 = \frac{P_1 V_1}{P_2} = \frac{(850)(261)}{(150)} = 1479 \text{ } in.^3$$

So, the equivalent gas volume at working pressure (150 psi) is 1,479 cubic inches, neglecting thermal deviations. In order to find out how many shots this volume of gas will give you in a single-acting system, where the cylinder is powered only on one side of the stroke, divide the gas volume by the volume of your cylinder. For a double-acting system, where you will be using gas to power *both* sides of the stroke (extend and retract), divide the gas volume by *double* the cylinder's volume. To account for the phase change from liquid to gas in CO_2, van der Waals forces, friction in the system, gas leaks, the rotation of the earth, Greenwich Mean Time, mutating gamma radiation, and other random variables, I would then multiply the number of shots by 80%, which will give you a more accurate picture of the capacity.

Note

In the above example, the temperature of the system was kept constant for comparison purposes. However, in reality, the faster your consumption rate, the more the temperature drops (as the phase change draws heat from its surroundings). This also causes the system pressure to drop, resulting in a lower volume of gas. So, if you need a rapid-fire system, then CO_2 might not be as ideal as HPA/N_2, which does not suffer from thermal effects.

Example 2: Gas Volume of an 88-Cubic-Inch HPA/N$_2$ Tank

The previous example was a bit involved because we had to calculate the volume of the CO_2 based on a liquid weight. This example is more straightforward, since HPA/N_2 tanks for paintball are rated by their actual physical volume. In this case, we can apply Boyle's Law directly given the following information:

P1 = 2,500 psi (competition limit)

V1 = 88 cubic inches

P2 = 150 psi (working pressure)

$$V_2 = \frac{P_1 V_1}{P_2} = \frac{(2500)(88)}{(150)} = 1467 \ in.^3$$

So, the equivalent gas volume at working pressure (150 psi) for this HPA/N_2 tank is 1467 cubic inches, neglecting thermal deviations. Again, for a single-acting system, divide the gas volume by the volume of your cylinder. For a double-acting system, divide the gas volume by *double* the cylinder's volume. To account for random variables, multiply the number of shots by 80%.

Example 3: Gas Volume of an 80-SCF Scuba Tank

Scuba tanks and other nonpaintball tanks are rated in SCF, or standard cubic feet, which is a measure of the volume of gas at standard temperature and pressure. In order to compare these tanks for use in our systems, we need the physical volume of the tank. In this case, we can apply Boyle's Law directly given the following information:

P1 = 14.7 psia (standard atmospheric pressure on the absolute psi scale)

V1 = 80 cubic feet = 138,240 cubic inches

P2 = 3,000 psig (tank rating) = 3014.7 psia

$$V_2 = \frac{P_1 V_1}{P_2} = \frac{(14.7)(138,240)}{(3014.7)} = 674 \ in.^3$$

So, the equivalent physical volume at the scuba tank's rated pressure (3,000 psi) is 674 cubic inches, neglecting thermal deviations, which is about the size of a regular scuba tank. Applying this new information to Boyle's Law again, we get:

$P1$ = 2,500 psi (competition limit)

$V1$ = 674 cubic inches (from above)

$P2$ = 150 psi (working pressure)

$$V =_2 \frac{P_1 V_1}{P_2} = \frac{(2500)(674)}{(150)} = 11,233 \ in.^3$$

So, the equivalent gas volume at working pressure (150 psi) for this scuba tank is 11,233 cubic inches, neglecting thermal deviations. Again, for a single-acting system, divide the gas volume by the volume of your cylinder. For a double-acting system, divide the gas volume by *double* the cylinder's volume. To account for random variables, multiply the number of shots by 80%.

Air Tank Basic Equipment

The *tank cutoff valve* isolates the tank from the rest of the system. This will allow you to fill the air tank for a match, while keeping the rest of the system depressurized (safe). Many competitions require that the weapon system only be pressurized in the arena, when the robot is ready to fight. Since you won't have much time to activate the robot before a match, your tank should be mounted so that the cutoff valve is externally accessible, or you should plan to have an access panel or secondary means to turn on the tank quickly and easily.

The *check valve* is what you use to refill an onboard air tank. It's a one-way valve that permits air to flow only into the tank for filling. Sometimes the check valve is spring-loaded (as in the case of a *pin valve* for CO_2), or it can be just a movable piece of steel that can slide back and forth (as in the case of a *fill nipple* for HPA/N_2).

The *burst disc* is a piece of copper that's usually installed directly on the tank. It's there to protect the tank from rupturing, and has the burst psi stamped on it. Excessive ambient heat (loaded tanks stored in a car trunk on a hot day, for example) can raise CO_2 pressure higher than the tank's ratings. The burst disc will blow and relieve the pressure inside the tank, saving the tank itself from blowing.

The *air pressure gauge* tells you how much gas is left in the tank. It must be externally visible for both CO_2 and HPA/N_2 systems, since this will tell you and/or arena personnel whether air is left in the system, or if the robot is safe.

Tanks for CO_2

Paintball tanks are the cheapest way to get CO_2 onboard the robot. CO_2 tanks are sized by the weight of liquid CO_2 that they can carry, since CO_2 gas goes into liquid form when pressurized. Standard aluminum CO_2 tank sizes are 9, 12, 16, and 20 ounces. Tanks range from 2.5 inches in diameter to 3.25 inches, and from 9.25 to 10.5 inches long. These tanks are relatively cheap, with prices under $40 for brand-new tanks.

As mentioned above, standard paintball CO_2 tanks have a *pin valve* on their output. This spring-loaded valve keeps the tank sealed until an *air system remote* pushes the pin to allow air to flow. The tank has threads that screw directly into the remote, which then connects to the rest of the air system. The remote should have a manual turn screw to depress the pin valve, which serves as the tank cutoff valve in a CO_2 system, as shown in Figure C.4. The remote and pin valve allow the user to easily disconnect the tank for filling, because CO_2 tanks are filled by weight, and it's easiest to weigh just the tank, and not the whole 400-pound robot. This means that CO_2 tanks should be mounted so that they're easily removable from the robot (yet still well secured). A CO_2 fill station has an adapter that screws onto the tank (like the remote) for filling through the pin valve.

Note   If your competition requires a tank-mounted cutoff valve, and the remote system is not adequate, then Smart Parts makes a cutoff valve that replaces the pin valve on the tank, and allows you to shut the tank off without the remote turn screw.

To help get the most shots out of a bottle, the paintball industry has developed something called an *anti-siphon tube*. This is a curved tube that points up away from the liquid, as shown in Figure C.5. This means that there is a definite *top side* to the horizontally mounted cylinder that's usually clearly marked by the paintball shop that installed the tube. Also, since the tube is usually fairly long, tanks equipped with an anti-siphon tube should never be mounted vertically. This will negate the anti-siphon effect, and cause you to pull liquid CO_2 all the time, which can damage seals and cause all other sorts of mayhem in your regulator and valves. If your robot design calls for a lot of inverted driving, then you may want to consider using HPA/N_2, which does not suffer from any difficulties related to the tank's particular orientation.

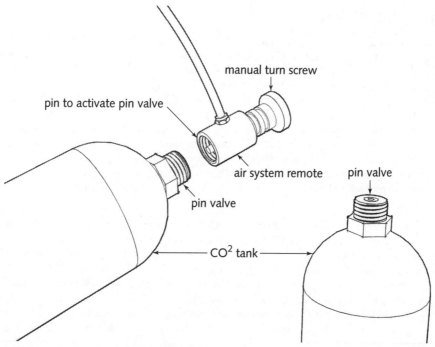

FIGURE C.4: Air system remote with turn screw.

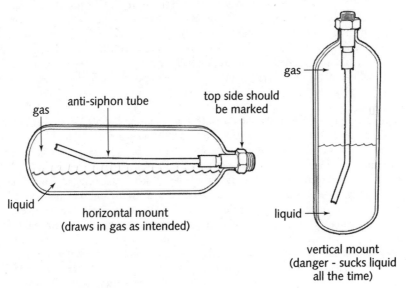

FIGURE **C.5**: Anti-siphon tube for a CO_2 tank.

Tanks for HPA/N$_2$

The paintball folks have also developed lightweight air tanks for use with HPA/N$_2$. These tanks can be steel, aluminum, or aluminum with a glass/carbon fiber wrap. They come in 48-, 68-, and 88-cubic-inch sizes, although you may still be able to find some 114-cubic-inch tanks. Note that all of these tank sizes share a *large* diameter compared to CO_2 tanks. They get shorter as you go to the lower capacities, but the diameter remains about 4.5 inches. This should be taken into account in your robot design. HPA/N$_2$ tanks can be mounted in virtually any position, as long as the tank pressure gauge is visible and the tank cutoff valve is easily accessible. HPA/N$_2$ tanks are usually sold along with the regulators mounted to them as a complete package. Prices of these systems tend to be quite high, as discussed in the regulator section. Since all competitions limit pressure below the maximum rating (usually around 2,500 psi), you won't need to bear the expense of a 4,500 psi tank. A cheaper 3,000 psi tank will be just fine.

A new trend in paintball is to sell HPA/N$_2$ tanks with a preregulator installed by the factory and set to an output pressure of 850 psi. The prereg's output threads match those of standard CO_2 tanks, and it has a standard CO_2 pin valve. The idea is that this tank/prereg system replaces a regular CO_2 bottle so that you can easily retrofit a CO_2 system for HPA/N$_2$. PMI, Crossfire, and a handful of other companies manufacture these systems. If you choose one of these setups (which range from about $150 to $250), you will need to use a separate CO_2 regulator setup, as described in the regulator section. Although they're meant to replace CO_2 tanks, these systems still use the fill nipple (as described below) and are filled by volume (pressure), not weight.

Siphon Tubes versus Anti-Siphon Tubes

As mentioned earlier, what you want is an anti-siphon tube to prevent liquid CO_2 from entering the regulator and valves. Unfortunately, there is also a technology in paintball that *relies* on using liquid CO_2 in the marker (gun). This technology uses a *siphon tube* or *dip tube*, which usually has a weight on the end so that the tube is always immersed in the liquid. This is the opposite of what you want. Make sure to avoid it.

All HPA/N$_2$ paintball systems have a *check valve* for filling the tank (also called a *fill nipple*) with a 1/8 NPT Foster-type FST Series quick-connect fitting, which is the universal standard in the paintball world for nonpermanent (quick-connect) high-pressure fittings. It has a locking ring that slips down and clicks to engage. Note that this is a *temporary* connection used only to fill the tanks. HPA/N$_2$ tanks are filled by monitoring the tank pressure gauge, and can be permanently mounted in the robot (no weighing involved like CO_2). Proper filling procedure is discussed at the end of this appendix.

As mentioned previously, the fill nipple has a movable piece of steel inside. When you're filling an empty tank, positive pressure pushes the steel piece in and allows the air to flow into the tank. When you've reached the pressure that you want, you turn off the air source (big supply tank) and *purge* the line. Purging the line means that you open a valve (called a *purge valve*) that's on the fill line. This makes the pressure in the line (which is now cut off from the supply tank) much lower than the pressure inside of the robot's air tank. The higher pressure in the tank pushes the check valve out and seals off the tank.

Tip

Make sure to get a little rubber cap for the fill valve when you buy the tank. Keeping the fill valve covered when you're not using it will help prevent random bits of dirt and metal from getting into the valve and being blown into the tank during a fill.

Other Tanks

Paintball tanks are the easiest, but by no means the only solution to an onboard air tank. Small fire extinguisher tanks are quite popular on the British circuit. Some modifications to the tanks are required.

The scuba tanks used by divers can be employed in some competitions (check the capacity limits), and SCBA (self-contained breathing apparatus) tanks have been used by a few teams. SCBA tanks are what firefighters and hazardous materials teams wear. They're smaller and lighter than a standard 80-cubic-foot scuba tank, since you don't have the buoyancy of the water to help you hold it up. Unfortunately, they are extremely expensive when purchased new, and can range from $500 to $1,200.

Air tanks are also available from the surplus market, but they tend to be made out of steel, which will add too much weight to your robot. Also, many of them are not equipped with standard NPT threads (described in the fittings section) and will require adapters. By the time you assemble all the necessary adapters and weigh in, you will find that you're not saving all that much money. More importantly, you will need current certifications, or you will have to pay to have these tanks tested, and they may fail anyway.

Tank Certification

All paintball tanks must have a current certification date for *hydrostatic testing*, which is usually printed on the side of the tank. Hydrostatic testing ensures that the tank can still handle its rated pressure. These dates expire, so you've got to make sure to stay on top of it. CO_2 tanks over 2 inches in diameter (12 oz or larger), must be recertified every 5 years, while tanks smaller than 2 inches in diameter are exempt from recertification. Steel and aluminum HPA/N_2 tanks must be recertified every 5 years. If your competition allows filament-wound HPA tanks, those must be recertified every 3 years, and have a 15-year overall lifespan. Annual visual inspections of the tanks are recommended, but not required.

Your local paintball shop should have a source for recertifying tanks. Also, scuba tanks must be recertified on a regular basis (and out-of-date tanks cannot be filled), so your local scuba shop may also have a lead on a recertification company. Make sure to allow 2 to 4 weeks for this service to be performed on your tanks. (Right before competition is obviously *not* a good time.) The fee is usually in the $20 to $50 range, and you should have a new DOT (Department of Transportation) sticker laminated to the tank. HPA/N_2 tanks may get a layer of epoxy, which may change the tank dimensions slightly right around the new label.

Rerouting the Check Valve

During the Las Vegas competition in November 2000, I was filling my air tanks and reached maximum pressure. I turned off the scuba tank and purged the air line, waiting for the check valve to engage again. It didn't. The robot was draining the air tanks through the purge valve on the fill line. Normally, after the pressure in the line is relieved, the pressure inside the tank is higher, and pushes the check valve out to shut off the air flow and seal the tank. This did not happen. Figuring that I had a sticky check valve, I performed a few more fill/purge cycles until it eventually engaged. This happened a few more times, but I managed to get through the competition—I carried three full scuba refill tanks at that time, but it wasted a lot of air. When I got home, I took apart the system and realized the problem. I had removed the check valve from the tank manifold and rerouted it so that I could add a line to another tank. The (extremely) important thing that I missed in my frenzied last-minute precompetition plumbing was that the check valve relies on having something behind it to run into when you fill the tank. (As described earlier, when filling the tank, you provide positive pressure, and this pushes the check valve back to open the seal to the tank and allow air to fill it.) I had put the check valve on a T-connector, and it was getting blown almost all the way to the other side. I was lucky that time, but I learned to avoid plumbing it into a T-connector. It will work with a 90-degree elbow just fine, but make sure to avoid those T's.

Tank Damage

If your tank is damaged during competition, your only choice is to remove it. If it has been cut through, then it's a nice trophy for your opponent, or it goes into the dumpster. If it looks salvageable, then it should not be used until you can have it recertified. If it passes recertification requirements, then it may be used again. If it fails, then it's a trip to the paintball store for a new tank.

Regulators

The *regulator* is the part of the system that takes the high pressure stored inside of the air tank and brings it down to the system pressure. Everything in your system that comes before the regulator is considered the *high side* (about 1,000 psi for CO_2 and 2,000 to 3,000 psi for HPA/N_2). Everything that comes after the regulator is considered the *low side* (150 to 250 psi for both CO_2 and HPA/N_2).

Regulators for CO_2

Since the paintball community has been developing CO_2 systems for years, this can be considered a mature technology. Light weight and reliability are big concerns for paintballers, and you can benefit from their advances. These regulators are designed to be mounted off the tank, with an air system remote, as discussed in the tank section. It's important to get a remote that has a manual screw to activate the tank's pin valve, which will allow you to fully attach the tank to the system and then activate it when you want (as a manual tank cutoff). Otherwise, screwing the remote onto the tank will immediately depress the pin valve and pressurize the system, which is a highly undesirable feature. As mentioned previously, if your competition requires a cutoff valve that is mounted to the tank, you can purchase a Smart Parts cutoff valve that can be installed by the paintball shop for a nominal fee.

One of the more popular CO_2 regulators for robot combat is the Palmer's Pursuit Stabilizer, which costs about $100 and weighs 4.5 ounces. The low-pressure version has a 0 to 400 psi output range. There are many regulators on the market, but finding one with a decent flow rate can be tricky. Consult your local paintball store and ask for recommendations for the high-flow regulators.

A non-paintball regulator that has gained popularity is the Victor SR 310 high-flow CO_2 regulator. This regulator is designed specifically for CO_2 with a maximum inlet pressure of 1,500 psi and an output range of 10 to 150 psi. It's equipped with fins to help prevent freezing from phase change in high-flow applications. Although it has a built-in self-resealing relief valve, the data sheet warns that it's not intended to protect downstream equipment. The drawback of this regulator is its weight of 2.5 pounds, which is much more than a standard paintball CO_2 regulator.

Note   Some competitions require a preapproval waiver for running a high-pressure system. This refers specifically to an unregulated CO_2 system. If you have a regulator, then you have a low-pressure pneumatic system. The pressure that they're talking about is what the actuator (cylinder) actually sees.

Regulators for HPA/N$_2$

Because the pressures involved in HPA/N$_2$ systems are so high, almost all paintball regulators are tank-mounted. This makes for a simple package and avoids having to deal with any high-pressure plumbing issues or adapters. I've used the Smart Parts Max Flow regulator for several years now with excellent results. I've also heard good things about the Nitro Duck Mega regulator. Both of these regulators are sold as complete systems along with the tank. Unfortunately, they cost upwards of $400.

As mentioned in the tank section, you can get a tank equipped with a preregulator (also called a *screw-on HPA tank*) that brings you down to about 850 psi. Since you're only working at CO$_2$ pressures, you can use brass and swivel fittings for plumbing. You'll need a separate CO$_2$ regulator as described earlier.

If you use a non-paintball tank such as a scuba or SCBA tank, then you will have to provide your own regulator. Make sure to observe all of the safety guidelines regarding high-pressure plumbing outlined in the fittings and tubing section that follows. Some competitors are using single-stage cartridge-type regulators from Tescom and GO Regulator. It should be noted that the GO Regulator Model PR-7 has a 1.1 C$_v$, which is quite impressive. Unfortunately, it weighs 3.2 pounds.

Accumulators

An *accumulator* (also called a *buffer tank* or *air reservoir*) is used to provide a reserve of air at the correct pressure, so that the flow of the regulator, which is usually pretty small, doesn't prevent the system from having a fast response. It's placed after the regulator and before the valve in the system. Basically, it's an air tank, but much smaller than the main air tank on the robot. Also, because it's on the low-pressure side of the regulator, it can be rated for a much lower pressure (usually around 200 to 250 psi) and made from thin-wall Type 304 stainless steel, much like a standard air cylinder.

In standard regulated CO$_2$ systems, liquid or vapor CO$_2$ can be sucked through the regulator and into the accumulator in a high-flow situation, where it can expand rapidly and create much higher pressures than the accumulator's rating. Therefore, these systems must have a downstream relief valve (also called a pop safety valve) on the low-pressure side of the regulator to protect the equipment and the operator from damage due to a vessel rupture. Relief valves are discussed in the "Safety Equipment" section later in this chapter.

Not all systems require an accumulator. It's just a way of compensating for a regulator that doesn't have a high enough flow to meet your speed requirements. If extremely high flow is your goal, then you will most likely require an accumulator, because very high C$_v$ (flow rate) values in regulators are pretty rare. The accumulator's volume should be based on the cylinder's volume and your rate of fire. For moderate rates of fire, a 1.5× volume is usually adequate, although as your rate of fire increases, you will need to use a larger reservoir.

Accumulators can be plumbed straight through or with a T-connector, as shown in Figure C.6. Personally, I believe that the through connection is a better configuration, if you have the space. I've used air reservoirs made by Bimba for my robots, but other manufacturers, such as Clippard, offer similar products.

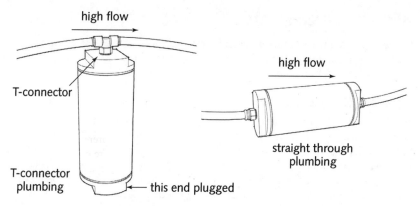

high flow

high flow

T-connector

straight through
plumbing

T-connector
plumbing

this end plugged

FIGURE **C.6: Accumulator plumbing options.**

Tubing and Fittings

Tubing and fittings are the means to connect all of your pneumatic components together. They are an important part of the system, and can be a bottleneck in the performance of a high-flow system if not properly sized. This section will introduce pipe threads and then discuss both high-side and low-side connectors and their usage and installation.

Pipe Threads and Sizes

Plumbing connectors in the United States use the National Pipe Thread, or *NPT*, standard (also called the American Standard Pipe Thread). These are different from regular screw threads because they're tapered along the length of the thread. This is done so that continued tightening of the threaded connector will increase radial pressure, creating a better seal to keep liquids and gases inside.

One of the biggest sources of confusion is that the size designations apparently have nothing to do with the actual dimensions of the fittings themselves. For example, a 1/8 NPT fitting is actually 0.38 inches in diameter. The nonintuitive feature that these sizes are actually based upon is the nominal *inside diameter* of a standard pipe that will accept this thread on the *outside*. (Oh, right. Obviously.) Much like machine screw threads (as discussed in Chapter 8, "Fasteners — Holding It All Together"), pipe threads are designated by the pipe size followed by the number of threads per inch (tpi). For example, the thread for a 1/8-inch inside diameter pipe with 27 tpi is called a 1/8-27 NPT thread. Unlike screw threads, pipe tapers have a single number of threads per inch for each pipe size. Since the number of threads per inch is redundant, it's usually dropped, and a 1/8-27 NPT thread is called a 1/8 NPT thread. A table of standard NPT thread sizes is included in Appendix F, "Tables and Charts."

The rule of thumb to get the maximum flow in a system is to use the largest diameter pipes and fittings that you can accommodate. Usually, the ports on the cylinder will be the limiting size factor of the system. Just make sure that everything else is as large as possible.

High-Pressure (Preregulator) Connectors

The connector choices for the robot builder are limited because of the pressures involved on the high side of the regulator. CO_2 and HPA/N_2 are discussed separately because of the different pressure ranges.

Connectors for CO_2 Systems

The paintball store is a good source for CO_2 fittings. All the brass fittings and adapters that you'll find there are acceptable. McMaster-Carr has a good selection of 1,200 psi fittings listed under "Extruded Brass Fittings." All components for the high side of a CO_2 system should be rated for at least 1,000 psi. (Team Inertia Labs also reports great success with the coated steel versions of these connectors from McMaster-Carr.) Teflon tape should be used on all connections. Proper application of the tape is described later in this chapter.

Usually, you can use the steel braided hose that you find at the paintball shop to bring the output of the remote into the regulator. Swivel connectors (made by Kick Ass Paintball Products and others) can also be used to make the plumbing easier.

Connectors for HPA/N_2 Systems

As mentioned previously, almost all paintball regulators for HPA/N_2 systems are tank-mounted, which means you don't have to worry about plumbing because the high tank pressure goes directly to the regulator. However, if you have a regulator that's not mounted to the tank, then you've got to deal with the higher pressures in HPA/N_2 systems, which require different plumbing than CO_2 systems. All components should be rated for 3,000 psi minimum. Anything less is risking disaster, not to mention not passing the stringent safety inspection. Nothing from the hardware store and few things from the paintball store will be able to help you here. The threads may fit, but those square nickel-plated brass paintball connectors (which are fine for CO_2) are not rated for HPA/N_2 pressures and should be avoided.

A rule of thumb is that regular brass fittings probably aren't going to work, with a precious few exceptions. If you want to be sure, then use steel or stainless-steel fittings. Parker has a series of industrial pipe fittings that meet or exceed all standards for HPA/N_2 systems. They're designed for hydraulic use and are the only type of plumbing that I can recommend for high pressure. They look different from the regular brass and plated brass connectors that you might find at a paintball store or automotive supplier because of their shape. There's an extra flange around the edge as shown in Figure C.7. Also, they have a P cast in the middle of the connector to identify them as Parker parts. Although this line does include brass fittings that are rated for pressures high enough to be used in an HPA/N_2 system, I would stick with the steel or stainless-steel versions. Also, on the high side of all HPA/N_2 systems, I like to use all 1/8 NPT fittings. On the advisement of various paintball sources, I have personally used medium-strength Loctite 242 (blue) threadlocker as a high-pressure sealant with great success, although Loctite 545 pipe sealant is recommended by hydraulic manufacturers. Make sure to clean the fitting and hole by removing all residual tape or old Loctite from the system before applying new sealant. Give it adequate time to cure (I know it's hard to wait) according to the instructions on the bottle.

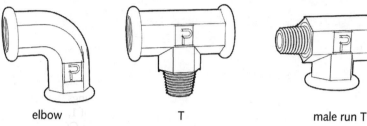

elbow   T   male run T

FIGURE **C.7: Parker high-pressure fittings.**

Note

McMaster-Carr carries a line of similarly rated components listed under "Steel High-Pressure Threaded Pipe Fittings." They have ratings in the 5,000 psi range. Actually, there is one line of brass fittings sold by McMaster-Carr that is rated for 2,900 psi. They are listed under "Brass Threaded Precision Pipe Fittings," and are also known as instrumentation fittings. As mentioned earlier, avoid the extruded and nickel-plated brass fittings, which only have a rating of 1,200 psi.

For high-pressure hoses, the paintball standard is a –4 size (5/16-inch OD) aircraft hose with an extruded Teflon hose with a 0.030-inch wall, a stainless-steel wire-braided cover, and 1/8 NPT male connectors on both ends. This hose has a minimum bend radius of 2 inches. Parker Parflex TFE 919 is a compatible hose rated for 3,000 psi. According to the manufacturer, all 91-series permanent (crimp) fittings are proofed up to the working pressure of the hose, regardless of whether they are brass, steel, or stainless steel. (The different material choices are provided to the end user for chemical compatibility.) You can also use custom-made hydraulic hose, but these tend to be much heavier than the steel braided hose.

Caution

The number one cause of failure with correctly rated steel braided hose is abrasion. Make sure to keep your hoses well isolated from moving chains and gear teeth, or anything else inside of the robot that might rub up against the hose.

Low-Pressure (Postregulator) Connectors

After the regulator, you have a lot more freedom and flexibility because almost everything works with 150 psi (or pressures in that range).

Quick-Connect Fittings

The easiest way to go is to use *push-to-connect* (also called *instant tube*) fittings. On one end, these fittings have standard NPT threads and tighten into the pneumatic system component. On the other end, they have a hole to accept compatible instant-fit tubing. Then you push the tubing into the fitting all the way until it stops. Presto! You're ready to go. In order to remove the tubing, you've got to push the outer ring, as shown in Figure C.8. This will release the fitting's grasp and allow you to remove the tubing without damaging it or ruining the fitting.

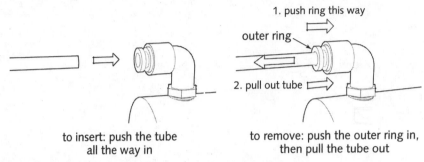

FIGURE C.8: Quick-connect tubing insertion and removal.

There are many types of push-to-connect fittings available to give you flexibility in your plumbing. They also swivel, which gives you the ability to tighten down the connector and then position the port where you need it, and they are reusable. I use glass-filled nylon instant tube fittings because they've got a high enough rating and they're available from McMaster-Carr. Straight connectors, 90-degree elbows, tees, and male run tees are all standard items, as shown in Figure C.9.

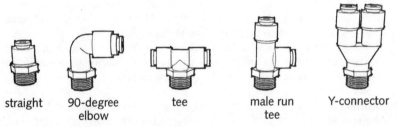

FIGURE C.9: Various quick-connect fittings.

Note In high-flow systems, you should avoid using elbows as much as possible because you will encounter a loss of speed due to air molecules colliding with the wall trying to get around the bend.

Note While instant tube fittings are the easiest to work with, they are by no means the *only* choice you have. Compression fittings (where the fitting is tightened down with a wrench and bites into the tubing) are available in higher psi ratings than instant tube fittings, and offer better resistance to extreme shock loads. (Although correctly installed instant tube fittings should be just fine.)

Speed Controls

Speed controls (also called *banjo valves*) are used to restrict the flow of air with a needle valve so that you get a slower action, which is desirable in some cases. They are usually available as 90-degree fittings with a locking thumb screw (as shown in Figure C.10) to adjust the speed. Even though they are designed to limit speed only in one direction (which is technically a needle valve with a bypass check), they severely limit flow in both directions and should be avoided in high-flow air pathways.

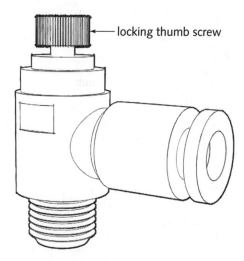

locking thumb screw

FIGURE **C.10: Speed control.**

Quick-Connect Tubing

The idea behind the push-to-connect fittings is that the tube pushes into the fitting and locks. This system allows you to use flexible tubing that can be cut easily with a sharp knife or tubing cutter. The important thing is not to mix inch and metric fittings. A metric hose used with an inch fitting (or vise versa) will probably blow out of its fitting when pressurized, or provide a leaky fit at best.

Both the quick-connect fittings and tubing are specified by the outside diameter of the tube. Nylon, polyurethane, and polyethylene are common tubing materials. One important thing to keep in mind when positioning pneumatic components is the minimum bend radius of the tubing. Since this is specified as a radius, remember that a full 180-degree bend will be twice this number. For example, tubing with a minimum bend radius of 2 inches bent into a half-circle needs a minimum of 4 inches from tip to tip. Also, higher psi tubing tends to be more rigid, and the more rigid the tubing is, the larger the minimum bend radius will be.

It's important to make sure that you cut the hose straight on the end, avoiding any ragged edges that may cause leaks when installed. Grainger has some excellent cutting tools for air hose. I use the #4HL90 handle for most jobs. If you're using stiffer tubing, you may want to consider the #4HL89 model.

Note I've seen some competitors, such as Team Mechanicus (creators of the Judge), plumb their air systems with copper tube instead of nylon, polyurethane, or polyethylene. For low-pressure (150 psi) plumbing, this offers a much, much larger diameter than you can get in standard tubing and fittings. Also, the junctions are as light as the tubing itself, the tubing has a high stiffness-to-weight ratio, and generally higher psi ratings than comparably sized plastic tubing. Team

Mechanicus recommends buying the hardened tubing (not the bendable annealed tubing) and using brazed-on rigid elbows to make turns. I haven't implemented this technique because I personally find brazing the copper a pain, but it works just fine.

Installing Fittings and Tubing

Plumbing can be a nightmare. When laying out tank, valve, and actuator positions, make sure to take into account the minimum bend radius of the tubing. I usually forget, and it turns the plumbing into a last-minute nightmare. You can bend the tubing a little bit tighter, and swiveling connectors certainly help, but in the end, the tubing will win out.

For high-flow systems, avoid sharp bends (such as 90-degree elbows) in the path between the regulator output, the valve, and the cylinder. Keep the tubing runs as short as possible, although make sure that you have enough slack to allow a cylinder to move without pulling on the hose. Instant tube fittings leak when the tube isn't seated correctly because it has too much sideways tension.

Teflon tape or another pipe-thread sealant is required. Some manufacturers have a dry-type sealant applied to the threads of their connectors. In this case, you don't need any tape. In all other cases, you should use Teflon tape (see proper wrapping in Figure C.11). You start with the fitting in your right hand with the threads facing away from you. Stretch out the Teflon tape across the threads and hold it down with your right thumb. The idea is that the tape should tighten when you tighten the fitting. Give it a few wraps, making sure not to get the tape too close to the tip of the fitting that goes inside of the thread. (You should try to leave two threads exposed at the tip.) After four or five wraps, pull the tape tight so that it breaks off and smooth it down. Then, the fitting's ready to be tightened.

1. Start with the threads facing away from you. Pinch the tape to the threads with your thumb. Don't start too close to the tip. Leave a few threads exposed.

2. Keeping the tape tight, give it four or five wraps around the threads making sure to stay away from the tip.

3. Hold onto the tape with your thumb and pull the roll until the tape snaps. Smooth the tape down onto the threads and you're ready to install the fitting.

FIGURE **C.11: Proper Teflon tape wrapping procedure.**

Fittings should be hand-tightened and then torqued with a wrench an additional two turns. The open end of a combination wrench is the correct tool for tightening pneumatic fittings. Adjustable wrenches may be used, but you risk damaging the flats on the fittings. Slip-joint pliers should never be used on pneumatic fittings.

Fixing Leaks

Identifying a leak may be difficult to do by eye. Most of the time it involves a slight hissing sound that's hard to pinpoint. One of the best ways to catch a leak is to prepare a cup full of soapy water and dab it around your pipe joints. A leak should immediately reveal itself by bubbling the water. Immediately depressurize the system completely to make it safe. Then, check to make sure that there is some sort of thread sealant in action. If you can see the edge of the Teflon tape, for example, then tighten down the fitting a little bit more. If you find no evidence of a thread sealant, then you must disassemble the unit and apply a thread sealant.

You may be asked to fill your pneumatic tanks well ahead of time and then line up and wait for a while. You can't predict how fast or slow the line will move. Tightening up your leaky pneumatic joints before competition will help you deal with long wait times. By fixing air leaks, you prevent your precious air supply from slowly trickling out, leaving you without a weapon in the arena.

Safety Equipment

In the old days of robot combat, safety equipment was virtually nonexistent. You had to be really, really careful. The items in this section are excellent additions to any pneumatic system and are required at many competitions.

Pressure Gauges

There are two air pressure gauges in a standard pneumatic system. The first gauge tells you how much air is left in the tank, and is either connected directly to the tank itself or is plumbed into the system before the regulator. The downstream pressure gauge tells you whether the weapon system itself is charged. Both gauges should be externally visible so that you can tell with a quick glance if the robot is safe or if it's still potentially dangerous.

Why do you need both? As long as there is enough air in the tank to run the regulator, the low-pressure gauge will continue to read the system pressure, which is usually 150 psi. This doesn't give you any indication as to what's left in the tank, however, which makes the high-pressure gauge useful.

The high-pressure gauge should have enough range to be able to read the tank pressure with reasonable accuracy. For CO_2 systems, you will need a 0 to 1,200 psi gauge, while HPA/N_2 systems need a 0 to 4,000 psi gauge. The low-pressure gauge should be 0 to 300 psi. Since air does not flow directly through them, you don't need huge orifice size. Usually 1/8 NPT is all you need. On the low-pressure side, you will need to create a branch for the gauge with a T-junction. You will also have to do this on the high-pressure side if the gauge isn't already equipped on the tank.

Relief Valve

The purpose of the *relief valve* (or *pop safety valve*) is to prevent excessive back pressure on the low side of the system. The valve is spring loaded, and the psi at which it activates is determined by the strength of the spring and should be stamped on the side of the valve, as shown in Figure C.12. They should be rated for no more than 30 percent over (1.3 times) the rating of the lowest-rated component in the system.

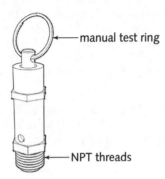

manual test ring

NPT threads

FIGURE **C.12: Relief valve.**

When a pressure higher than the activation psi is present in the low side of the system, the relief valve kicks in and opens an orifice to bleed the excess system pressure. Because it's spring loaded, once the excess pressure is relieved, it self-seals and allows the system to continue doing its job as before. An excess pressure situation can happen if a cylinder experiences a force that's larger than the one that it's pushing with. The larger force wins out and the cylinder is compressed. This creates a larger psi in the whole low-pressure system.

By the way, this all happens extremely quickly, and it sounds like a loud "pop." If the overpressure situation persists, such as a sticky regulator, it makes a loud "blat" sound. If this happens, make sure to clean regulator thoroughly and check the system pressure before continuing.

Note

Many competitions call specifically for an ASME-certified valve. ASME stands for American Society of Mechanical Engineers, and this certification guarantees that the valve complies with ASME Code Section VIII for unfired pressure vessels. Although some relief valves are adjustable, the ASME valves are fixed and must be ordered for the psi that you need. They also have some means to test the system by a lever or pull ring.

Purge Valve

The *purge valve* (also called a *plug valve*) is usually a ball-type turn valve, as shown in Figure C.13 below. It is used to drain (purge) the low side of the system of air pressure, rendering it safe. It is closed (no flow) when the low side is pressurized for battle (making the weapon active) and is opened (full flow) at the end of a match to make the robot safe for transport. The purge valve doesn't need to have a high pressure rating since it will be on the low side of the regulator, only looking at about 150 psi. It also does not need to have an especially large orifice, either, since it is not involved in the high-speed air flow path. It's usually plumbed to the side of a T-junction, so that the high-flow air has a straight through path. The purge valve is only really used at the conclusion of a match to make the system safe.

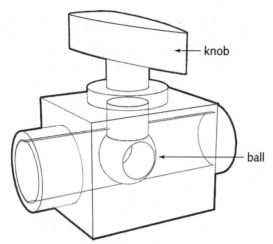

turning the knob rotates the ball,
allowing air to flow through the valve

FIGURE C.13: Ball-type purge valve.

Valves

The job of the *air valve* (also called a *pneumatic solenoid valve*) is to route air into and out of the cylinder or other actuator, which determines the direction of travel for the cylinder. In this section, I'll go through each of the options that you need to consider when selecting an air valve, with special attention paid to high-flow systems.

Flow Rate (C_v Factor)

The commonly accepted measure for the flow rate of a system component is the C-sub-v factor (designated as C_v). The C_v factor is critical in optimizing your system for high-speed, high-flow operation. This section will help you calculate the necessary C_v factor for your valve using available information.

Calculating Required C_v Factor

So by now, you've probably selected the cylinder that you want based on the amount of force that you want. You've selected the tank size to give you an adequate number of shots. The next step is to calculate the C_v factor of your system and make sure that the valve has enough flow to handle the actuation speed that you desire. The C_v factor is the most critical performance factor in selecting a valve. Basically, higher speeds mean higher air flow, which means a higher C_v factor.

1. Find the internal volume of the cylinder in cubic inches, as described in the cylinder section. This is done by multiplying the area of the piston (in square inches) by the stroke length (also in inches).

$$Volume = \pi r^2 \times stroke = \pi \left(\frac{bore}{2}\right)^2 \times stroke$$

2. Find the cylinder pressure Pc. This is usually 150 psi, but is essentially whatever system pressure you're running. You should run the highest system pressure that your components will allow, because this helps determine the force of the cylinder as shown next.

$$Force = \pi r^2 \times Pc = \pi \left(\frac{bore}{2}\right)^2 \times Pc$$

3. Find the compression ratio, which compares the cylinder pressure Pc to atmospheric pressure Pa, which is 14.7 psia.

$$CompRatio = \frac{Pc + Pa}{Pa} = \frac{Pc + (14.7)}{(14.7)}$$

4. Find the CFM (cubic feet per minute) of flow, which relies on the cylinder volume (in cubic inches), compression ratio, and time (in seconds) to achieve the full stroke travel. The 28.8 in the denominator is a constant.

$$CFM = \frac{V \times CompRatio}{time \times 28.8}$$

5. Multiply the CFM by the "A" factor from Table C.1, which takes into account several constants to simplify the C_v formula. I usually use a 2-psi drop for my calculations. (The *drop* refers to the loss in pressure that you're willing to accept.)

$$C_v = CFM \times A$$

Table C.1   "A" Factor for Various System Pressures

Inlet Pressure (psig)	Compression Factor	"A" Constant 2 psi drop	"A" Constant 5 psi drop	"A" Constant 10 psi drop
10	1.6	.155	.102	
20	2.3	.129	.083	.066
30	3.0	.113	.072	.055
40	3.7	.097	.064	.048
50	4.4	.091	.059	.043
60	5.1	.084	.054	.040
70	5.7	.079	.050	.037
80	6.4	.075	.048	.035
90	7.1	.071	.045	.033
100	7.8	.068	.043	.031
110	8.5	.065	.041	.030

Inlet Pressure (psig)	Compression Factor	"A" Constant 2 psi drop	"A" Constant 5 psi drop	"A" Constant 10 psi drop
120	9.2	.062	.039	.039
130	9.9	.060	.038	.028
140	10.6	.058	.037	.027
150	11.2	.056	.036	.026
160	11.9	.055	.035	.026
170	12.6	.054	.034	.025
180	13.3	.052	.033	.024
190	14.0	.051	.032	.024
200	14.7	.050	.031	.023

Source: Parker Industrial Pneumatic Technology Bulletin 0275-B1.

Example: Calculating C_v

This example will use a 3-inch bore cylinder with a 4-inch stroke, being run at 150 psi. The required time of actuation is 0.1 seconds.

$$Volume = \pi\left(\frac{bore}{2}\right)^2 \times stroke = (3.14)\left(\frac{3\ in.}{2}\right)^2 \times (4\ in.) = 28.274\ in.^3$$

$$Force = \pi\left(\frac{bore}{2}\right)^2 \times Pc = (3.14)\left(\frac{3\ in.}{2}\right)^2 \times (150\ psi) = 1,060\ lbs.\ of\ force$$

$$CompRatio = \frac{Pc + Pa}{Pa} = \frac{(150) + (14.7)}{(14.7)} = 11.2$$

$$CFM = \frac{V \times CompRatio}{time \times 28.8} = \frac{(28.274\ in.^3) \times (11.2)}{(0.1\ sec) \times 28.8} = 110$$

$$C_V = CFM \times A = (110) \times (0.56) = 6.16$$

So, for the above cylinder to fire in the given time, a valve with a C_v of 6.16 or greater will be needed. The cylinder will consume 28.274 cubic inches of air on each activation (double that if powered in both directions), and will have a resulting force of 1,060 pounds at a system pressure of 150 psi.

Ports and Ways

Technically, *ways* are paths for air to flow through the valve, and *ports* are physical holes on the valve body that connect to these paths. However, they are often confused for two reasons: the nonobvious way to count the number of ways through a valve, and the fact that, for the most part, the number of ports corresponds to the number of ways. Also included in this section are examples to help demonstrate some of the options available in selecting a valve and as a way of selecting which strategy best suits your weapon.

Open/Closed Valve Conventions

Those readers who are familiar with electronics may find the open/closed conventions for pneumatic valves counterintuitive. Why? In electronics, normally-closed switches and relays *allow* the flow of current until activated. In pneumatics, normally-closed valves *prevent* the flow of air until activated, as shown in Figure C.14.

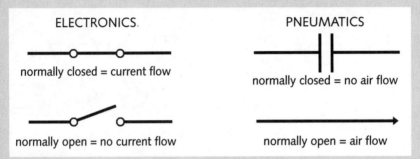

FIGURE C.14: **Electronic switches versus pneumatic valves.**

To complicate matters, you will usually interface your pneumatic valve to the control system with an electrical servo switch. The servo switch will control the voltage that goes to activate the valve's coil. In this case, the normal configuration is for a normally-open electrical switch to control a normally-closed pneumatic valve.

The Two-Way Valve

A two-way valve has two ports: an inlet and an output. This is used to provide an on/off switching function. These valves can be purchased in either NO (normally-open) or NC (normally-closed) configurations. A normally-closed valve will only allow air to flow when it's energized. A normally-open valve closes off air flow when it's energized. This type of valve cannot repeatedly fire a cylinder by itself, because there is no way to exhaust the cylinder with a single switch. However, when used in pairs (as in the following example), two-way valves can control single-acting cylinders.

Example: Dual Two-Way Valves (Single-Acting Cylinder)

In this application example, two normally-closed two-way valves are used: one for positive pressure, and one for exhaust, as shown in Figure C.15. Valve A connects the air source to the cylinder. Energizing this valve will fire the cylinder. One end of valve B is connected to the cylinder and the other end is left open. Energizing this valve will exhaust the cylinder. Since the cylinder is powered in one direction only (single acting), it needs a spring or gravity return to retract it and reset the system for the next activation. This will give you the fastest response for a single-acting situation because you can get larger C_v rates on individual valves than you can with a three- or four-way valve. The switching is a little complicated because you will need to fire the solenoids individually. Usually, cylinders are used in the single-acting configuration for lifters and launchers, where power only needs to be applied in one direction.

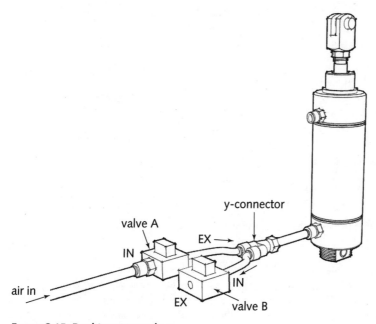

FIGURE C.15: Dual two-way valves.

The Three-Way Valve

A three-way valve usually has three ports: inlet, output, and exhaust. This type of valve can be used to fire and retract a single-acting cylinder on its own. When used in pairs, they can control a double-acting cylinder.

Example: Single Three-Way, Two-Position Valve (Single-Acting Cylinder)

This method uses a single three-way valve with the output port connected to a single-acting cylinder as shown in Figure C.16. This valve takes the place of the dual two-way valves in the first example. This valve routes air into one end of the cylinder to provide powered motion in one direction only. It also exhausts the cylinder as in the previous example, requiring a gravity or spring return. The normally-closed version of this valve provides a simple single-valve solution for launchers and lifters. Energize the valve to fire the cylinder. Remove power and it exhausts.

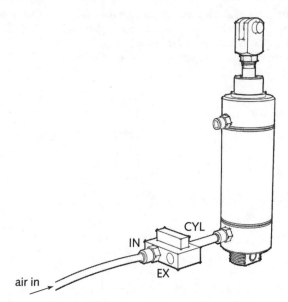

FIGURE C.16: Single three-way, two-position valve.

Example: Dual Three-Way, Two-Position Valves (Double-Acting Cylinder)

Unlike the previous example, this coupling of two three-way valves (shown in Figure C.17) allows powered motion in *both* directions of travel with a double-acting cylinder. A normally-closed valve is placed on either port of the cylinder, and one valve is energized to extend the cylinder, while the other valve is energized to retract the cylinder. By putting a speed control on the valve that exhausts while retracting the cylinder, you can get a slower action that's much more gentle on your robot. When neither valve is powered, the cylinder will move freely. This method is used in a situation where it's desirable to actively pull a weapon back in, which might be the case with retracting a hammer, for instance. Because the cylinder can move freely when neither valve is powered, the rest position for the arm should be stable or spring-loaded.

The Four-Way Valve

A four-way valve may have four or five ports, as shown in Figure C.18. In the four-port configuration, it has an inlet, a cylinder A output, a cylinder B output, and a common exhaust. In the five-port configuration, the valve has the same inlet and output ports, but separate exhausts for A and B. Since A and B will only be connected to their own exhaust ports, the five-port configuration is still technically a four-way valve. Why would having separate exhausts be a useful thing? So that you can retract a cylinder much more slowly than you extend it using a speed control, which is a commonly desirable trait when firing a hammer. It would be better to pull the hammer back in more slowly so you don't pound yourself.

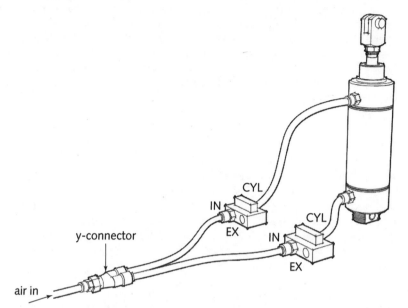

FIGURE C.17: Dual three-way, two-position valves.

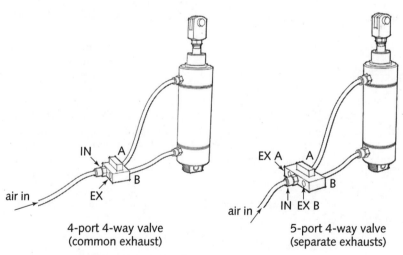

4-port 4-way valve
(common exhaust)

5-port 4-way valve
(separate exhausts)

FIGURE C.18: Four-way valve connection.

A four-way valve can control a double-acting cylinder by itself by routing positive pressure at the inlet to one output port while exhausting the other. When the valve is energized, it switches inlet and exhaust directions, which reverses the travel of the cylinder. Although this is an easy configuration to work with, to get maximum flow, you might consider dual three-way valves to control a double-acting cylinder as discussed previously. Four-way valves are further classified by the number of *positions* they have. These are possible switching positions for spool, which actually does the routing of the air inside of the valve.

Note   Some manufacturers have five-way valves, but these are technically four-way valves with five ports, as described previously.

Four-Way, Two-Position Valve (Double-Acting Cylinder)

A *two-position valve* has only two states. The cylinder is either fully extended or fully retracted. This is the easiest type of valve for the pneumatics novice to implement, because the control is easy, and can be interfaced with a single servo switch. Apply power, and the cylinder moves to one end. Remove the power, and it goes back to the original position. This type of valve uses a single solenoid (described next). An example of a weapon employing this type of valve would be a striking weapon with an unstable neutral position (head held aloft).

Note   If you're keeping something aloft (like a hammer head), you might be better off with a three-position open-center valve and a spring to hold the arm in place, rather than a two-position valve. The reason is that you want the ability to bring the arm back under power, but for the maximum speed in your stroke, the cylinder must be completely exhausted before attempting to fire. The back pressure will slow you down. You probably don't need a huge amount of force to keep the arm aloft.

Four-Way, Three-Position, Closed-Center Valve

A *three-position, closed-center* valve allows you to stop the cylinder at any time in the travel, and it will stay in that position. Closed-center means that when the valve is in the stopped (center) position, there is no exhaust to either side of the cylinder. This doesn't mean that the cylinder stops in the exact center of the stroke. It can stop anywhere and hold — midstroke, near either end, or at the end — it doesn't matter where. As long as the solenoids are not energized, there is no air flow. No air flow means that the cylinder maintains whatever position it's at. When you reenergize the solenoid, the cylinder moves in the direction of air flow. This type of valve can be used with a lifting arm and speed controls. The control for this type of system is a bit more complicated and requires two servo switches for running a double-solenoid pilot, as described later. An example of a weapon using this type of valve is a lifting arm that can be stopped at any point in the lift.

Four-Way, Three-Position, Pressure-Center Valve

A *three-position, pressure-center* valve allows you to stop the cylinder at any time in the travel and it will stay in that position, exactly like the closed-center valve above. However, pressure-center means that instead of cutting off the flow of air completely to immobilize the cylinder, both sides of the cylinder are connected to positive pressure. This type of valve behaves exactly as the closed-center version does, although some manufacturers warn that because of minor pressure inequities, the cylinder may creep to the fully extended position, and recommend the closed-center valve for stopping the cylinder mid-stroke.

Four-Way, Three-Position, Open-Center Valve

A *three-position, open-center valve* connects exhaust to both sides of the cylinder when it's in the rest (center) position. This type of valve is best used in a high-speed system. In some weapon systems, speed counts for a lot. Even a small amount of speed difference can mean a large amount of force lost. In the case of an open-center valve, you get the maximum speed response because you're not trying to exhaust a chamber full of pressurized air at the same time that you're filling the other side. Having air present on the other side of the piston will slow down the response because of the need to push it out at the same time that you're pushing air in. This also means that since the cylinder is exhausted on both sides at rest, it's free to flop around unrestrained. If this cylinder is running an arm, then it's best to have the arm fold back into the body, with a spring return or just using gravity. The control for an open-center valve also requires two servo switches for running a double-solenoid pilot, as described later in this chapter. An example of a weapon using this type of valve would be a hammer with a stable neutral position (folds back into the body).

Solenoid Pilots

The *pilot* is the part of the valve that controls the switching. Most of the time, you will use a *solenoid* (electrical) pilot, which is a lot like an electrical relay because it creates a magnetic field that causes a physical motion. In this case, the field moves a spool that changes the direction of air flow. You can choose the solenoid activation voltage. Commonly available voltages are 12 volts DC, 24 volts DC, and 120 volts AC. You should choose the DC voltage that works best with the voltages available in your robot.

Caution

You have a choice of which is the default condition when no power is applied to the solenoid. This should be made so that the weapon arm is in the safest position when the system has no power. Also remember that you should have a motion restraint to prevent the arm from accidentally firing during your power up sequence in the arena.

Single-Solenoid Pilot

All of the two-position valve examples described earlier (two-way, three-way, and four-way valves) use a single-solenoid configuration, which uses a spool with a spring return. The valve is in one position when deenergized. To switch positions, you energize the solenoid. To return to the original position, you deenergize the solenoid, and the spring pulls the spool back. This type of valve can be controlled with a single switch or relay.

Double-Solenoid Pilot

The four-way, three-position valve examples described earlier employ a double-solenoid configuration, which uses a spool that is pushed back and forth, with a spring that returns it to the center. Energizing a solenoid at either end of the spool changes the position. When both solenoids are deenergized, the spool returns to the center. This type of valve requires a servo switch for each solenoid.

Air Pilots

Some high-flow valves require an air pilot, instead of a solenoid pilot, for activation. In place of a magnetic field to move the spool, air power is used. The user is provided with a pilot port, where the air goes in. In order to interface this to the robot, however, you will still need a solenoid (albeit a much smaller one) to route air into the pilot port. Because this adds an extra component to the system (and yet another part that can fail), air pilots should only be used when necessary. It isn't generally worth the extra weight and cost.

Valve Manufacturers

I've personally used MAC four-way and Norgren Prospector valves in my robots, although there are many different manufacturers to choose from, including SMC, Parker-Skinner, Festo, and KIP to name a few.

The Fill System

Getting air into the system safely is a major concern. This can be as dangerous as the entire pneumatic system itself. Flying hoses can slap you in the face or hands. Ruptured connections can send a cloud of CO_2 flying into your face. You should make sure that your filling apparatus is properly rated for the type of gas you are using, and that it is in good working order.

CO_2 Fill Stations

The most common source for big CO_2 supply tanks is your local welding supply shop. They stock CO_2 tanks and refill the tanks. As far as the actual filling apparatus goes for paintball tanks, the paintball shop should have an off-the-shelf fill station that attaches directly to the supply tank, as shown in Figure C.19. It's important that the fill station has an air system remote that has a manual screw to activate the tank's pin valve, which will allow you to fully attach the tank to the fill station and then activate it when you want. Otherwise, screwing the remote onto the tank will immediately depress the pin valve and pressurize the line, which is a highly undesirable feature. If you have any questions, the people at the paintball shop are usually more than happy to tell you all about the equipment and how to use it.

You'll also need a scale that has the appropriate resolution for your tank size to make sure that you're not overfilling your CO_2 tanks. They should only be filled to their rated capacity. For example, a 20-oz CO_2 tank should be placed on the scale and zeroed. Then you should add CO_2 until the scale reads 20 ounces. No more.

Caution

Don't overfill your CO_2 tanks. You run the risk of stretching the burst disc, which is there to prevent tank rupture in an overpressure situation. Compromising this safety device, whether it's intentional or not, is unacceptable.

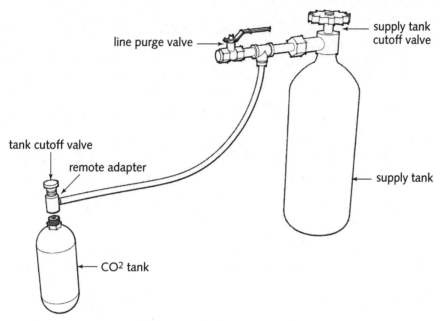

line purge valve

supply tank
cutoff valve

tank cutoff valve

remote adapter

supply tank

CO^2 tank

FIGURE **C.19: Typical paintball-type CO$_2$ fill station.**

HPA/N$_2$ Fill Stations

The most popular source for an HPA supply tank is your local scuba shop. As mentioned before, standard scuba tanks are rated for 3,000 psi and have a capacity of 80 cubic feet, which is enough to refill smaller paintball tanks numerous times. Some of the larger competitions provide air-tank filling for free to competitors, monitored by a sponsoring welding supplier, usually for safety and insurance reasons. You should find out ahead of time if you will be required to provide your own air supply during the competition. If this is the case, you may need more than one tank to get you through a weekend.

Don't walk into a scuba shop thinking that you have the right to purchase any equipment that you want and use it however you wish, especially if you're not a certified diver. You don't. They have a legal responsibility to make sure that the equipment they sell is properly used for the purpose for which it was intended (which, by the way, is not robot combat). However, it's not unheard of to use scuba tanks for purposes other than diving. Airbrush artists use them as a quiet alternative to a noisy compressor. More recently, paintball players have been using them to fill their HPA tanks. The key is to talk to the manager or owner of the shop to see if he or she is willing to help you out. You may be required to sign a waiver that says that you won't be using the supply tanks for their intended purpose. Make sure to get any and all training that is offered, and make sure to ask questions so that you have a full understanding of all of the dangers involved.

I would avoid trying to make an HPA/N$_2$ fill rig using individual parts from a scuba shop. It's possible, but a whole lot easier to get something premade. Your best bet for a fill rig is the local paintball shop. Many shops now carry fill rig assemblies that fit scuba tanks. The Java SCUBA fill station (shown in Figure C.20) clamps directly to the standard K-valve and gives you a high pressure gauge, which does not come standard with scuba tanks. It retails for about $60. The only drawback is that the fill nipple is directly on the fill station, which is fine for most paintball applications, but several-hundred-pound robots would like to stay on the floor while you fill them. To remedy this, you can go a few different ways. You can purchase a coiled remote from the paintball shop that's rated for 3,000 psi. (Don't be surprised if the coil stiffens during a fill.) You can get a custom hydraulic hose from a hydraulic shop (look wherever there are bridges and trucks). Finally, you can order a premade Teflon hose with a stainless-steel braided cover from McMaster-Carr (#4468K201) in a 24- to 36-inch length with male 1/8 NPT connectors on both ends for about $15.

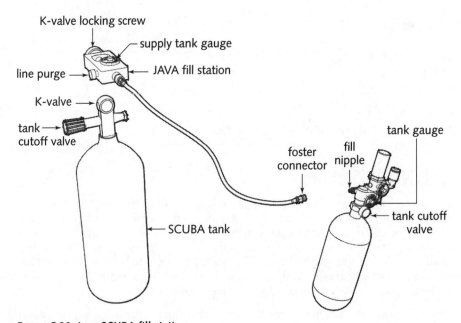

FIGURE **C.20: Java SCUBA fill station.**

Nitrogen tanks, much like CO$_2$ tanks, can easily be acquired from your local welding supply shop. Unfortunately, since the use of nitrogen has only recently gained popularity, paintball shops have not yet started stocking fill stations specifically for nitrogen. The CO$_2$ fill stations are not rated for the same pressure, and use different fittings. You can, however, make your own fill rig with adapters from a welding supply store and appropriately rated hose and Foster-type adapters, as described in the fittings section. It is important to add an appropriately rated high-pressure purge valve to the line so that you can depressurize the line, allowing the check valve to engage, as described later in this chapter.

Proper Tank-Filling Procedures

Outlined next are the proper filling procedures for CO_2 and HPA/N_2 systems. These are general guidelines and should be customized to whatever additional safety procedures are deemed necessary by your fill station manufacturer and supply tank provider. The best training comes from the source. Wherever you buy your pieces of filling equipment, you should ask the shop to give you a brief rundown of the system's operation. Don't be afraid to ask questions until you feel completely comfortable with the equipment.

 Caution Proper eye protection should be worn at all times while filling and handling pneumatic tanks. A full-face shield is recommended for the filling procedure.

CO$_2$ Systems

The following procedure applies to standard paintball CO_2 tanks with a pin valve. Although screw-on type preregulated HPA tanks have a pin valve, you should use the procedure for HPA/N_2 tanks for them.

1. Remove the CO_2 tank from the robot and place it on the scale.

2. Make sure that the manual screw on the fill station's air system remote is unscrewed all the way so that it cannot accidentally activate the pin valve.

3. Screw the fill station's air system remote onto the air tank's pin valve.

4. Close the fill station line purge valve.

5. Zero the digital scale, or record the current weight.

6. Slowly turn the manual screw on the fill station's air system remote clockwise to activate the CO_2 tank's pin valve.

7. Slowly open the supply tank valve, and allow the tank to fill at a slow rate.

8. It may take several minutes to reach the desired weight of CO_2. Filling a tank too quickly will generate a lot of heat, and may damage the tank.

9. When you reach the desired weight, turn off the supply tank valve.

10. Turn the manual screw on the fill station's air system remote counterclockwise to deactivate the CO_2 tank's pin valve.

11. Open the fill station line purge valve.

12. Unscrew the fill station's air system remote from the air tank's pin valve.

13. This tank is now filled and ready for combat.

14. Make sure that the robot's purge valve is open.

15. Make sure that all weapon safety restraints and covers are in place.

16. Make sure that the manual screw on the robot's air system remote is unscrewed all the way so that it cannot accidentally activate the pin valve.

17. Screw the robot's air system remote onto the air tank's pin valve.

18. Secure the air tank in its holder. The robot is now ready for combat. Do not remove the weapon safety restraints or covers until the robot is in the arena.

Don't overfill your CO_2 tanks. You run the risk of stretching the burst disc, which is there to prevent tank rupture in an overpressure situation. Compromising this safety device, whether it's intentional or not, is unacceptable.

HPA/N$_2$ Systems

The following procedure applies to all HPA/N$_2$ systems, including the newer screw-on-type preregulated HPA tanks that are intended to replace CO_2 tanks.

1. Make sure that the robot is in a stable position and cannot move. All weapon safety restraints and covers should be in place.

2. Make sure the Foster connector is pushed all the way down onto the fill nipple, and that the locking ring clicks before applying any air pressure of any kind.

3. Close the robot's purge valve.

4. Close the fill station line purge valve.

5. Slowly open the tank cutoff valve. (With screw-on type preregulated HPA tanks, skip this step. They don't need to open the tank cutoff valve.)

6. Make sure that the tank pressure gauge is clearly visible.

7. Slowly open the supply tank valve a quarter turn, and listen for the click of the check valve. Once you hear the click, turn the supply tank valve a little bit more and then hold at that position allowing the tank to fill at a slow rate. If you do not hear a click, stop immediately and make sure that the tank cutoff valve is open.

8. It may take several minutes to reach your desired tank pressure. At a recommended rate of no higher than 600 psi per minute, this means a *minimum* fill time of 4 minutes and 10 seconds to reach 2,500 psi. Filling a tank too quickly will generate a lot of heat, and may damage the tank. It also increases the risk of dieseling, as described in the safety section.

9. When you reach the desired pressure, turn off the supply tank valve.

10. Close the tank cutoff valve. (With screw-on type pre-regulated HPA tanks, skip this step.)

11. Open the fill station line purge valve. The HPA/N$_2$ should vent, and the check valve should close off the flow of air from the tank.

12. Open the robot's purge valve.

13. Disconnect the Foster connector from the fill nipple.

14. The robot is now ready for combat. Do not remove the weapon safety restraints or covers until the robot is in the arena.

Keep a close eye on the robot's pressure gauge. Some nitrogen and scuba tanks can provide fills above the pressure ratings of paintball tanks. Do not overfill the tank, or damage may occur.

Online Resources

In the past few years, the amount of information and resources available to builders has exploded. Here are many of the most popular online resources for combat robot builders, sorted by category.

Table D.1   Competitions

Web Site	Notes
BattleBots http://www.battlebots.com	The largest competition in the United Sates. Negotiations are underway to bring this competition back onto TV. Check the Web site for updates.
Robot Fighting League (RFL) http://www.botleague.com	A growing collective of local fighting robot events. There is a unified rule set that can be scaled to the size of the available arena. This is currently your best bet for participating in a competition in the United States.
Robot Wars http://www.robotwars.com	Hugely popular in England, this competition is televised weekly there, and is the only remaining show currently on TV.

Table D.2   Metals

Web Site	Notes
Aircraft Spruce http://www.aircraftspruce.com	Carries 4130 Chromoly tubing.
DC Waterjet http://www.dcwaterjet.com	Not a metal supplier, but a vendor that specializes in cutting you custom patterns in metal with an abrasive waterjet.
Metals Depot http://www.metalsdepot.com	Specializing in small quantities and good prices on aluminum, steel, tool steel, brass, and stainless steel.
Online Metals http://www.onlinemetals.com	All the usual varieties of aluminum and steel, as well as more exotic materials such as titanium and tool steel. Nice metal packs help you stock up your shop economically.
President Titanium http://www.presidenttitanium.com	Serious inquiries only. See the "Materials" section for a discussion of titanium and average prices.
Tico Titanium http://www.ticotitanium.com	Serious inquiries only. See the "Materials" section for a discussion of titanium and average prices.
Titanium Joe http://www.titaniumjoe.com	Specializes in good prices on surplus titanium. Order by e-mail and PayPal.
Wick's Aircraft Supply http://www.wicksaircraft.com	Carries 4130 Chromoly tubing.

Table D.3   Motors

Web Site	Notes
EV Parts http://www.evparts.com	Electric vehicle parts. Etek and AME motors.
NPC Robotics http://www.npcrobotics.com	Home of the most popular gearmotors in robot combat, as well as custom gear kits that you can easily assemble yourself.
Robot Marketplace http://www.robotmarketplace.com	Carries just about every motor used in robot combat, including EV Warriors, MagMotors, and NPC gearmotors.
Team Delta Engineering http://www.teamdelta.com	Home of the DeWalt drill motor mounting solutions engineered by Dan Danknick and Peter Abrahamson. Also carries Bosch GPA equivalents and Etek motors.
Team Whyachi http://www.teamwhyachi.com	Custom-engineered gearboxes for the high-power MagMotors.

Note: See also the surplus suppliers listed in Appendix D.

Table D.4 Wheels and Hub Solutions

Web Site	Notes
American Airless http://www.americanairless.com	Foam-filled wheels. They can also custom fill your wheel/hub combo, no matter where it comes from. Excellent service and quick turnaround.
Azusa Engineering http://www.azusaeng.com	Cool aftermarket hubs and wheels for various applications. Has aluminum sprockets and can do custom work.
Go Kart Supply http://www.gokartsupply.com	Lots of hubs, rims, and wheels.
Kart World http://www.kartworld.com	Lots of hubs, rims, and wheels.
NPC Robotics http://www.npcrobotics.com	Home of the flat-proof series of tires. Custom hub solutions to match their motors. Also caster wheels.
Robot Marketplace http://www.robotmarketplace.com	Just about every type of wheel used in robot combat, including NPC flat-proof tires and Carefree BattleTreads.
Team Delta Engineering http://www.teamdelta.com	Colson caster wheels and custom-engineered hubs.

Table D.5 Control Systems

Web Site	Notes
IFIrobotics http://www.ifirobotics.com	Home of the Isaac 16 and Isaac 32 control systems. Full line of accessories, including Spike servo switches and signal amplifiers. Extensive documentation and other resources available for free download.
Robot Marketplace http://www.robotmarketplace.com	Carries Futaba and JR radio packages, as well as various servos and accessories, including Deans antennas. Also carries the IFIrobotics Isaac control system.
Tower Hobbies http://www.towerhobbies.com	Carries various R/C packages, as well as servos and accessories, including Deans antennas and servo extensions.

Table D.6   PBasic Programming Resources

Web Site	Notes
FIRST http://www.usfirst.org	Check out the workshop archives for tutorial presentations on various topics.
How Stuff Works http://electronics.howstuffworks.com	Look under "Programming the Basic Stamp" for an excellent and detailed tutorial.
Parallax, Inc. http://www.parallax.com	Originators of the PBasic programming language, which is used to program the IFIrobotics control systems. Excellent "getting started" section.

Note: Also try searching under "PBasic programming" or "basic stamp programming" on Google.com.

Table D.7   Speed Controls

Web Site	Notes
4QD http://www.4qd.co.uk	High-power speed controls. Very popular among UK competitors.
IFIrobotics http://www.ifirobotics.com	Bulletproof speed controls that are both readily available and affordable.
NPC Robotics http://www.npcrobotics.com	Source for the Roboteq speed control.
Robot Marketplace http://www.robotmarketplace.com	Carries just about every speed control used in robot combat, including IFIrobotics, Vantec, Roboteq, and OSMC controls.
Robot Power http://www.robot-power.com	A source for the Open-Source Motor Controller (OSMC).
Robotic Sporting Goods http://roboticsportinggoods.com	Dual-channel speed controls.
Team Delta Engineering http://www.teamdelta.com	Carries IFIrobotics speed controls and custom-engineered signal amplifiers.
Vantec http://www.vantec.com	Single- and dual-channel speed controls. One of the first companies to embrace robot combat.

Table D.8   Batteries and Wiring Supplies

Web Site	Notes
Ballistic Batteries http://www.ballisticbatteries.com	R/C racing batteries. Also has a line of NiCad batteries for combat robots.
NPC Robotics http://www.npcrobotics.com	Connectors and SLA batteries.
Powerwerx http://www.powerwerx.com	Connectors and wiring supplies.
Rebco Performance Products http://www.rebcoperformance.com	Racing products. Source for high-power disconnect switches.
Robot Marketplace http://www.robotmarketplace.com	Wiring supplies and high-power switches. Also carries BattlePacks and SLA batteries.
Robotic Power Solutions http://www.battlepack.com	Makers of the popular BattlePack custom battery packs designed specifically for fighting robots. Specializes in NiCad and NiMH packs. Also carries wiring supplies.
Team Delta Engineering http://www.teamdelta.com	Wiring supplies and high power switches.
Tower Hobbies http://www.towerhobbies.com	R/C racing batteries and wiring supplies (Deans Ultra Wire and Wet Noodle wire).
West Marine http://www.westmarine.com	Marine supplier. Source for high-power disconnect switches.

Table D.9   Specialty R/C Switches and Mixers

Web Site	Notes
RobotLogic http://www.robotlogic.com	Custom-engineered radio mixers that allow you to control inverted driving. Also carries R/C interfaces and other types of mixers.
Robot Marketplace http://www.robotmarketplace.com	Carries the RobotLogic mixers as well as R/C gyros.
Team Delta Engineering http://www.teamdelta.com	Home of the popular D-switch R/C interface switch (with many useful variations), a custom-engineered solution by Dan Danknick.
Tower Hobbies http://www.towerhobbies.com	Carries R/C helicopter gyros.

Table D.10   Pneumatics

Web Site	Notes
Compressed Gas Association http://www.cganet.com	Organization that helps set industry standards for the safe use of compressed gases.
Fluid Power Educational Foundation http://www.fpef.org	Organization that's committed to educating people in fluid power. Lots of downloadable learning resources.
National Fluid Power Association http://www.nfpa.com	Products locator has lots and lots of links for pneumatic and hydraulic supplies.
FIRST http://www.usfirst.org	Check out the excellent pneumatics manual, which should be under competition documents.

Note: Also see the builder's Web sites in Table D.11 for detailed pneumatics information.

Table D.11   Tips from Other Builders

Web Site	Notes
BattleBots Builder's Forum http://forums.delphiforums.com/ BattleBot_Tech	The single greatest resource of collective knowledge about combat robotics on the internet. It can literally take you weeks to comb through all of the subjects. Of particular usefulness are the "Ultimate Guides."
Team Automatum http://www.automatum.com	Detailed build reports with plenty of pictures and excellent trip reports.
Team Biohazard http://www.robotbooks.com	Tips and resources from a master builder.
Team Carnage Robotics http://www.carnagerobotics.com	Detailed build reports.
Team Coolrobots http://www.coolrobots.com	Builder's corner has excellent tips and tutorials. Database of robots and builders.
Team DaVinci http://www.teamdavinci.com	Many tips for builders. Excellent pneumatics section. Lots of CAD files.
Team Deadblow http://www.deadblow.net	Detailed build reports.
Team Delta http://www.teamdelta.com	Detailed build reports and how-to sections on DeWalt drill motor timing and modifications, noise reduction, and other topics.

Web Site	Notes
Team Hazard http://www.legalword.com/wod.html	Lots of tips and information for builders.
Team Hurtz http://www.teamhurtz.com	Detailed how-to sections on pneumatics, Bosch GPA motor modifications, and other topics.
Team Inertia Labs http://www.interialabs.com	Lots of builder tips on various subjects. Innovators in coaxing manufacturers into modifying their products for robot combat and bringing them to our market.
Team Infernolabs http://www.infernolab.com	Lots of build reports with a ton of pictures. Excellent wheel comparison guide.
Team Mechanicus http://www.machinegod.com	Build reports from a pneumatics master.
Team Minus Zero http://www.tmz.com	Detailed build reports. Lots of battery info.
Team Nightmare http://www.robotcombat.com	Trip reports, the history of robot combat, and builder's database. The most up-to-date news on the sport.
Team Puppetmaster Robotics http://puppetmaster-robotics.com	Lots of general (and hard to find) information.
Team Redneck Robotics http://www.redneckrobots.com	EV Warrior modifications.
Team Robot Dojo http://www.robotdojo.com/	Lots of tutorials and a ton of pictures.
Team Robot Village http://www.robot-village.com	How-to sections for EV Warrior modifications. Good trip reports and build reports.
Team Rolling Thunder http://www.teamrollingthunder.com	How-to section for Bosch motor modifications. Lots of pictures and detailed build reports.
Team Storm http://www.stormrobot.com	How-to sections and build reports. Nice section on hydraulics.
Team Stupid http://www.stupidrobots.com	Lots of builder's tips. Nice pneumatics section.
Team Toad http://www.lazytoad.com/teamtoad/	Tutorials and thoughtful, well-documented experiments.

Catalogs

A s you saw in Appendix D, you can get almost everything you need to build your robot online. However, there are a few good mail-order catalogs that you should have on your bookshelf.

Younger builders should note that one of the most important tools for robot building is a credit card. When you have a credit card, ordering parts for the companies listed in this section — online or on the phone — is a snap. If you don't have a credit card, then you may have to place your order COD, where you pay when the shipment arrives. Not all companies offer this, however, and it's best to contact each company and find out its policy.

Note Additional handling fees may apply to shipments made outside the United States. Minimum order amounts may also apply at some of these vendors. Make sure to contact the vendor and ask first thing.

Industrial Suppliers

The suppliers in Table E.1 are the biggest suppliers on the block. They generally carry everything, and it's usually in stock. From mechanical items to raw materials, these stores have the complete range. They also stock tools and accessories. The bulk of your non-robot-specific parts and materials will come from these suppliers.

Table E.1   Industrial Suppliers

Supplier	Notes
McMaster-Carr P.O. Box 54960 Los Angeles, CA 90054-0960 (562) 692-5911 (562) 695-2323 FAX http://www.mcmaster.com	Lots of mechanical components and raw materials from titanium to neoprene. Also a good source for stocking your toolbox.
MSC Industrial Supply, Inc. 75 Maxess Rd. Melville, NY 11747-3151 (800) 645-7270 (713) 862-8665 (outside USA & Canada) (800) 255-5067 FAX http://www.mscdirect.com	Their catalog is so massive, they actually call it *The Big Book*.
Grainger Industrial Supply Huge network of stores throughout the USA, Canada, and Puerto Rico. Check the Web site for a local branch. (888) 361-8649 http://www.grainger.com	Good stock of tools and chain drive components. General-purpose hardware. Business-to-business sales only. If you're a consumer with no business of your own, then you'll have to get someone with a business to help you purchase products.

Specialty Suppliers

These suppliers target more-specific interest groups. They are included in Table E.2 because many of the tools and materials used in robot building come from various other disciplines.

Table E.2   Specialty Suppliers

Supplier	Notes
Aircraft Spruce 225 Airport Circle Corona, CA 92880 (877) 477-7823 (909) 372-9555 (outside of USA & Canada) (909) 372-0555 http://www.aircraftspruce.com	Home-built aircraft supplies. Good source for 4130 Chromoly tubing.

Supplier	Notes
Eastwood 263 Shoemaker Rd. Pottstown, PA 19464 (800) 345-1178 (610) 323-2200 (outside USA & Canada) (610) 323-6269 FAX http://www.eastwoodcompany.com	Automotive parts and tool catalog. Large selection of welding and auto-body repair equipment and accessories.
Micro-Mark 340 Snyder Ave. Berkeley Heights, NJ 07922-1595 (800) 225-1066 (908) 464-2984 (outside USA & Canada) (908) 665-9383 FAX http://www.micromark.com	Modelmaker tools and supplies. Small metalworking tools, and small vertical mill and lathe.
Northern Tool & Equipment Company P.O. Box 1499 Burnsville, MN 55337-0499 (800) 533-5545 (952) 882-6927 FAX http://www.NorthernTool.com	Farm equipment catalog. Huge selection of hydraulic components. Also some useful go-kart supplies.
Small Parts, Inc. 13980 N.W. 58th Court P.O. Box 4650 Miami Lakes, FL 33014-0650 (800) 220-4242 (305) 557-7955 (outside USA & Canada) (800) 423-9009 FAX http://www.smallparts.com	They have a little bit of everything, but specialize in, well, small parts. Good selection of exotic materials. Lots of tools and small mechanical parts.
West Marine 500 Westridge Dr. Watsonville, CA 95076 (800) 262-8464 (831) 761-4800 (outside USA & Canada) (831) 761-4421 FAX http://www.westmarine.com	Marine supply catalog. Of interest for the selection of high-power switches.
Wick's Aircraft Supply 410 Pine St. Highland, IL 62249 (800) 221-9425 (618) 654-7447 (outside USA & Canada) (618) 654-6253 http://www.wicksaircraft.com	Home-built aircraft supplies. Good source for 4130 Chromoly tubing.

Surplus Suppliers

One of my favorite things to do is check out the surplus catalogs for new and interesting stuff. Popular motors such as the AME D-Pack and EV Warrior were discovered in surplus catalogs, listed in Table E.3. Who knows what you might find?

Table E.3 Surplus Suppliers

Supplier	Notes
American Science & Surplus P.O. Box 1030 Skokie, IL 60076 (847) 647-0011 http://www.sciplus.com	Truly a strange and wonderful catalog full of interesting science-related items. Random selection of electronic and mechanical parts.
C&H Sales 2176 East Colorado Blvd. Pasadena, CA 91107 (800) 325-9465 (626) 796-2628 (outside USA & Canada) (626) 796-4875 FAX http://www.candhsales.com	One of my old stand-bys for sourcing surplus mechanical parts and electronics (big transformers, power supplies). Also has a big selection of pneumatic components.
Herbach & Rademan 353 Crider Ave. Moorestown, NJ 08057 (800) 848-8001 (856) 802-0422 (outside USA & Canada) (856) 802-0465 FAX http://www.herbach.com	Another one of my favorites. Similar selection to C&H.
Mendelson Electronic Company, Inc. (MECI) 340 E. First St. Dayton, OH 45042 (800) 344-4465 (800) 344-6324 FAX http://www.meci.com	Wide selection of motors and electronic stuff. Where the EV Warrior was discovered.
Servo Systems Company 115 Main Rd. P.O. Box 97 Montville, NJ 07045-0097 (800) 922-1103 (973) 335-1007 (outside USA & Canada) (973) 335-1661 http://www.servosystems.com	Interesting catalog with lots of industrial robots that I can't afford. They sell components for making your own motion-control systems, and are the source for the AMC amplifier mentioned in the weapons section.

Supplier	Notes
Surplus Center 1015 West "O" St. P.O. Box 82209 Lincoln, NE 68501-2209 (800) 488-3407 (402) 474-4055 (outside USA & Canada) (402) 474-5198 FAX	Cool catalog full of pneumatics, hydraulics, and motors, as well as other random equipment.

Radio-Control Hobby Suppliers

The stores listed in Table E.4 come out with regular catalogs that have all the hottest R/C airplanes and cars. They are also a good source for radios and servos, often beating local hobby shops.

Table E.4   Radio-Control Hobby Suppliers

Supplier	Notes
Tower Hobbies P.O. Box 9078 Champaign, IL 91826-9078 (800) 637-4989 (217) 398-3636 (outside USA & Canada) (800) 637-6050 FAX http://www.towerhobbies.com	The biggest and the best online hobby supplier. Excellent catalog.
Hobby Lobby 5614 Franklin Pike Circle Brentwood, TN 37027 (615) 373-1444 (615) 377-6948 FAX http://www.hobby-lobby.com	Another option if Tower is out of stock.

Electronic Parts

While I don't recommend making your own speed controls or other custom circuits, you may need electronic parts for a custom control box for your IFI system. The suppliers in Table E.5 are all reliable and well known.

Table E.5   Electronic Parts

Supplier	Notes
All Electronics P.O. Box 567 Van Nuys, CA 91408-0567 (888) 826-5432 (818) 904-0524 (outside USA & Canada) (818) 781-2653 FAX http://www.allelectronics.com	This is a southern California surplus supplier that carries not only electronic parts, but also a random assortment of mechanical items as well.
Digi-Key Corporation 701 Brooks Avenue South Thief River Falls, MN 56701 (800) 344-4539 (218) 681-6674 (outside USA & Canada) (218) 681-3380 FAX http://www.digikey.com	Easily the most reliable and professional of the electronic parts suppliers listed in this table, although you will also be paying slightly more than the others.
Marlin P. Jones P.O. Box 530400 Lake Park, FL 33403 (800) 652-6733 (561) 848-8236 (outside USA & Canada) (800) 432-9937 FAX http://www.mpja.com	Lots of surplus electronic and computer equipment and bargain prices.
Mouser Electronics 1000 North Main St. Mansfield, TX 76063 (800) 346-6873 (817) 804-3888 (outside USA & Canada) (817) 804-3899 FAX http://www.mouser.com	Good prices and service. Runs a close second to Digi-Key.

Machine and Power Tools

Eventually, you may decide that you need to expand your power-tool collection to more than the hardware-store variety. The suppliers listed in Table E.6 can help you out. You can also find more-specific metalworking tools in their catalogs.

Table E.6 Machine and Power Tools

Supplier	Notes
Airgas/Rutland Tool 2225 Workman Mill Rd. Whittier, CA 90601-1437 (800) 289-4787 (408) 436-8220 (outside USA & Canada) (800) 444-4787 FAX http://www.rutlandtool.com	Hand tools for professional machinists as well as machine tools (mills and lathes). Big selection.
Enco 400 Nevada Pacific Hwy. Fernley, NV 89408 (800) 873-3626 (800) 965-5857 http://www.use-enco.com	Hand tools for professional machinists as well as machine tools (mills and lathes). Good prices.
Grizzly Industrial, Inc. P.O. Box 3110 Bellingham, WA 98227-3110 (800) 523-4777 (570) 546-9663 (outside USA & Canada) (800) 438-5901 FAX http://www.grizzly.com	Power tools such as big drill presses, bandsaws, mills, and lathes.
Harbor Freight Tools 3491 Mission Oaks Blvd. Camarillo, CA 93011-6010 (800) 423-2567 (800) 905-5220 FAX http://www.harborfreight.com	Their catalog is full of suspicious knock-off tools made in China, but for these prices, who cares? (They're almost disposable.) There are also refurbished tools and random equipment, all at good prices.
Penn Tool Company 1776 Springfield Ave. Maplewood, NJ 07040 (800) 526-4956 (973) 761-1494 FAX http://www.penntoolco.com	Hand tools for professional machinists.

Mechanical Parts

The companies in Table E.7 manufacture mechanical components such as gears and racks. If you can't find it at McMaster-Carr, then you can go to the source. Note that this tends to be a pretty expensive option. These catalogs don't generally have prices because you have to call the manufacturer for a quote. However, each of their catalogs is full of useful engineering information.

Table E.7   Mechanical Parts

Supplier	Notes
PIC Design 86 Benson Rd. P.O. Box 1004 Middlebury, CT 06762 (800) 243-6125 (203) 758-8272 (outside USA & Canada) (203) 758-8271 FAX http://www.pic-design.com	All types of gears and other mechanical components. Also, special belt drives and precision positioning equipment. Their catalog is a great reference. You can get pricing and place an order online with a credit card (minimum order).
Stock Drive Products/Sterling Instruments 2101 Jericho Turnpike Box 5416 New Hyde Park, NY 11042-5416 (800) 819-8900 (516) 328-3300 (outside USA & Canada) (516) 326-8827 FAX http://www.sdp-si.com	Excellent multipart catalog that's a good mechanical reference. You can get pricing and place orders online with a credit card.
W. M. Berg, Inc. 499 Ocean Ave. East Rockaway, NY 11518 (800) 232-2374 (516) 599-5010 (outside USA & Canada) (800) 455-2374 FAX http://www.wmberg.com	Good reference catalog. It appears that you can place orders online with a credit card.

Tables and Charts

This appendix is your resource for helpful tables and charts to aid you in your design, and to supplement the information in the chapters.

Table F.1   Tap and Body Drill Sizes for American Standard Screw Threads

Screw Size	Dec. Size	Threads per Inch	Tap Drill	Dec. Size	Close Fit Drill Size	Dec. Size	Free Fit Drill Size	Dec. Size
#0	0.060"	80	#56	.0465"	#52	.0635"	#50	.0700
#1	0.073"	64	#53	.0595"	#48	.0760"	#46	.0810
#1	0.074"	72	#53	.0595"	#48	.0760"	#46	.0810
#2	0.086"	56	#50	.0700"	#43	.0890"	#41	.0960
#2	0.086"	64	#50	.0700"	#43	.0890"	#41	.0960
#3	0.099"	48	#47	.0785"	#37	.1040"	#35	.1100
#3	0.099"	56	#45	.0820"	#37	.1040"	#35	.1100
#4	0.112"	40	#43	.0890"	#32	.1160"	#30	.1285
#4	0.112"	48	#42	.0935"	#32	.1160"	#30	.1285
#5	0.125"	40	#38	.1015"	#30	.1285"	#29	.1360
#5	0.125"	44	#37	.1040"	#30	.1285"	#29	.1360
#6	0.138"	32	#36	.1065"	#27	.1440"	#25	.1495
#6	0.138"	40	#33	.1130"	#27	.1440"	#25	.1495
#8	0.164"	32	#29	.1360"	#18	.1695"	#16	.1770
#8	0.164"	36	#29	.1360"	#18	.1695"	#16	.1770
#10	0.190"	24	#25	.1495"	#9	.1960"	#7	.2010
#10	0.190"	32	#21	.1590"	#9	.1960"	#7	.2010
#12	0.216"	24	#16	.1770"	#2	.2210"	I	.2280
#12	0.216"	28	#14	.1820"	#2	.2210"	I	.2280
1/4	0.250"	20	#7	.2010"	F	.2570"	H	.2660
1/4	0.250"	28	#3	.2130"	F	.2570"	H	.2660
5/16	0.3125"	18	F	.2570"	P	.3230"	Q	.3320
5/16	0.3125"	24	I	.2720"	P	.3230"	Q	.3320
3/8	0.375"	16	5/16"	.3125"	W	.3860"	X	.3970
3/8	0.375"	24	Q	.3320"	W	.3860"	X	.3970
7/16	0.4375"	14	U	.3680"	29/64"	.4531"	15/32"	.4687
7/16	0.4375"	20	25/64"	.3906"	29/64"	.4531"	15/32"	.4687
1/2	0.500"	13	27/64"	.4219"	33/64"	.5156"	17/32"	.5312
1/2	0.500"	20	29/64"	.4531"	33/64"	.5156"	17/32"	.5312

Source: *Machinery's Handbook* (ANSI/ASME B94.11M-1993)

Table F.2   Standard Drill Sizes

Drill Size	Decimal	Drill Size	Decimal	Drill Size	Decimal
#80	.0135	1/16"	.0625	#27	.1440
#79	.0145	#52	.0635	#26	.1470
1/64"	.0156	#51	.0670	#25	.1495
#78	.0160	#50	.0700	#24	.1520
#77	.0180	#49	.0730	#23	.1540
#76	.0200	#48	.0760	5/32"	.1562
#75	.0210	5/64"	.0781	#22	.1570
#74	.0225	#47	.0785	#21	.1590
#73	.0240	#46	.0810	#20	.1610
#72	.0250	#45	.0820	#19	.1660
#71	.0260	#44	.0860	#18	.1695
#70	.0280	#43	.0890	11/64"	.1719
#69	.0295	#42	.0935	#17	.1730
#68	.0310	3/32"	.0938	#16	.1770
1/32"	.0312	#41	.0960	#15	.1800
#67	.0320	#40	.0980	#14	.1820
#66	.0330	#39	.0995	#13	.1850
#65	.0350	#38	.1015	3/16"	.1875
#64	.0360	#37	.1040	#12	.1890
#63	.0370	#36	.1065	#11	.1910
#62	.0380	7/64"	.1094	#10	.1935
#61	.0390	#35	.1100	#9	.1980
#60	.0400	#34	.1110	#8	.1990
#59	.0410	#33	.1130	#7	.2010
#58	.0420	#32	.1160	13/64"	.2031
#57	.0430	#31	.1200	#6	.2040
#56	.0465	1/8"	.1250	#5	.2055
3/64"	.0469	#30	.1285	#4	.2090
#55	.0520	#29	.1360	#3	.2130
#54	.0550	#28	.1405	7/32"	.2188
#53	.0595	9/64"	.1406	#2	.2210

Continued

Table F.2 (continued)

Drill Size	Decimal	Drill Size	Decimal	Drill Size	Decimal
#1	.2280	9/32"	.2812	3/8"	.3750
A	.2340	L	.2900	V	.3770
15/64"	.2344	M	.2950	W	.3860
B	.2380	19/64"	.2969	25/64"	.3906
C	.2420	N	.3020	X	.3970
D	.2460	5/16"	.3125	Y	.4040
E	.2500	O	.3160	13/32"	.4062
1/4"	.2500	P	.3230	Z	.4130
F	.2570	Q	.3320	27/64"	.4219
G	.2610	R	.3390	7/16"	.4375
17/64"	.2656	11/32"	.3438	29/64"	.4531
H	.2660	S	.3480	15/32"	.4688
I	.2720	T	.3580	31/64"	.4844
J	.2770	23/64"	.3694	1/2"	.5000
K	.2810	U	.3680		

Source: *Machinery's Handbook* (ANSI/ASME B94.11M-1993)

Table F.3 Basic Dimensions of American National Standard Pipe Threads, NPT

Pipe Size	Actual Pipe OD	Threads per Inch
1/8"	0.405"	27
1/4"	0.540"	18
3/8"	0.675"	18
1/2"	0.840"	14
3/4"	1.050"	14
1"	1.315"	11.5
1-1/4"	1.660"	11.5
1-1/2"	1.900"	11.5
2"	2.375"	11.5
2-1/2"	2.875"	8
3"	3.500"	8

Source: *Machinery's Handbook* (ANSI/ASME B1.20.1-1983[R1992])

Table F.4   Standard Dimensions for Cap Screws

Screw Size	Dec. Size	Socket-Head			Flat-Head			Button-Head		
		Hex Key	Head Height	Head Dia.	Hex Key	Head Height	Head Dia.	Hex Key	Head Height	Head Dia.
#0	0.060"	0.050"	0.060"	0.096"	0.035"	0.044"	0.138"	0.035"	0.032"	0.114"
#1	0.073"	1/16"	0.073"	0.118"	0.050"	0.054"	0.168"	0.050"	0.039"	0.139"
#2	0.086"	5/64"	0.086"	0.140"	0.050"	0.064"	0.197"	0.050"	0.046"	0.164"
#3	0.099"	5/64"	0.099"	0.161"	1/16"	0.073"	0.226"	1/16"	0.052"	0.188"
#4	0.112"	3/32"	0.112"	0.183"	1/16"	0.083"	0.255"	1/16"	0.059"	0.213"
#5	0.125"	3/32"	0.125"	0.205"	5/64"	0.090"	0.281"	5/64"	0.066"	0.238"
#6	0.138"	7/64"	0.138"	0.226"	5/64"	0.097"	0.307"	5/64"	0.073"	0.262"
#8	0.164"	9/64"	0.164"	0.270"	3/32"	0.112"	0.359"	3/32"	0.087"	0.312"
#10	0.190"	5/32"	0.190"	0.312"	1/8"	0.127"	0.411"	1/8"	0.101"	0.361"
1/4"	0.250"	3/16"	0.250"	0.375"	5/32"	0.161"	0.531"	5/32"	0.132"	0.437"
5/16"	0.3125"	1/4"	0.3125"	0.469"	3/16"	0.198"	0.656"	3/16"	0.166"	0.547"
3/8"	0.375"	5/16"	0.375"	0.563"	7/32"	0.234"	0.781"	7/32"	0.199"	0.656"
7/16"	0.4375"	3/8"	0.4375"	0.656"	1/4"	0.234"	0.844"	-	-	-
1/2"	0.500"	3/8"	0.500"	0.750"	5/16"	0.251"	0.938"	5/16"	0.265"	0.875"

Note: The head diameter data can be used to size counterbores and countersinks.

Source: *Machinery's Handbook* (ANSI/ASME B18.3–1998)

Table F.5 Keyless Bushing Dimensional Data

Shaft Dia.	Required Sprocket Bore	Overall Length	Max Torque (in-lbs)	Max Thrust (in-lbs)	Installation Torque (in-lbs)	Weight (oz)
3/16"	5/8"	3/4"	100	700	125	0.5
1/4"	5/8"	3/4"	150	790	125	0.5
5/16"	3/4"	7/8"	200	890	150	1
3/8"	3/4"	7/8"	250	925	150	1
7/16"	7/8"	1"	300	950	175	1.5
1/2"	7/8"	1"	350	980	175	1.5
9/16"	1"	1-1/8"	400	990	200	2
5/8"	1"	1-1/8"	450	1000	200	2
5/8"	1-1/2"	1-1/2"	1750	3300	1200	8
3/4"	1-1/2"	1-1/2"	2500	4400	1200	8
13/16"	1-3/4"	1-7/8"	2600	4950	1500	8
7/8"	1-3/4"	1-7/8"	2800	5500	1500	11
15/16"	1-3/4"	1-7/8"	3100	6050	1500	11
1"	1-3/4"	1-7/8"	3500	6600	1500	11
1-1/16"	2"	2-1/4"	4000	7000	2000	11
1-1/8"	2"	2-1/4"	4600	7500	2000	16
1-3/16"	2"	2-1/4"	5200	8000	2000	16
1-1/4"	2"	2-1/4"	6000	8500	2000	16
1-3/8"	2-3/8"	2-3/4"	6400	9500	2300	27
1-7/16"	2-3/8"	2-3/4"	6700	10000	2300	27
1-1/2"	2-3/8"	2-3/4"	7000	10500	2300	27
1-5/8"	2-5/8"	3-1/8"	8500	11750	2800	37
1-3/4"	2-5/8"	3-1/8"	10000	12750	2800	37
1-7/8"	2-7/8"	3-9/16"	11750	14000	3900	48
1-15/16"	2-7/8"	3-9/16"	12750	14500	3900	48
2"	2-7/8"	3-9/16"	14000	15000	3900	48

Note: Shafts and hubs must have +/- 0.0015" tolerance for 5/8" shaft (1" sprocket bore) and smaller, and +/- 0.003" tolerance for 5/8" shaft diameter (1-1/2" sprocket bore) and larger.

Source: Fenner Mannheim/Trantorque

Table F.6 Dimensional Data for #35 ANSI Sprockets

Number of Teeth	Outside Diameter	Thickness	Hub Diameter
9	1.26"	3/4"	27/32"
10	1.38"	3/4"	31/32"
11	1.5"	3/4"	1-1/16"
12	1.63"	3/4"	1-7/32"
13	1.75"	3/4"	1-1/4"
14	1.87"	3/4"	1-1/4"
15	1.99"	3/4"	1-11/32"
16	2.11"	3/4"	1-15/32"
17	2.23"	3/4"	1-19/32"
18	2.35"	3/4"	1-23/32"
19	2.47"	3/4"	1-27/32"
20	2.59"	3/4"	1-15/16"
21	2.71"	3/4"	2"
22	2.83"	3/4"	2"
23	2.95"	3/4"	2"
24	3.07"	3/4"	2"
25	3.19"	3/4"	2"
26	3.31"	3/4"	2"
28	3.55"	7/8"	2"
30	3.79"	7/8"	2"
32	4.03"	7/8"	2"
35	4.39"	1"	2-1/4"
36	4.51"	1"	2-1/4"
40	4.99"	1"	2-1/4"
42	5.23"	1"	2-1/4"
45	5.59"	1"	2-1/4"
48	5.95"	1"	2-1/4"
60	7.38"	1"	2-1/4"

Table F.7   Standard Keyway and Setscrew Data for ANSI Sprockets

Shaft Diameter	Key Size	Setscrew Sizes
5/16"	3/32"	#8-32
3/8"	3/32"	#8-32
1/2"	1/8"	#10-32
5/8"	3/16"	1/4"
3/4"	3/16"	1/4"
7/8"	3/16"	1/4"
1"	1/4"	5/16"
1-1/8"	1/4"	5/16"
1-3/16"	1/4"	5/16"
1-1/4"	1/4"	5/16"
1-3/8"	5/16"	5/16"
1-7/16"	3/8"	3/8"
1-1/2"	3/8"	3/8"
1-5/8"	3/8"	3/8"
1-3/4"	3/8"	3/8"
1-15/16"	1/2"	1/2"
2"	1/2"	1/2"
2-3/16"	1/2"	1/2"
2-7/16"	5/8"	5/8"
2-15/16"	3/4"	3/4"

Note: Length of setscrew may vary depending on the hub diameter and bore.

Note: Depth of keyway is half the key size.

Note: Setscrews are usually placed one above the keyway and the other at 90 degrees from the keyway.

Table F.8   Radio-Control Frequencies and Channels

27 MHz Band		50 MHz Band		72 MHz Band		75 MHz Band	
Ch	Freq (Hz)	Ch	Freq (Hz)	Ch	Freq (Hz)	Ch	Freq (Hz)
A1	26.995	00	50.800	11	72.010	61	75.410
A2	27.045	01	50.820	12	72.030	62	75.430
A3	27.095	02	50.840	13	72.050	63	75.450
A4	27.145	03	50.860	14	72.070	64	75.470
A5	27.195	04	50.880	15	72.090	65	75.490
A6	27.255	05	50.900	16	72.110	66	75.510
		06	50.920	17	72.130	67	75.530
		07	50.940	18	72.150	68	75.550
		08	50.960	19	72.170	69	75.570
		09	50.980	20	72.190	70	75.590
				21	72.210	71	75.610
				22	72.230	72	75.630
				23	72.250	73	75.650
				24	72.270	74	75.670
				25	72.290	75	75.690
				26	72.310	76	75.710
				27	72.330	77	75.730
				28	72.350	78	75.750
				29	72.370	79	75.770
				30	72.390	80	75.790
				31	72.410	81	75.810
				32	72.430	82	75.830
				33	72.450	83	75.850
				34	72.470	84	75.870
				35	72.490	85	75.890
				36	72.510	86	75.910
				37	72.530	87	75.930
				38	72.550	88	75.950

Continued

Table F.8 *(continued)*

27 MHz Band		50 MHz Band		72 MHz Band		75 MHz Band	
Ch	Freq (Hz)	Ch	Freq (Hz)	Ch	Freq (Hz)	Ch	Freq (Hz)
				39	72.570	89	75.970
				40	72.590	90	75.990
				41	72.610		
				42	72.630		
				43	72.650		
				44	72.670		
				45	72.690		
				46	72.710		
				47	72.730		
				48	72.750		
				49	72.770		
				50	72.790		
				51	72.810		
				52	72.830		
				53	72.850		
				54	72.870		
				55	72.890		
				56	72.910		
				57	72.930		
				58	72.950		
				59	72.970		
				60	72.990		

Note: The 72 MHz band may ONLY be used for aircraft models. The 75 MHz band may ONLY be used with surface models. Use of the 50 MHz band requires an Amateur Radio Operator License from the FCC.

Table F.9   Stall Current and Torque Constant of Various Motors

Motor	Stall Current (amps)	Torque Constant Kt (oz-in/amp)
AME D-Pack	1240	2.78
Andrus 5015-1	220	6.7
Andrus 5017-1	132	7
Andrus 5019-1	436	6.5
Astroflight Cobalt 15	100	0.42
Astroflight Cobalt 40	100	2
Astroflight Cobalt 60	100	3.6
Astroflight Cobalt 90	100	5.3
Bosch GPA	180	8.7
DeWalt 18V (gearbox in high)	110	21.8
Dustin Motor (36V hi speed)	180	16.5
EV Warrior	130	6.6
MagMotor mini (S28-150)	385	5.25
MagMotor 3" (S28-400)	571	6.57
MagMotor 4" (C40-300)	480	8.05
NPC 1200	107	9.5
NPC 2212 (12V only)	80	36
NPC 2423	34.2	10.3
NPC Black Max (Scott)	705	9.56
NPC R81/R82	105	136.5
NPC T64	154	121.6
NPC T74	210	112.76
NPC 02446	81.4	8.4
NPC 41250	112	73.2
Sullivan Hi-Tork	41	5.4
ThinGap TG3200-35	267	0.72
ThinGap TG3200-42	141	0.87

Table F.10   Unit Conversions

$1 \text{ in} = 2.54 \times 10^{-2} \text{ m} = 2.540 \text{ cm} = 8.333 \times 10^{-2} \text{ ft}$

$1 \text{ in}^2 = 6.452 \times 10^{-4} \text{ m}^2 = 6.452 \text{ cm}^2 = 9.94 \times 10^{-3} \text{ ft}^3$

$1 \text{ in}^3 = 5.79 \times 10^{-4} \text{ ft}^3 = 1.639 \times 10^{-2} \text{ L} = 1.64 \ 10^{-5} \text{ m}^3 = 1.64 \times 10^{-2} \text{ cm}^3$

$1 \text{ lb} = 16 \text{ oz.} = 0.4536 \text{ kg} = 453.6 \text{ g} = 4.448 \text{ N}$

$1 \text{ lb/in}^3 = 2.768 \times 10^4 \text{ kg/m}^3 = 27.68 \text{ g/cm}^3$

$1 \text{ in-lb} = 16 \text{ oz-in} = 8.333 \times 10^{-2} \text{ ft-lbs} = 1.152 \times 10^{-2} \text{ kg-m} = 1.152 \times 10^3 \text{ g-cm} = 0.113 \text{ N-m} = 1.129 \times 10^6 \text{ dyne-cm}$

$1 \text{ W} = 1 \text{ J/sec} = 1.341 \times 10^{-3} \text{ HP} = 0.238 \text{ cal/sec} = 0.7376 \text{ ft-lb/sec} = 84.5 \text{ in-lb·RPM}$

$1 \text{ RPM} = 6 \text{ degrees/sec} = 0.1047 \text{ rad/sec}$

$1 \text{ ft/sec} = 12 \text{ in/sec} = 0.6818 \text{ MPH} = 0.3048 \text{ m/sec}$

$1 \text{ psi} = 6.895 \times 10^{-2} \text{ bar} = 6895 \text{ Pa} = 7.031 \times 10^{-2} \text{ kg/cm}^2 = 6.805 \times 10^{-2} \text{ atm}$

$R \text{ (univ. gas const.)} = 8.31451 \text{ J/mol·K} = 8.21 \times 10^{-2} \text{ L·atm/mol·K} = 1.987 \text{ cal/mol·K}$

Table F.11   Sample Tool Checklist for Competition

Safety

    Safety glasses

    Full face shield

Power tools

    Jigsaw (with spare blades)

    Angle grinder

    Cordless drill (with charger and spare battery)

    Dremel tool (assorted bits)

Hand tools

    Screwdrivers (assorted flat-head and Phillips)

    Allen wrenches

    Adjustable (crescent) wrenches

    Combination wrenches

    Socket set

    Needle nose pliers

    Slip-joint pliers

    Measuring tape

    Ruler (12" steel)

    Combination square

    Machinist's square

    Hammer (ball-peen)

    Hacksaw

    C-clamps

    Files (assorted)

    Deburring tool

    Hobby (X-acto) knife

    Chain tools (if necessary)

    Shop-vac (small 1HP)

    Extension cords (2)

    Power strip

    Flashlight

    Dental mirror

    Telescoping magnet

    Automatic center punch

    Drill index

    Tap wrench (assorted taps)

    Soldering iron

    Solder

    Desoldering tool

    Wire crimper

    Wire cutter

    Wire stripper (manual and automatic)

    Digital multimeter (DMM)

Chemicals

    Cutting lubricant (WD-40)

    Solvent (isopropyl alcohol)

    Loctite (various formulations)

Supplies

    Rags

    Markers (ultra fine tip)

    Masking tape

    Electrical tape (regular and liquid)

    Cable (zip) ties

    Velcro

Administrative

    Pad of paper

    Construction/design notes

    Documentation for pneumatic/hydraulic parts (for safety inspectors)

    Competition rulebook

Index

Continued

Continued

Continued

Continued